PROTEOMICS APPLICATIONS IN FOOD SCIENCE

PROTEOMICS APPLICATIONS IN FOOD SCIENCE

Edited by:

VALERIA SEVERINO

Delve
PUBLISHING

www.delvepublishing.com

Proteomics Applications in Food Science

VALERIA SEVERINO

Delve Publishing
2010 Winston Park Drive,
2nd Floor
Oakville, ON L6H 5R7
Canada
www.delvepublishing.com
Tel: 001-289-291-7705
001-905-616-2116
Fax: 001-289-291-7601
Email: orders@arclereducation.com

ISBN: 978-1-77361-136-5 (Hardcover)

DECLARATION

Some content or chapters in this book are open access copyright free published research work, which is published under Creative Commons License and are indicated with the citation. We are thankful to the publishers and authors of the content and chapters as without them this book wouldn't have been possible.

ABOUT THE EDITOR

VALERIA SEVERINO

Valeria Severino obtained her PhD in Biological Processes and Biomolecules from the Second University of Naples in 2009. Her scientific interests concern the structural and functional characterization of proteins and peptides by using biochemical approaches. Fields of expertise comprises food chemistry, cellular biology, natural products and related applications. She published more than 40 papers on international peer-reviewed journals.

TABLE OF CONTENT

SECTION III: PROTEOMICS AND FOOD ALLERGIES

LIST OF CONTRIBUTORS

Valeria Severino

Peter Feist
Department of Chemistry and Biochemistry, Integrated Biomedical Sciences Program, and the Harper Cancer Research Institute, 251 Nieuwland Science Hall, University of Notre Dame, Notre Dame, IN 46556, USA

Amanda B. Hummon
Department of Chemistry and Biochemistry, Integrated Biomedical Sciences Program, and the Harper Cancer Research Institute, 251 Nieuwland Science Hall, University of Notre Dame, Notre Dame, IN 46556, USA

Yasset Perez-Riverol
European Molecular Biology Laboratory, European Bioinformatics Institute (EMBL-EBI), Wellcome Trust Genome Campus, Hinxton, Cambridge, UK

Emanuele Alpi
European Molecular Biology Laboratory, European Bioinformatics Institute (EMBL-EBI), Wellcome Trust Genome Campus, Hinxton, Cambridge, UK

Rui Wang
European Molecular Biology Laboratory, European Bioinformatics Institute (EMBL-EBI), Wellcome Trust Genome Campus, Hinxton, Cambridge, UK

Henning Hermjakob
European Molecular Biology Laboratory, European Bioinformatics Institute (EMBL-EBI), Wellcome Trust Genome Campus, Hinxton, Cambridge, UK

Juan Antonio Vizca´ıno
European Molecular Biology Laboratory, European Bioinformatics Institute (EMBL-EBI), Wellcome Trust Genome Campus, Hinxton, Cambridge, UK

Gerhard Mayer
Medizinisches Proteom Center (MPC), Ruhr-Universität Bochum, D-44801 Bochum, Germany

Andrew R. Jones
Institute of Integrative Biology, University of Liverpool, Liverpool L69 7ZB, UK

Pierre-Alain Binz
SIB Swiss Institute of Bioinformatics, Swiss-Prot group, Rue Michel-Servet 1, CH-1211 Geneva 4, Switzerland

Eric W. Deutsch
Institute for Systems Biology, 401 Terry Avenue North, Seattle, WA 98109, USA

Sandra Orchard
EMBL-EBI, Wellcome Trust Genome Campus, Hinxton, Cambridge, CB10 1SD, UK

Luisa Montecchi-Palazzi
EMBL-EBI, Wellcome Trust Genome Campus, Hinxton, Cambridge, CB10 1SD, UK

Juan Antonio Vizcaíno
EMBL-EBI, Wellcome Trust Genome Campus, Hinxton, Cambridge, CB10 1SD, UK

Henning Hermjakob
EMBL-EBI, Wellcome Trust Genome Campus, Hinxton, Cambridge, CB10 1SD, UK

David Oveillero
EMBL-EBI, Wellcome Trust Genome Campus, Hinxton, Cambridge, CB10 1SD, UK

Randall Julian
Indigo BioSystems, Indianapolis, IN 46240, USA

Christian Stephan
Medizinisches Proteom Center (MPC), Ruhr-Universität Bochum, D-44801 Bochum, Germany
Kairos GmbH, Universitätsstraße 136, D-44799 Bochum, Germany

Helmut E. Meyer
Medizinisches Proteom Center (MPC), Ruhr-Universität Bochum, D-44801 Bochum, Germany

Martin Eisenacher
Medizinisches Proteom Center (MPC), Ruhr-Universität Bochum, D-44801 Bochum, Germany

Sameh Magdeldin
Department of Structural Pathology, Institute of Nephrology, Graduate School of Medical and Dental Sciences, Niigata University, 1-757 Asahimachi-dori, Niigata, Japan.
Department of Physiology, Faculty of Veterinary Medicine, Suez Canal University, Ismailia, Egypt.

Shymaa Enany
Department of Structural Pathology, Institute of Nephrology, Graduate School of Medical and Dental Sciences, Niigata University, 1-757 Asahimachi-dori, Niigata, Japan.
Department of Microbiology and Immunology, Faculty of Pharmacy, Suez Canal University, Ismailia, Egypt.

Yutaka Yoshida
Department of Structural Pathology, Institute of Nephrology, Graduate School of Medical and Dental Sciences, Niigata University, 1-757 Asahimachi-dori, Niigata, Japan.

Bo Xu
Department of Structural Pathology, Institute of Nephrology, Graduate School of Medical and Dental Sciences, Niigata University, 1-757 Asahimachi-dori, Niigata, Japan.

Ying Zhang
Department of Structural Pathology, Institute of Nephrology, Graduate School of Medical and Dental Sciences, Niigata University, 1-757 Asahimachi-dori, Niigata, Japan.

Zam Zureena
UKM Medical Centre, Kuala Lumpur, Malaysia.

Ilambarthi Lokamani
UKM Medical Centre, Kuala Lumpur, Malaysia.

Eishin Yaoita
Department of Structural Pathology, Institute of Nephrology, Graduate School of Medical and Dental Sciences, Niigata University, 1-757 Asahimachi-dori, Niigata, Japan.

Tadashi Yamamoto
Department of Structural Pathology, Institute of Nephrology, Graduate School of Medical and Dental Sciences, Niigata University, 1-757 Asahimachi-dori, Niigata, Japan.

Ignacio Ortea, Gavin O'Connor, Alain Maquet
European Commission, DG Joint Research Centre (JRC), Directorate F. Health, Consumers and Reference Materials, Retieseweg 111, 2440 Geel, Belgium

Piotr Minkiewicz
Department of Food Biochemistry, University of Warmia and Mazury in Olsztyn, Plac Cieszyński 1, Olsztyn-Kortowo 10-726, Poland

Małgorzata Darewicz
Department of Food Biochemistry, University of Warmia and Mazury in Olsztyn, Plac Cieszyński 1, Olsztyn-Kortowo 10-726, Poland

Anna Iwaniak
Department of Food Biochemistry, University of Warmia and Mazury in Olsztyn, Plac Cieszyński 1, Olsztyn-Kortowo 10-726, Poland

Jolanta Sokołowska
Department of Food Biochemistry, University of Warmia and Mazury in Olsztyn, Plac Cieszyński 1, Olsztyn-Kortowo 10-726, Poland

Piotr Starowicz
Department of Food Biochemistry, University of Warmia and Mazury in Olsztyn, Plac Cieszyński 1, Olsztyn-Kortowo 10-726, Poland

Justyna Bucholska
Department of Food Biochemistry, University of Warmia and Mazury in Olsztyn, Plac Cieszyński 1, Olsztyn-Kortowo 10-726, Poland

Monika Hrynkiewicz
Department of Food Biochemistry, University of Warmia and Mazury in Olsztyn, Plac Cieszyński 1, Olsztyn-Kortowo 10-726, Poland

Chuan Hong
Laboratory of Human Evolution, Institute of Vertebrate Paleontology and Paleoanthropology, CAS, Beijing, China,
Department of Scientific History and Archaeometry, Graduate University, Beijing, China,

Hongen Jiang
Laboratory of Human Evolution, Institute of Vertebrate Paleontology and Paleoanthropology, CAS, Beijing, China,
Department of Scientific History and Archaeometry, Graduate University, Beijing, China,

Enguo Lu
Xinjiang Institute of Archaeology, Xinjiang Uygur Autonomous Region, Urumchi, China,

Yunfei Wu
College of Life Science, Graduate University, CAS, Beijing, China,

Lihai Guo
Asia Pacific Application Support Center, AB SCIEX, Shanghai, China

Yongming Xie
Asia Pacific Application Support Center, AB SCIEX, Shanghai, China

Changsui Wang
Laboratory of Human Evolution, Institute of Vertebrate Paleontology and Paleoanthropology, CAS, Beijing, China,
Department of Scientific History and Archaeometry, Graduate University, Beijing, China,

Yimin Yang
Laboratory of Human Evolution, Institute of Vertebrate Paleontology and Paleoanthropology, CAS, Beijing, China,
Department of Scientific History and Archaeometry, Graduate University, Beijing, China,

Francesco Di Girolamo
Department of Laboratory Medicine, Bambino Gesù Children's Hospital, IRCCS, Piazza Sant'Onofrio 4, Rome 00165, Italy

Andrea Masotti
Gene Expression-Microarrays Laboratory, Bambino Gesù Children's Hospital, IRCCS, Piazza Sant'Onofrio 4, Rome 00165, Italy

Guglielmo Salvatori
Neonatal Intensive Care Unit, Department of Medical and Surgical Neonatology, Bambino Gesù Children's Hospital, IRCCS, Piazza Sant'Onofrio 4, Rome 00165, Italy

Margherita Scapaticci
Department of Laboratory Medicine, San Camillo Hospital, Viale Vittorio Veneto 18, Treviso 31100, Italy

Maurizio Muraca
Department of Laboratory Medicine, Bambino Gesù Children's Hospital, IRCCS, Piazza Sant'Onofrio 4, Rome 00165, Italy

Lorenza Putignani
Parasitology Unit, Bambino Gesù Children's Hospital, IRCCS, Piazza Sant'Onofrio 4, Rome 00165, Italy
Metagenomics Unit, Bambino Gesù Children's Hospital, IRCCS, Piazza Sant'Onofrio 4, Rome 00165, Italy

Romina Pedreschi
European Commission (EC), DG Joint Research Centre (JRC), Institute for Reference Materials and Measurements (IRMM), Geel 2440, Belgium

Jørgen Nørgaard
European Commission (EC), DG Joint Research Centre (JRC), Institute for Reference Materials and Measurements (IRMM), Geel 2440, Belgium

Alain Maquet
European Commission (EC), DG Joint Research Centre (JRC), Institute for Reference Materials and Measurements (IRMM), Geel 2440, Belgium

Bindukumar Nair
Division of Allergy, Immunology, and Rheumatology, Department of Medicine, University at Buffalo, 640 Ellicott Street, Room 444 Innovation Center, Buffalo Niagara Medical Campus, 640 Ellicott Street, Buffalo, NY 14203, USA

John C. Wheeler
Division of Allergy, Immunology, and Rheumatology, Department of Medicine, University at Buffalo, 640 Ellicott Street, Room 444 Innovation Center, Buffalo Niagara Medical Campus, 640 Ellicott Street, Buffalo, NY 14203, USA

Donald E. Sykes
Division of Allergy, Immunology, and Rheumatology, Department of Medicine, University at Buffalo, 640 Ellicott Street, Room 444 Innovation Center, Buffalo Niagara Medical Campus, 640 Ellicott Street, Buffalo, NY 14203, USA

Paula Brown
Division of Allergy, Immunology, and Rheumatology, Department of Medicine, University at Buffalo, 640 Ellicott Street, Room 444 Innovation Center, Buffalo Niagara Medical Campus, 640 Ellicott Street, Buffalo, NY 14203, USA

Jessica L. Reynolds
Division of Allergy, Immunology, and Rheumatology, Department of Medicine, University at Buffalo, 640 Ellicott Street, Room 444 Innovation Center, Buffalo Niagara Medical Campus, 640 Ellicott Street, Buffalo, NY 14203, USA

Ravikumar Aalinkeel
Division of Allergy, Immunology, and Rheumatology, Department of Medicine, University at Buffalo, 640 Ellicott Street, Room 444 Innovation Center, Buffalo Niagara Medical Campus, 640 Ellicott Street, Buffalo, NY 14203, USA

Supriya D. Mahajan
Division of Allergy, Immunology, and Rheumatology, Department of Medicine, University at Buffalo, 640 Ellicott Street, Room 444 Innovation Center, Buffalo Niagara Medical Campus, 640 Ellicott Street, Buffalo, NY 14203, USA

Stanley A. Schwartz
Division of Allergy, Immunology, and Rheumatology, Department of Medicine, University at Buffalo, 640 Ellicott Street, Room 444 Innovation Center, Buffalo Niagara Medical Campus, 640 Ellicott Street, Buffalo, NY 14203, USA

Nuria Prieto
Genetics Department, Biology Faculty, Complutense University of Madrid, Madrid 28040, Spain

Carmen Burbano
Food Technology Department, National Institute for Agricultural and Food Research and Technology (INIA), Ctra. La Coruña Km. 7.5, Madrid 28040, Spain

Elisa Iniesto
Genetics Department, Biology Faculty, Complutense University of Madrid, Madrid 28040, Spain

Julia Rodríguez
Allergy Service, Research Institute Hospital 12 de Octubre (i+12), Avenida de Córdoba s/n, Madrid 28041, Spain

Beatriz Cabanillas
Allergy Service, Research Institute Hospital 12 de Octubre (i+12), Avenida de Córdoba s/n, Madrid 28041, Spain

Jesus F. Crespo
Allergy Service, Research Institute Hospital 12 de Octubre (i+12), Avenida de Córdoba s/n, Madrid 28041, Spain

Mercedes M. Pedrosa
Food Technology Department, National Institute for Agricultural and Food Research and Technology (INIA), Ctra. La Coruña Km. 7.5, Madrid 28040, Spain

Mercedes Muzquiz
Food Technology Department, National Institute for Agricultural and Food Research and Technology (INIA), Ctra. La Coruña Km. 7.5, Madrid 28040, Spain

Juan Carlos del Pozo
Center for Biotechnology and Plant Genomic, Polytechnic University of Madrid-National Institute for Agricultural and Food Research and Technology (UPM-INIA), Montegancedo Campus, Boadilla del Monte, Madrid 28660, Spain

Rosario Linacero
Genetics Department, Biology Faculty, Complutense University of Madrid, Madrid 28040, Spain

Carmen Cuadrado
Food Technology Department, National Institute for Agricultural and Food Research and Technology (INIA), Ctra. La Coruña Km. 7.5, Madrid 28040, Spain

PREFACE

In the last decades, proteomics has rapidly advanced with novel developments and cutting-edge approaches, resulting in reliable platforms widely used for several purposes. Among others, food science represents one of the most dynamic and growing research areas in which proteomics is extensively applied. Food proteomics is commonly used to characterize raw materials, determine the food composition and evaluate the relationship between the food protein content and the nutritional properties. Moreover, quality control during food production can be monitored with proteomics, along with food traceability, quality and safety, in particular by detecting the presence of allergens or contaminants. Considering the complexity of food matrices and the elevated variability of food-associated proteins and peptides, it is hard to choose a single proteomic strategy. Often, a combination of multiple techniques achieves the better results, with proteomic approaches well adapted to each problem.

This book covers several aspects of proteomics in food science, highlighting a range of methods and applications for analyzing the food composition, the presence of specific allergens or microorganisms, and for evaluating food identity and quality. To this aim, the book is organized into three Sections.

In Section 1 are reviewed technical aspects at the basis of proteomics, focusing on concepts and challenges of proteomic analyses, sample preparation and techniques commonly used.

In Section 2 are reported proteomic applications in food science, emphasizing on food authentication, for example producing a fingerprint typical of each food or food component. Moreover, this Section contains some interesting applications such as the identification of milk components in ancient food and the use of proteomics to evaluate ex vivo protein modifications and their effects on foods.

Finally, the Section 3 focuses on proteomics and food allergies, a particularly dynamic field also called food allergenomics. Here are reported some particular proteomic applications to assess the physicochemical properties of food allergens.

Editor

INTRODUCTION

Dr. Valeria Severino

The term proteomics refers to the study of the entire protein content, also known as "proteome", of a biological system. Proteome was coined by Marc Wilkins in 1994 as the crasis of protein and genome, in order to indicate the protein complement of the genome. Nowadays, proteomics has a wide broad meaning, indicating both the large-scale characterization of proteins in a sample and targeted studies on specific proteins of interest. This latter aim has been explored especially by using mass spectrometry (MS)- or immune-based proteomics.

In the last decade, the use of proteomics in food science has been widely exploited by developing appropriate methods and instruments. However, at the beginning proteomic approaches in food research and biotechnology were rather limited. The main difficulty encountered was related to the use of plant material, because the genome sequences of many vegetal species were uncompleted or unknown. Moreover, instruments and equipment were expansive and bulky. Fortunately, proteomics has become a routine technique in several fields, including food science, mainly due to the rapid sequencing of many genomes, including plant species commonly used in human nutrition (e.g. rice, potato, wheat, maize, soybean and other crops) and technical improvements.

Nowadays proteomics is commonly used in food industry for bioprocess improvement, validation and quality control. Moreover, it is helpful for basic research in food science, providing both large-scale and detailed information about proteins and peptides present in food.

Proteomic analysis can be challenging: samples are often complex with very different chemical species. As an example, the same protein can be present as isoform or with modifications. Moreover, the dynamic range of the cellular proteome is extremely wide, with about seven orders of magnitude: this means that proteins can be expressed at the same time from one copy per cell to ten million copies per cell [1]. It is thus necessary to reduce the complexity of the sample in order to improve proteomic analysis, and this can be achieved by associating classical separation techniques. At the protein level, those most widely used in proteomics are summarized below.

SDS-PAGE (sodium dodecyl sulphate-polyacrylamide gel electrophoresis). In SDS-PAGE proteins are dissolved in a loading buffer containing the anionic detergent SDS, thus forming negatively charged adducts that are loaded onto a polyacrylamide gel of a defined pore size. Proteins are then separated by electrophoretic migration according to their molecular weight (Fig. 1). SDS-PAGE is a technique compatible with a wide range of biochemical experimental approaches [2].

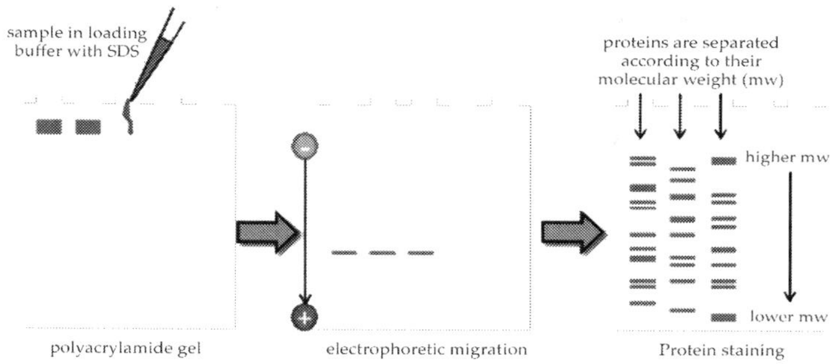

Figure 1. Schematization of protein separation by SDS-PAGE experiment.

2D-PAGE (two-dimensional polyacrylamide gel electrophoresis). In SDS-PAGE proteins are separated according to their molecular weight, thus according to one dimension. Their charge is uniformly negative due to the effect of SDS. In 2D-PAGE proteins are separated according to two dimensions, first by their charge (correctly, their isoelectric point, pI) and then by their molecular weight (Fig. 2).

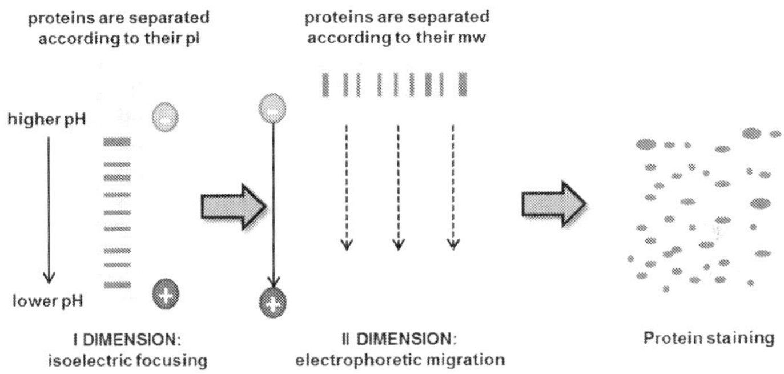

Figure 2. Schematization of protein separation by 2D-PAGE experiment.

For several years high-resolution 2D-PAGE has been one of the principal techniques used in proteomics allowing separating up to 10,000 protein spots along with isoforms and post-translational modified species [3, 4]. 2D fluorescence difference gel electrophoresis (2D-DIGE) is a variant of 2D-PAGE that uses fluorescent internal standard so that samples can be easily compared and quantitated [5].

Chromatographic techniques. Classical chromatographic approaches allow separating proteins according to several parameters, such as the size (size exclusion chromatography), charge (ionic exchange chromatography, cation or anionic), hydrophobicity (reverse phase chromatography) or affinity to particular ligands (affinity chromatography). Currently, highly efficient combined chromatographic approaches have been developed to improve proteomic analyses [6, 7].

Protein arrays. The array technology represents a high-throughput powerful method to characterize enriched subsets of proteins. Typically, arrays are constituted by immobilized molecules, such as antibodies, able to only bind the protein of interest. Then the array works likely ELISA technique: another conjugate antibody is used to detect by chemiluminescence the linked protein (Fig. 3).

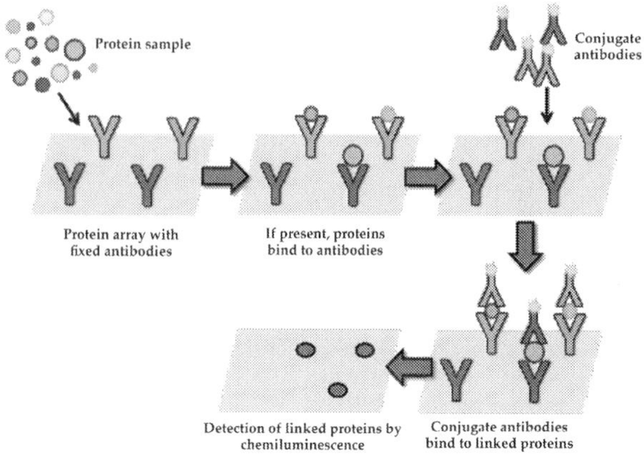

Figure 3. Schematization of working mechanism at the basis of protein arrays.

Protein arrays or microarrays (also known as protein chip) can detect at the same time thousands of proteins, including kinases, proteases, membrane receptors and many others. A variant of classical arrays include the possibility to fix antigens instead of antibodies, in order to detect in the sample of interest the presence of specific antibodies. These arrays have been widely used to evaluate food allergies [8].

Proteomics in food science

Proteomic approaches have been successfully applied in several fields of food science, including the analysis of food composition, the assessment of food authentication, quality and safety, the identification of food allergens, the evaluation of industrial processes on food properties and others.

Proteomics has contributed to gather detailed information about many foods, such as milk and dairy products, wine, meat, and fish.

Milk and dairy products are very common foods in human nutrition. Many products are based on bovine milk that has an enormous value worldwide. The detailed knowledge of bovine milk composition, along with milk from other species, is a reference for people working in the dairy sector allowing the prevention of food frauds and ensuring milk quality. Recently, the knowledge of milk proteome has been expanded by applying extensive fractionation steps including acidification, filtration, and centrifugation. With this approach milk proteome resulted increased by 366 new proteins [9]. Non-targeted proteomics has been also used to analyze differences present in the breast milk proteome between allergic and non-allergic mothers with the aim to clarify the relation breast milk composition-allergy [10]. Beyond its nutritive properties, milk proteins can act as an allergen. Cow's milk proteins (CMPs) are among the most common food allergens, affecting about 3% of children until three years of age [11]. In particular, α-casein and β-lactoglobulin are the most allergenic proteins in cow's milk. To reduce allergenic impact of milk several alternative products have been considered and their proteome has been compared even with the aim to found a valuable alternative for breast milk [12]. Proteomic methodologies have been also applied to monitor dairy products authentication, preventing frauds such as the undeclared addition of chemical substances or milk from other species.

Wine proteome is the entire set of proteins deriving both from grapevine and microorganisms typical of wine, such as yeasts and bacteria, thus resulting in a very complex system. Proteomics has already been used to analyze wine itself but also grape berries under different conditions, and wine yeast and bacteria. Various approaches have been adopted for analyzing white or red wine, accordingly to their different composition. For example, white wine proteomes have been extensively studied with combinatorial peptide ligand libraries (CPLL) able to distinguish proteins belonging to *Vitis vinifera*, as well as proteins from *S. cerevisiae,* commonly used in the fermentation process [13]. Moreover, CPLL have been used for harvesting and identifying low-amount fining agents such as casein in white wines [14]. *S. cerevisiae* is one of the most important industrially used yeast species in wine production. As a consequence, wine quality is strictly related to strains employed for fermentation, and a reliable method to compare and differentiate *S. cerevisiae* strains is a useful tool. Proteomic approach based on matrix-assisted-laser-desorption/ionization time-of-flight mass spectrometry (MALDI-TOF MS) can be particularly adapt, producing fingerprint easily comparable, as performed on 33 *S. cerevisiae* strains [15].

The use of proteomics to verify meat quality and safety has gained always more robustness and accuracy. For example, proteomics has been used monitor *in vivo* performances of livestock animals, but also to investigate molecular processes underlying meat quality, mainly related to the muscular post mortem events. These alterations in muscle proteome are extremely complex, reflecting several processes interconnected also able to affect meat tenderness, a meat feature crucial in determining quality. Simplifying, meat tenderness is mainly attributed to endogenous proteolytic enzymes interacting with other proteins such as glycolytic enzymes, heat shock proteins and structural component of muscles. Proteomic studies have demonstrated that meat tenderness in some species is associated with the accumulation of actin and myosin [16], but also mitochondrial proteins such as prohibitin, mitofilin and others [17]. In addition, proteomics is useful in industrial production of meat and cooked products. Recently, scandals of meat adulterations have been raised, and more attention has been devoted to technical improvements able to ensure meat quality. In this framework, besides classical approaches, proteomics has been used to monitor modified proteins, such as phosphorylated myosin regulatory light chain-2, alpha actin-1, triosephosphate isomerase and 14-3-3 protein epsilon, useful for evaluating beef quality [18].

Proteomics for food safety

In food biotechnology and processes it is of utmost importance to carefully monitor food safety in terms of microbial contaminations in the final product as well as during the whole production process. Detection, confirmation, and quantification of bacteria, other microorganisms and toxins in food are crucial analytical problems. There are several common dangerous bacteria and reliable and sensitive methods for their detection, even at low levels, are urgently required. Proteomics and genomics approaches have enough features to be considered as main sensitive and specific methods of choice.

Among commonly used products, seafood represents an important part able to affect people's health. Assessment of seafood authenticity and origin is crucial to prevent frauds and guarantee consumer's confidence in a transparent market. For example, the labeling process with intentional or unintentional mislabeling can be a sensible step for fraudulent substitution of fish species or undeclared addition of chemicals. With the discovery of food biomarker, proteomics can provide powerful tools to monitor food quality and safety. Several methodologies have been proposed to ensure authentication of fish species by using proteomics [19].

Allergenomics

The identification of food protein allergens by using proteomics-based approaches is named allergenomics. In recent years, proteomic analysis of food allergens has become an increasing field of interest mainly because food allergies represent a relevant problem in clinical medicine. Recently, several allergenomics studies and profiles have been proposed, including processed fish and seafood [20], plant [21], wheat [22], peanuts [23] and others.

have recently been performed, aiming at better understanding the cause of sensitization to cow's milk in breastfed infants and at assessing both the safety of food (e.g. transgenic) and in particular the allergenic properties.

The possibility to perform quantitative target analysis such as single- or multiple-reaction monitoring (SRM and MRM, respectively) MS with the use of isotope-labeled internal standard, offers absolute quantitation of food allergens even present in very low amounts, thus representing a novel proteomic strategy that could better manage food allergies [24].

This chapter belongs to a book illustrating principal proteomic applications in food science, reporting first basic principles of proteomics, and subsequently some cutting-edge instrumental developments, data treatment, and applications. It has been conceived as a compendium of the knowledge of proteomic applications in food science, and for a broad range of readers, including students, researchers, chemists and new-comers into the field.

REFERENCES

1. Zubarev, R.A., The challenge of the proteome dynamic range and its implications for in-depth proteomics. Proteomics, 2013. 13(5): p. 723-6.
2. Brunelle, J.L. and R. Green, One-dimensional SDS-polyacrylamide gel electrophoresis (1D SDS-PAGE). Methods Enzymol, 2014. 541: p. 151-9.
3. Klose J., Kobalz U. Two-dimensional electrophoresis of proteins: An updated protocol and implications for a functional analysis of the genome. Electrophoresis. 1995;16:1034–59.

4. Lohnes, K., et al., Combining high-throughput MALDI-TOF mass spectrometry and isoelectric focusing gel electrophoresis for virtual 2D gel-based proteomics. Methods, 2016. 104: p. 163-9.
5. Alban, A., David, S. O., Bjorkesten, L., Andersson, C. et al. A novel experimental design for comparative two-dimensional gel analysis: Two-dimensional difference gel electrophoresis incorporating a pooled internal standard. Proteomics 2003, 3, 36-44.
6. Valeja SG, Xiu L, Gregorich ZR et al. Three dimensional liquid chromatography coupling ion exchange chromatography/ hydrophobic interaction chromatography/reverse phase chromatography for effective protein separation in top-down proteomics. Anal Chem. 2015;87(10):5363-71.
7. Binai NA, Marino F, Soendergaard P et al. Rapid analyses of proteomes and interactomes using an integrated solid-phase extraction-liquid chromatography-MS/MS system. J Proteome Res. 2015 Feb 6;14(2):977-85.
8. Alcocer, MJC, Ares, SC, Lopez-Calleja I. Recent advances in food allergy. Braz. J. Food Technol. 2016, vol.19.
9. Nissen A, Bendixen E, Ingvartsen KL et al. Expanding the bovine milk proteome through extensive fractionation. J Dairy Sci. 2013;96(12):7854-66.
10. Hettinga KA, Reina FM, Boeren S et al. Difference in the breast milk proteome between allergic and non-allergic mothers. PLoS One. 2015 Mar 23;10(3):e0122234.
11. Sampson HA. Update on food allergy. J Allergy Clin Immunol, 113 (2004) 805-819.
12. D'Auria E, Agostoni C, Giovannini M et al. Proteomic evaluation of milk from different mammalian species as a substitute for breast milk. Acta Pædiatrica, 2005; 94: 1708–1713.
13. D'Amato A, Fasoli E, Kravchuk AV et al. Mehercules, adhuc Bacchus! The debate on wine proteomics continues. J Proteome Res. 2011 Aug 5;10(8):3789-801.
14. Cereda A, Kravchuka AV, D'Amato A et al. Proteomics of wine additives: Mining for the invisible via combinatorial peptide ligand libraries. J Proteom 2010; 73(9), 1732–1739.

15. Usbeck JC, Wilde C, Bertrand D et al. Wine yeast typing by MALDI-TOF MS. Appl Microbiol Biotechnol. 2014 Apr;98(8):3737-52.
16. Lametsch R, Karlsson A, Rosenvold K et al. Postmortem proteome changes of porcine muscle related to tenderness. J Agric Food Chem. 2003; 51, 6992-6997.
17. Laville E, Sayd T, Morzel M et al. Proteome changes during meat aging in tough and tender beef suggesting the importance of apoptosis and protein solubility for beef aging and tenderization. J Agric Food Chem. 2009; 57, 10755-10764.
18. Rodrigues RT de S, Chizzotti ML, Vital CE, et al. Differences in Beef Quality between Angus (*Bos taurus* taurus) and Nellore (*Bos taurus* indicus) Cattle through a Proteomic and Phosphoproteomic Approach. Loor JJ, ed. PLoS ONE. 2017;12(1):e0170294. doi:10.1371/journal.pone.0170294.
19. Mazzeo MF, Siciliano RA. Proteomics for the authentication of fish species. J Proteomics. 2016 Sep 16;147:119-24.
20. Abdel Rahman AM, Kamath SD, Gagné S et al. Comprehensive proteomics approach in characterizing and quantifying allergenic proteins from northern shrimp: toward better occupational asthma prevention. J Proteome Res. 2013 Feb 1;12(2):647-56.
21. Nakamura R, Teshima R. Proteomics-based allergen analysis in plants. J Proteomics. 2013 Nov 20;93:40-9.
22. Akagawa M, Handoyo T, Ishii T et al. Proteomic analysis of wheat flour allergens. J Agric Food Chem. 2007 Aug 22;55(17):6863-70.
23. Johnson PE, Sayers RL, Gethings LA et al. Quantitative Proteomic Profiling of Peanut Allergens in Food Ingredients Used for Oral Food Challenges. Anal Chem. 2016 Jun 7;88(11):5689-95.
24. Ahsan N, Rao RS, Gruppuso PA et al. Targeted proteomics: Current status and future perspectives for quantification of food allergens. J Proteomics. 2016 Jun 30;143:15-23.

SECTION I: PRINCIPLES OF PROTEOMICS

CHAPTER

1

PROTEOMIC CHALLENGES: SAMPLE PREPARATION TECHNIQUES FOR MICROGRAM-QUANTITY PROTEIN ANALYSIS FROM BIOLOGICAL SAMPLES

Peter Feist and Amanda B. Hummon

Department of Chemistry and Biochemistry, Integrated Biomedical Sciences Program, and the Harper Cancer Research Institute, 251 Nieuwland Science Hall, University of Notre Dame, Notre Dame, IN 46556, USA

ABSTRACT

Proteins regulate many cellular functions and analyzing the presence and abundance of proteins in biological samples are central focuses in proteomics. The discovery and validation of biomarkers, pathways, and drug targets for various diseases can be accomplished using mass spectrometry-based proteomics. However, with mass-limited samples like tumor biopsies, it can be challenging to obtain sufficient amounts of protcins to generate high-quality mass spectrometric data. Techniques developed for macroscale quantities recover sufficient amounts of protein from milligram quantities of starting material, but sample losses become crippling with these techniques when only microgram amounts of material are available. To combat this challenge, proteomicists have developed micro-scale techniques that are compatible with decreased sample size (100 μg or lower) and still enable excellent proteome coverage. Extraction, contaminant removal, protein quantitation, and sample handling techniques for the microgram protein range are reviewed here, with an emphasis on liquid chromatography and bottom-up mass spectrometry-compatible techniques. Also, a range of biological specimens, including mammalian tissues and model cell culture systems, are discussed.

Keywords: proteins; mass-spectrometry; mass-limited samples; microgram samples

INTRODUCTION

Proteins are essential cellular machinery, performing and enabling tasks within biological systems. The variety of proteins is extensive, and the role they occupy in biology is deep and complex; life depends on proteins. Each step of cellular generation, from replication of genetic material to cell senescence and death, relies on the correct function of several distinct proteins. The precision of cellular machinery can be disrupted, however, resulting in disease. Because much of the machinery essential to cell health and survival remains unknown, studying proteins is of great interest and importance. The field of proteomics is the large-scale study of proteins and the proteome and encompasses many techniques, such as immunoassays [1] and two-dimensional differential gel electrophoresis (2-D DIGE) [2,3]. Another group of methodologies that are growing in popularity for protein discovery and analyses are mass spectrometry-based approaches.

There are two main approaches for mass spectrometry-based proteomics, top–down and bottom–up analyses. Top–down methods analyze whole proteins; bottom–up approaches investigate the peptides from digested proteins. There are unique methods of analysis that each group has developed, but they share in common their mode of analysis. In mass spectrometric analysis, the mass-to-charge ratios (*m*/*z*) of molecular species are determined. By collecting this data, compounds in the sample can be identified by comparing against standard databases of compounds and molecules with known masses. From whole protein analysis in top–down proteomics, to peptide analysis in bottom–up proteomics, each particle measured has an m/z signature detectable by the mass spectrometer. By pairing mass analyzers and detectors, adding equipment in different configurations, and coupling separations and mass spectrometers together, there are virtually limitless possibilities, functionalities, and speeds of data acquisition for mass spectrometry-based proteomic analysis.

Mass spectrometry-based proteomics has advanced rapidly since the advent of "soft" ionization techniques, namely electrospray ionization (ESI) and matrix-assisted laser desorption ionization (MALDI) [4,5,6] in the late 1980's. Methods of detection previously used for organic chemicals and other sample identifications were adapted for proteomics. Several combinations of ionization sources and mass analyzers are available commercially, each with merits for specific applications. Time-of-flight (TOF) instruments are often coupled with MALDI instruments [7,8]. ESI-TOF instruments can provide high-speed, continuous measurements without compromising resolution. High-resolution Fourier Transform Ion Cyclotron Resonance (FT-ICR) mass spectrometers are costly, but provide unsurpassed data collection power using both electrospray and MALDI ion sources [9]. The invention and commercial distribution of the Orbitrap mass analyzer by Makarov and Thermo greatly increased the proteomic capabilities of mass spectrometry, combining higher sensitivity and improved mass resolution with lightweight benchtop instruments [10]. The high-field compact Orbitrap, introduced in 2011, provided a high mass resolution instrument at a lower cost than Fourier Transform Ion Cyclotron Resonance (FT-ICR) [11]. The most current Orbitrap instrument, the Orbitrap Fusion, has shown promising results in sensitivity and scan speed [12]. With high mass accuracy and specificity, combined with the selectivity of ion traps and quadrupoles, analysis of complex samples, even in small quantities, can be completed with increasing ease and confidence.

Many mass spectrometry methods are used to analyze proteins. Some methods are discovery-based, where samples are analyzed to determine what proteins are present in the sample. Often a high-resolution mass spectrometer is used for this purpose, as the false-discovery rate of protein identifications from peptides rely on highly accurate mass-to-charge measurements [13]. Some methods are targeted, focusing on single proteins of interest and quantifying them in different samples or sample fractions. Highly selective methods using ion traps and quadrupoles are ideal for targeted analysis [14,15,16]. Proteins and peptides can be fragmented in the mass spectrometer for tandem mass spectrometry in a variety of ways, and those fragments are analyzed for *de novo* peptide sequencing [17,18] or peptide mass fingerprinting [19,20,21,22]. The vast majority of peptide identifications are accomplished with fragmentation followed by protein database searches of the resulting fragments. Electron capture dissociation (ECD) [23], electron transfer dissociation (ETD) [24,25], higher energy collisional dissociation (HCD) [26,27], collision-induced dissociation (CID) [9,28,29], and a host of other fragmentation methods are available [30,31,32], each with recommended applications [33]. Furthermore, mass spectrometry methods are often customized within software.

The mass spectrometer is a critical aspect in proteomic experiments; however, the results obtained from the mass spectrometer are limited by the sample. Regardless of the analysis approach used, a high quality sample is critical for a successful experiment. Proteomic analyses depend on the sample containing proteins to analyze. Sample preparation approaches that are time-consuming, or worse, incur massive sample losses, are intolerable. This review will focus on sample preparation strategies for bottom–up mass spectrometry-based proteomics, with brief focus on obtaining samples for analysis.

MACROSCALE *VERSUS* MICROSCALE TECHNIQUES

Sample preparation for mass spectrometry-based proteomics has many options available when the sample is not limited. To grasp the need for microproteomics, an understanding of large-scale sample preparation is necessary. By understanding macroscale techniques, proteomicists can adapt their methods to suit microscale techniques. All of the macroscale techniques described here involve quantities of sample in the milligram to 100 microgram range, which do not necessitate as much care and precision as microscale samples, generally defined as samples less than 100 μg. Sample loss in the macroscale is not as pressing, for several reasons. The concentration of the sample is easier to maintain with larger quantities, leading to fewer losses in the processing steps. The overall percentage of lost volumes is lower, because milligram quantities of sample require larger volumes to fully solubilize. Milligram quantities can also be divided more easily while maintaining the majority of protein in the sample, including low-abundance species. Large quantities of protein from complex samples are also generally more diverse in their protein composition [8].

Obtaining a Sample for Proteomic Analysis

In order to obtain protein from a biological specimen, the sample must first be harvested from the organism, culture, or patient. Samples can be obtained by several methods. Traditional dissection from animal species, biopsies, blood draws, and additional methods can deliver adequate protein for analysis. The ethics of obtaining samples are not reviewed here, but guidelines are in place for human, animal, and cell culture procedures [34,35,36].

The proteins in the sample must be made readily accessible via lysis and extraction from the cells in the sample. There are numerous methods for lysis and extraction; lysis buffers and mechanical disruption strategies have been reviewed in the past [37,38]. Most common methods involve a chemical lysis and extraction agent, along with some mechanical stimulus that physically breaks apart the cell, allowing the chemical agent to solubilize the available protein [39]. A list of common detergents, including critical micellar concentration (CMC), can be found in Table 1 [40,41].

Table 1. Common detergents for mass spectrometry and cell lysis, organized by type. Adapted from Swiderek *et al.* [41] and Thermo-Pierce documentation.

Detergent Name	Type	Molecular Weight	CMC, mM	Mol. Weight (Micelle)	Suggested Removal
Triton X-100	Nonionic	647	0.24	90,000	TCA/Acetone
NP-40	Nonionic	617	0.29	90,000	Acetone
Tween 20	Nonionic	1228	0.06		Acetone
Tween 80	Nonionic	1310	0.01	76,000	Acetone

Octyl Glucoside	Nonionic	292	23–24	8000	Ethyl acetate
Octyl thioglucoside	Nonionic	308	9		Ethyl Acetate
Big CHAP	Nonionic	878	3–4	8781	Filtration
Deoxycholate	Anionic	415	2–6	2000	Acetone, TCA
Sodium Dodecyl Sulfate	Anionic	288	6–8	17,887	Filtration/FASP
CHAPS	Zwitter-ionic	615	8–10	6149	Filtration
CHAPSO	Zwitter-ionic	631	8–10	7000	Filtration

Highly efficient lysis and extraction has been achieved using several lysis buffer formulations, with mechanical perturbations ranging from gentle rocking or cell scraping to sonication or French pressing. The choice of lysis buffer and mechanical disruption depends on the protein target of extraction, sample size, and experience in preparation methods.

Lysis buffers can differ in critical micelle concentration (CMC). The CMC is the concentration at which the detergent forms micelles spontaneously, which can affect their efficacy and removal in different environments. Above this point, the detergent forms micelles, and detergent added will move directly into micelles. Higher CMC values are associated with weaker hydrophobic binding to monomers. Thus, higher CMC detergents tend to be more easily removed by buffer exchange and dialysis. Solutions with lower CMC values form micelles more easily, and generally require less detergent to effectively solubilize protein. Another factor that can affect a lysis buffer is the micelle molecular weight (MMW). Lower-weight micelles are more easily removed than higher-weight micelles. Making use of CMC and MMW, one can more easily determine the best course for the experiment. Choosing a lysis buffer depends greatly on these detergent factors. Most of the detergents listed are incompatible with downstream mass spectrometry analysis, and must be removed. There is no absolute "best way" to lyse a sample, as illustrated by the number of lysis and extraction protocols that exist in the literature; however, advantage lies with speed, sample retention, and cleanliness of preparation.

Lysis, extraction, and denaturation of protein can occur in the same step with certain procedures, such as with sodium dodecyl sulfate SDS while boiling and agitating the sample. Gutstein, *et al.* published an excellent review on microproteomic techniques describing methods for harvesting and lysis [8]. Bodzon-Kulakowska [42], Visser [43], and Hilbrig [44] published comprehensive reviews on sample preparation techniques with methods of cell disruption; Bodzon-Kulakowska has the most recent, large-scale review containing detailed analyses on the extraction of protein and subsequent removal of nearly all types of contaminants [42]. Visser includes classification of different techniques into "hard" conditions, which differ significantly from physiological conditions, and "soft" lysis, which are often used when the biological activity of the analyte needs preservation [43]. Hilbrig focuses on affinity precipitation of proteins [44].

Macroscale proteomic lysis techniques can efficiently extract protein from large amounts of sample. The issue with broadly applying a certain technique to a microscale sample is the removal of detergents and contaminants present in the lysate, which interfere with later steps. Catastrophic losses can occur from indiscriminate application of lysis techniques without proper planning and care. Microproteomic techniques focus on the efficient lysis of sample and removal of contaminants while retaining maximum sample. For example, the French Press has a large surface area where proteins can adsorb and be lost. This mode of lysis is not ideal with microgram quantities of sample. Instead, sonication in microcentrifuge tubes is be a better choice, so as to efficiently lyse cells, maximize sample concentration of protein, and thus minimize loss to surfaces.

Contaminant Removal

Once protein is extracted, removal of contaminants and detergents is necessary. Some detergents will interfere with enzymatic digestion, and most will interfere with reverse-phase separations and mass spectrometry, sometimes damaging instruments and irreversibly ruining columns [45]. Removal of unwanted cellular material, such as lipids and genomic DNA, prevents signal suppression, chromatographic interference, and presents a much cleaner, clearer spectrum from which to obtain protein identification data. One common approach for contaminant removal is precipitation. The uses of precipitations vary based on three main factors: The detergent or contaminants for removal; whether the proteins must be kept in a native or denatured state; and the post-processing analysis. Here, the focus is on bottom–up mass spectrometry-based proteomics; thus, detergent removal is a must, and the denaturation of the protein has little to no bearing. The post-processing analysis is singular, but there are numerous added steps to consider. Some common detergents can be removed using acetone precipitation, the classic precipitation technique in many proteomics and biochemical procedures [46]. Others can be removed using trichloroacetic acid precipitation [47,48,49]; chloroform-methanol mixture [50] or ethyl acetate [51,52] can remove contaminants with high yield, and most interferents can be eliminated using molecular weight cutoff filters to capture the protein in the sample. There are various other precipitation techniques that have been developed and compared for recovery and post-precipitation analysis in gels or for in-solution digestion (Table 2) [53,54,55]. TCA, chloroform-methanol, ethyl acetate, and acetone precipitation have similar efficiency for a wide variety of samples; chloroform-methanol has been found to work better for membrane proteins [49,56], while acetone precipitation sequesters mostly water-soluble proteins.

Table 2. Approaches for contaminant and detergent removal in proteomic samples.

Approach	Description	Reference
"Salting out"	Precipitation uses saturation of salt to precipitate protein from solution. Most commonly an ammonium sulfate precipitation, but also uses sodium sulfate.	[57,58,59]
Ultrafiltration	Centrifugation at high speed using molecular weight cutoff filter to remove contaminants; prominent in Filter-Aided Sample Preparation (FASP).	[57,60]
Polyethyleneimine (PEI)	Cationic polymer precipitates nucleic acids in 1 M NaCl, leaving proteins in the supernatant. PEI must be removed before further analysis.	[57,58,61,62]
Isoelectric Point (PI)	The pH of solution is adjusted with mineral acid to the isoelectric point of most proteins (pH 4–6). Neutral proteins will aggregate and precipitate.	[57,63,64]
Thermal	Cell extracts are denatured using heat; denatured proteins aggregate and precipitate, but stability is enhanced.	[57,65,66]
Nonionic polymer Polyethylene glycol (PEG)	Concentration of PEG unique to the protein mixture is added. Proteins precipitate based on an excluded volume principle. Centrifugation pellets the precipitated protein. PEG must be removed before mass spectrometry analysis.	[57,67,68,69]

Acetone precipitation is simple to perform. The standard procedure for acetone precipitation involves the addition of cold (0 to −20 °C) acetone to aqueous sample mixtures to a composition of 80%. Crowell, *et al.* examined the varying reports for acetone precipitation efficiency [70], which ranged from 50% to 100% in the literature [71,72,73,74,75]. After observing that removal of SDS from a protein solution of bovine serum albumin (BSA) caused a precipitous drop in efficiency, despite increasing concentration of BSA, acetone, and non-ionic detergent, they optimized conditions to improve yields from acetone precipitation (Figure 1). They found that the initial concentration of protein obtained during cell lysis is a major factor, as well as the percentage of acetone used and the ionic strength of the pre-precipitation solution. The ionic strength correlated with protein charge and dielectric conditions in the protein sample. Ideal conditions for near-quantitative to quantitative yield of protein were established to include addition of 80% acetone to nearly any protein mixture containing 1 to 100 mM of NaCl or similar salt. The salt concentration necessary for complex mixtures can vary widely, requiring some optimization for particular cell lysates.

Another mature precipitation technique is the methanol-chloroform precipitation. Wessel and Flugge introduced this efficient method of precipitation and concentration for dilute samples, especially those containing membrane proteins. They describe the procedure using a 4:1:3 ratio of methanol:chloroform:water, with an additional three volumes of methanol added to pellet the protein. Proteins that are only slightly soluble in methanol-chloroform collect on the water-organic interface, and are then pelleted by the additional methanol and centrifugation. The efficiency of this procedure approaches 100% for a variety of protein concentrations and detergent solutions. Methanol-chloroform precipitation has uses for bottom–up and top–down proteomics [58]. Ideally, precipitation would be avoided in microproteomics, because of the possibility of total sample loss. However, several factors discussed above that govern precipitation efficiency can be altered to produce successful precipitations, even in the microscale. These manipulations, especially as they relate to protein concentration, will be discussed in detail later in this manuscript.

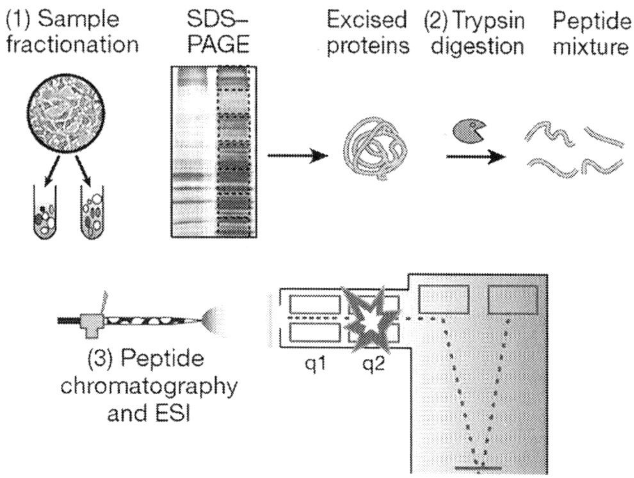

Figure 1. Acetone precipitation efficiency varies with salt concentration. The protein yield from an acetone precipitation in a complex mixture correlates with protein concentration, dielectric strength of the solution, and intrinsic protein charge. The overall model proposed is an ion-pairing model for organic solvents. With this protocol, efficiencies of 100% can be achieved for acetone precipitation. Figure adapted from Crowell, *et al.* (2013) [70].

Dissolving precipitated samples can be a challenge; as the protein precipitate is usually compacted into a pellet by centrifugation steps and the pellet is often dried to some extent. Exposure of the whole pellet to the dissolution buffer is necessary. Part of the precipitate can easily be left behind if thorough examination of the solution and container is not performed [76,77]. Strategies for fully suspending the pellet in any lysis buffer of choice include vigorous vortexing [58,59,78], sonication [58,79], shaking, and even two-step, on-pellet trypsin digestion [80]. In microscale, it is best to avoid vigorous agitation, because adsorption and loss occurs with increased exposure to surfaces; sonication avoids much of the splash-up that vortexing or shaking involves, but efficiently exposes the pellet to buffer.

Digestion Strategies

In-solution and in-gel digestion are two well-used approaches to prepare a bottom–up proteomic sample. In-solution digestion is the one of simplest and most powerful of the macroscale techniques in shotgun proteomics to perform. In-solution digestion involves denaturing, reducing, alkylating, and digesting the protein sample in the liquid phase, as opposed to in a gel or on a filter. In-solution digestion is extremely common and has been used with a variety of samples [81,82,83,84]. In-solution digestion can be performed using single-tube approaches, eliminating much of the sample loss that occurs during solution transfer between different vessels. Generally in-solution digests are fractionated after digestion, but fractionation can be performed previous to digestion using different forms of chromatography, including, but not limited to, strong and weak ion exchange, reverse-phase, and size exclusion chromatography.

Gel-based mass spectrometry analysis is widely used [85,86,87] as a first method of separation prior to LC–MS/MS analysis. Different methods of gel electrophoresis have been discussed in several reviews by Herbert [88], Lilley [3], Görg [89], Rabilloud [90,91], and others. Before the digestion, separation of the protein is performed using a gel. Basic overviews of gel-based mass spectrometry protocols can be seen in Figure 2 [9]. In one of the most common proteomic sample preparation strategies, a denaturing gel (sodium dodecyl sulfate in a polyacrylamide gel, SDS-PAGE) is used for bottom–up proteomics, as the protein will be cleaved into peptides in later steps [9,92].

The stained gel pieces are then excised from the gel, destained, and the protein within the gel piece is subjected to digestion. Several strategies are commonly used before eventual in-gel digestion, and there are many variations on one- and two-dimensional gel separations. One-dimensional gels are excellent for simple fractionation; proteins often have specific molecular weights, allowing for a semi-targeted approach. Less interference is involved in this approach, as the sample is dramatically simplified, so that only isobaric proteins will exist in the digest. SDS-PAGE is possibly the most widely used proteomic technique today [93], owing to the ability to separate thousands of proteins in a single sample from a complex mixture [94]. SDS-PAGE is also a relatively simple procedure, and the gel provides a good vehicle for the safe storage of proteins for future analyses [93].

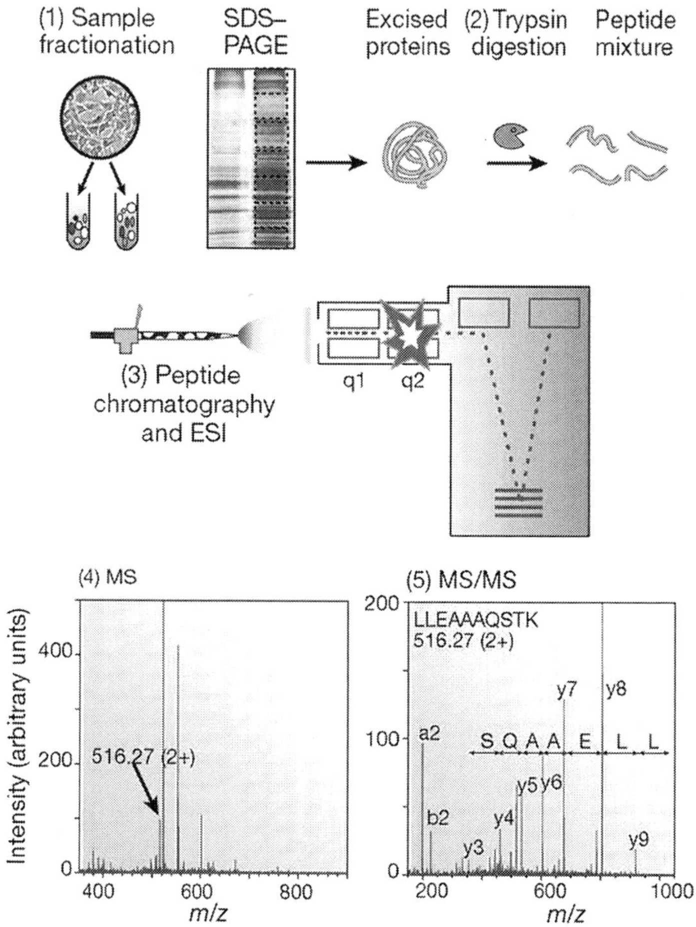

Figure 2. Typical workflow for gel-based mass spectrometry analysis. The gel is used to separate whole protein in one or two dimensions. After destaining, the proteins are excised from the gel and subjected to enzymatic proteolysis. Peptides can then be analyzed via mass spectrometry. Figure adapted from Aebersold *et al.*(2003) [9].

Two-dimensional (2D) gels add time to the protocol, but increase the selectivity of the process. 2D gels can use more than one orthogonal property of the protein for separation purposes; for example, isoelectric point (P_I) and molecular weight. One of the main differences between 1D and 2D gels is the fact that separation by P_I requires non-denaturing conditions. This condition means that common solubilization procedures using SDS, as well as use of other charge-coating detergents and running buffers, cannot be used in a first dimension isoelectric point focusing. Sample preparation then requires more thought and care. Rabilloud and Lelong have published a tutorial on 2D gel electrophoresis [95], covering the process, limitations, and advantages of the technique. 2D gels are run with different, orthogonal linear coordinates; the gel axes of the first dimension are rotated 90°, so that the proteins migrate on what was previously the horizontal axis, onto a second gel. 2D gel electrophoresis has been used for the fractionation and identification of many thousands of proteins in milligram sample amounts, as well as in the analysis of laser capture microdissection (LCM) samples [94,96].

In-gel digestion has several advantages over in-solution digestion. Using a gel, samples of greater complexity can be analyzed via mass spectrometry, because fractionation, and thus simplification, of the sample occurs on the gel. In-solution digests do not reduce the complexity of the sample without further pre-fractionation. Rabilloud and Lelong generated a cost-benefit comparison of gels *versus* mass spectrometry time [95]. Normal shotgun analysis on a mass spectrometer has a limited dynamic range and can only handle so much complexity, requiring more injections and mass spectrometry time to identify proteins of interest. Gels, used as a pre-fractionation procedure, can decrease the amount of mass spectrometry time needed to obtain the same or better information. Time on the mass spectrometer is generally more costly than even a 2D gel set-up. Gels are also useful for easing the removal of contaminants from the sample that may interfere with mass spectrometry analysis [97].

In-solution digests have the advantage of simplicity; fewer things can go wrong in the straightforward methods of in-solution digests. In-solution digests require less protein, because sample loss is often more severe for gel-based analysis. Sample recovery for gel-based procedures is estimated to be 70%–80% by Granvogl *et al.* [92]; Speicher *et al.* have a similar estimate [98]; extraction of peptides from a gel is inherently less efficient than an in-solution digestion [99], with estimates of 70%–80% of in-solution digest efficiency [93]. Sample complexity can be compensated using pre- or post-digestion fractionation for in-solution digests. Contaminants can be removed, at the expense of some sample loss. Concentration can be roughly controlled in solution, while the quantitative amount of protein from a gel is more difficult to ascertain; a rough estimate is possible based on staining [100].

The advantages of in-solution digestion and in-gel digestion have been combined in the form of spin-filter aided digestion protocols. Filter-aided digestion was originally developed by Manza, *et al.* [101,102] in 2005, and the acronym FASP (for Filter-Aided Sample Preparation) was coined by the Mann group in 2009 [103]. The process has now been widely derivatized for many mass spectrometry protocols and sample compositions [104,105,106]. As discussed by Wiśniewski and colleagues, the proteins in the sample are trapped in a high-molecular weight cutoff filter. Salts and low-molecular weight compounds flow through the filter and can be discarded. One key feature of FASP is the effective removal of detergents and contaminants from samples. The use of urea buffer enables the removal of nearly all (≥99.9%) of SDS from the sample; this is one of the unique identifiers separating Wiśniewski's FASP from Manza's spin-filter protocol [107]. After carboamidomethylation of the proteins, trypsin is added directly to the filter. The filter acts as a reactor for the trypsin digestion. Once on-filter trypsin digestion is accomplished, the peptides can be eluted in whatever volume of buffer necessary, and the filter retains any high-molecular weight interferents. FASP is a single-tube protocol, substituting aqueous- or organic-phase digestion for a solid-phase, reactor-based protocol. Sample losses have been variably reported [101,102,108,109]; however, whether these are technical variations or downfalls in the protocol are unknown. Nevertheless, the FASP protocol has been improved upon in recent years, with evidence of efficient low-microgram sample analysis [110]. In a comparison of bottom–up approaches by Weston *et al.*, FASP outperformed in-solution digestion techniques with a 66.2% identification rate, as opposed to 47.8% for the in-solution digest [111]. In our experience, individual protein bands become difficult to visualize and excise with traditional Coomassie Blue staining, especially with sample amounts less than forty micrograms of complex protein mixture. Silver staining methods have arisen that alleviate this problem [95]. The peptide recovery of in-gel digestion also affects its use in microscale, *versus* the higher recovery efficiency of FASP or in-solution digests.

Fractionation and Separation of Proteomic Mixtures

Various mass spectrometry and fractionation combinations have been developed or refined with the intent of delving deeply into the proteome of organisms and model systems. For the yeast proteome, the multi-dimensional protein identification technology was developed by the Yates research group [112]. Accurate mass tags (AMT) were developed in order to decrease the need for tandem mass spectrometry while providing more sensitive measurements and greater dynamic range [113]. High-performance liquid chromatography (HPLC) is a very common separation technique with a wide variety of stationary phases. With the advent of ultra-performance liquid chromatography (UPLC), chromatographic separations have increased both in resolving power and speed of separation. HPLC and UPLC function on the same principles. UPLC columns generally offer smaller particle sizes, resulting in decreased analyte path length and higher column pressures (10,000 pounds per square inch (PSI) or greater in maximum). Liquid Chromatography is often coupled with electrospray ionization and tandem mass spectrometers for both top–down and bottom–up proteomic studies. LC has been used for a staggering number of analyses [114,115]; LC–MS is a proteomic workhorse [116]. Liquid chromatography is robust, customizable based on the functionality of the stationary particles in the separation column.

For bottom–up proteomic analysis, the most common HPLC/UPLC stationary phase is the C18 reverse-phase column. The reverse-phase column uses the hydrophobicity of peptides for separation, utilizing a gradient from low to high organic-phase solvent. Acidified methanol and acetonitrile are commonly used as organic-phase, also known as "B" or "strong" solvents because of their miscibility with aqueous solutions. Acidified water is most often the "weak" solvent, also known as "A". Both buffers are acidified with the same acid, generally with formic acid or trifluoroacetic acid (TFA) at 0.1% or 0.01%, respectively. Formic acid is preferred over TFA, as TFA tends to form adducts and suppress signal [117,118]. While reverse-phase columns are very popular, many stationary phases are in use for proteomic work in both one- and two-dimensional separations, online and off-line. A separation strategy known as electrostatic repulsion hydrophilic interaction chromatography (ERLIC) has gained popularity for phosphoproteomic work, using adjustments in pH and volatile salts for gradient separations. As the name suggests, ERLIC uses the charge and hydrophilicity of peptides as a basis for separation. Typically ERLIC begins with a low-organic, high-pH gradient, moving to high organic and low-pH as the separation moves on. In this way, ERLIC elutes peptides in order of increasing hydrophobicity and acidity. ERLIC has proven effective at separating and identifying modified and unmodified proteins [119,120,121,122]. Smaller-diameter columns with lower loading capacities and smaller stationary phase particles offer an advantage in microproteomics. By increasing the local concentration of peptide and decreasing eddy diffusion, sample loading amounts can be minimized and still provide adequate peptide signal; the chromatographic resolution necessary for complex sample separation is not compromised.

MICROPROTEOMICS

Sample size and complexity are challenges in mass spectrometry. The sub-field of microproteomics focuses on improved analysis of microgram-quantities of samples through careful sample handling and increasing efficiency of processing. Depending on extraction techniques, tissue type, and cell density, 1 to 100 micrograms of protein can be obtained, processed with microproteomic techniques, and analyzed, with tissue remaining for use in other techniques. Alternatively, laser capture microdissection can be used to obtain samples for microproteomics post-microscopy analysis, providing thousands of cells for diagnostic analysis [94,96,123,124,125]. Complex samples are processed in a different manner than simple protein mixtures. Cells and tissue have a vast range of protein amounts, from 100 or less copies per cell to 10^6 copies for yeast [126]; it is conceivable that the dynamic range is greater in mammals [127,128]. Complex mixtures require fractionation prior to mass spectrometry analysis, while simple protein mixtures can be separated in line with the mass spectrometer. With micrograms of sample available, fractionation becomes a greater challenge. Successive sample losses can be crippling; dilution of sample in complex mixtures can mean losing low-abundance protein species below signal-to-noise cutoffs. Microproteomics techniques help ensure the maximum amount of sample possible, hopefully allowing for detection and identification of even low-abundance proteins in the sample.

As with macroscale analyses, obtaining sample is the first defining step of microproteomics analysis. Microproteomic techniques often rely on precise excision of tissue as much as preparing the sample with care from start to finish. Laser capture microdissection allows for the precise removal of several thousand cells or less from a fixed, microscopy sample. Other methods for obtaining sample are also viable. Wang *et al.* used flow cytometry to obtain 500 to 5000 cancer cells in their circulating tumor cell simulation [129]. Sun *et al.* used *Xenopus* embryos and ova, measuring roughly 1.2 mm in diameter [130]. Smaller organisms or parts of organisms can be homogenized or dissected to provide microproteomic samples as well. The lower limit on complex proteomic sample sizes has not been firmly established; descriptions of nanogram analyses have been reported [94,123,129].

Sensitivity and Microscale Analysis

Advances in instrumentation benefit the field of microproteomics, lowering the threshold of detection for less abundant proteins in samples, which would normally be lost with less sensitive instrumentation. Although Fourier Transform Ion Cyclotron Resonance (FT-ICR) instruments, which provide unmatched mass resolution and mass accuracy, are powerful tools, they are not ideal for microproteomics. These instruments are expensive and compromise scan speed as mass resolution is increased; the hybrid Orbitrap family is more routinely used [131]. The advent of the Orbitrap mass analyzer provided a sensitive benchtop alternative with speed and selectivity, furthering the utility of microproteomics and decreasing sample size requirements, in some cases down to single cell measurements [9]. Comparisons between the LTQ Orbitrap Velos, a linear ion trap-Orbitrap hybrid mass spectrometer, and the Q-Exactive, a hybrid quadrupole-Orbitrap system, show increases in protein identification and the superiority of each instrument in different modes of fragmentation on a reasonably complex proteomic sample. Sun *et al.* determined that the Q-Exactive outperformed the Orbitrap Velos over a range of loading amounts from 1000 to 1 ng on a Waters NanoAcquity UPLC system. For this system, it is also notable that the CID fragmentation mode was out-performed by HCD fragmentation on the Orbitrap Velos [132].

Increases in sensitivity are not always due to advances in technology. Meyer *et al.* reported an increase in electrospray sensitivity with the addition of low percentages (~5%) of dimethyl sulfoxide (DMSO) to separation buffer systems [133]. Upon investigation with an Orbitrap Elite system, Hahne *et al.* determined that an increase in electrospray efficiency and charge state reduction, due in part to the high proton affinity of DMSO in the gas phase, were major factors. With a charge-state reduction experiment, DMSO strips multiply charged species of protons, leaving more doubly charged peptides for detection, thus increasing signal and decreasing ion injection times [134]. While appealing, there are conflicting views on the use of DMSO in mass spectrometry [133,134,135,136], possibly due to conjugation product that form when DMSO sits for long periods of time, as well as the high boiling point of DMSO itself (189 °C). One adduct, dimethylsulfone, has a very high boiling point (238 °C) and thus does not ionize to gas phase in electrospray ionization. Accumulation of dimethyl sulfone as a jelly-like substance will damage a mass spectrometer or column, so freshly distilled DMSO or no DMSO may be preferable, despite possible gains.

Clean Sample Preparation

Microproteomics focuses on the separation, preparation and analysis of protein samples under 100 micrograms, which are very sensitive to losses in the proteomic workflow. With each transfer and processing step, protein can be lost. Reproducibility of sample preparation is key for reducing and troubleshooting sample loss in microproteomics. The concentration of protein plays a large role in sample loss. More concentrated samples tend to have less catastrophic losses in precipitation, desalting, and resuspension, due to the concentration-dependent adsorption maxima of many proteins. Select proteins in the sample will adsorb less to surfaces, such as beads, at higher concentrations compared to lower concentrations [137,138]. Higher concentration minimizes catastrophic sample loss by surpassing the adsorption maxima of proteins in the sample. Microproteomic techniques help ensure minimal sample loss before analysis, and maximal identification of peptides and protein groups during analysis, all while adhering to the strict purity standards necessary for mass spectrometric analysis. For example, some techniques use detergent-free methods of digestion, which lessens the cleanup steps involved in the protocol

[139,140,141]. Acetonitrile, ammonium bicarbonate, and Rapigest© have successfully been used in digestion protocols with microgram quantities of protein. Others have developed removal methods that eliminate the protein from solution without precipitation steps, such as the Filter-Aided Sample Preparation protocol. Although many microproteomicists prefer not to precipitate low quantities of samples due to risk of total loss, precipitation remains a viable and effective microproteomics technique.

Because precipitation does not provide 100% recovery in most cases, two main types of precipitations are favored for their ease of use and high recovery: Acetone precipitation and trichloroacetic acid (TCA) precipitation in deoxycholate. Acetone precipitation is a concentration-based procedure; higher concentrations of protein in the original solution result in higher recovery [142]. In an acetone precipitation, efficiency ranges from 50% to 100% recovery, with dependency on ionic strength of the solution as well as initial protein concentration. If acetone precipitation is to be used for microproteomics, it is ideal with lysis reagents, such as Nonidet P-40 (NP-40), which obtain good protein yields during extraction [40,129]. Even nanoscale analyses can be performed using acetone precipitation. Wang, *et al.* analyzed the proteome of 500 to 5000 cancer cells using NP-40 lysis buffer and acetone precipitation [129]. Using a commercial reverse-phase liquid chromatography separation, electrospray ionization, and quadrupole-time-of-flight (Q-TOF) mass spectrometer, an average of 619 ± 59 proteins were identified from 1.4 μg peptides obtained from 5000 cells; an average of 167 ± 21 proteins were identified from an estimated 140 ng peptides obtained from 500 cells. The interference from NP-40's polyethylene glycol units was not observed. Although protein losses occurred during the precipitation, significant numbers of proteins were identified from even the smallest sample amount. With the advances in sensitivity and throughput from newer mass spectrometers, these numbers could improve. Acetone precipitation also works relatively well for SDS removal, with about 100-fold removal [143]. Care must be taken, however, to remove enough SDS to prevent problems with chromatography, as small amounts of SDS can interfere chromatographically with reverse-phase separations [144]. This can be especially troublesome with low-quantity samples.

Quantitative protein precipitation using TCA is specific to deoxycholate and certain other detergents in solution; TCA can be used alone for precipitation, but the pellet is not readily dissolved, causing sample losses [59,145]. In a typical TCA-deoxycholate (DOC) precipitation, the protein solution is mixed with a dilute sodium deoxycholate buffer. The protein is intercalated by deoxycholate and then precipitated from solution with TCA. Acidification of the deoxycholate can cause a change in solubility [49,146], causing aggregation and precipitation. The deoxycholate can then be preferentially removed using acetone, leaving near-pure protein behind [47,48]. TCA-DOC precipitation is remarkably efficient, even in dilute samples, with recovery values from 90% to 100% of total protein [48].

Both the TCA-DOC and acetone precipitation methods have been adapted to the microscale and can recover the majority of protein input. Acetone precipitation is inherently more difficult to perform in the microscale compared to the macroscale, due to the dependence of its efficiency on protein concentration [70]. In either case, the protein is denatured and also separated from materials that interfere with liquid chromatography-mass spectrometry analysis, such as lipids, cellular debris, and non-compatible detergents. The pellet left behind can be dissolved in a buffer of choice. Care must be taken in dissolving precipitated samples, particularly when using TCA, as these pellets can be extremely difficult to re-suspend if over-dried, or if the proteins have limited solubility in the buffer of choice. Often dissolution is performed in a denaturing buffer or chaotropic agent compatible with mass spectrometry, or an agent easily diluted in the sample for the prevention of interference. An 8 M urea or 6 M urea with 2 M thiourea mixture, Rapigest©, ammonium bicarbonate, and 80% acetonitrile are common buffers for resuspension. Usually resuspension of a microproteomic sample is performed in small volumes, to keep protein concentration high and prevent losses, so resuspension techniques such as shaking or vortexing are discouraged in plastic tubes. The exposure of protein to the plastic walls may result in irreversible loss. Thus, sonication or on-pellet digestion are better approaches for dissolving precipitation pellets.

Filter-aided sample preparation (FASP) has also been adapted to microscale analysis. Sample preparation is similar to the macroscale techniques described. FASP is considered efficient for a wide range of proteins and preparations, and shows great efficacy for removal of detergents and contaminants without sample losses. FASP has been used for small numbers of cells (3000 and under) from formalin-fixed paraffin-embedded samples with LCM. Wisenewski *et al.*reported 2055 protein identifications from 3000 cells, containing nanograms worth of protein, with no interference from formalin or paraffin [110]. Protein can also be concentrated and desalted in few steps. FASP can be combined with other methods, such as with isobaric tags for relative and absolute quantitation (iTRAQ) [147], a multiplexed approach to protein quantitation, and antibody affinity selection [148].

FASP adapts well to the microscale due to the increase in local concentration of protein in relation to sample volume. Because the filter traps the proteins, it artificially increases the concentration. Recent advances by Erde, *et al.* suggests that the addition of a carrier such as Tween-20 to the molecular-weight cutoff filter before protein addition reduces losses 300-fold [104] (Figure 3). Also, instead of loading urea-solubilized peptides onto the filter unit, 0.2% deoxycholic acid can be used. This technique increases trypsin digestion efficiency, but necessitates cleanup by an additional method; the use of TCA-DOC precipitation post-FASP could be implemented for this use.

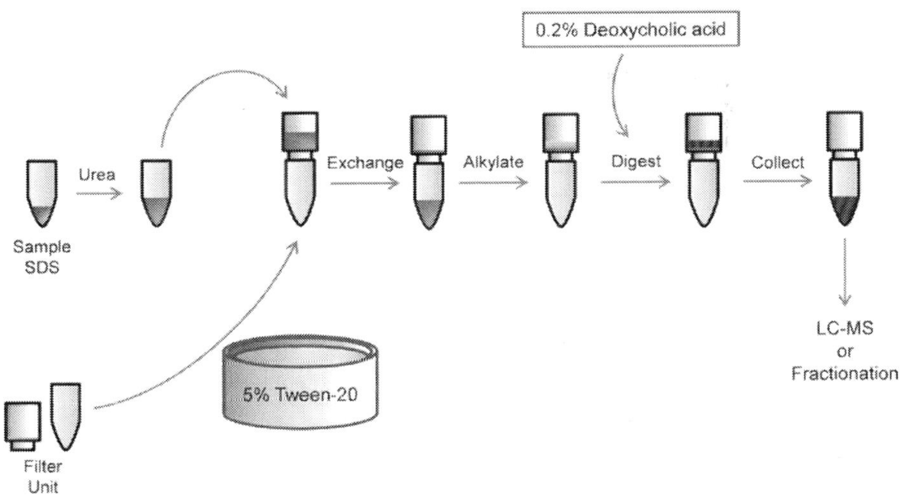

Figure 3. Enhanced filter-aided sample preparation (FASP) workflow. Samples are prepared in 4% sodium dodecyl sulfate (SDS) and diluted in 8 M urea to dissociate SDS from the proteins. Filter units are passivated overnight with 5% Tween-20, followed by thorough washing in MS-grade water. Diluted samples are applied to the filter units for buffer exchanges, eliminating contaminants. Proteins are alkylated with urea present, followed by successive buffer

exchanges. Proteins are digested with surfactants present, then liberated with centrifugation. Extraction with organic solvent leaves behind pure peptides for LC–MS analysis. This procedure reduces losses about 300-fold. Figure adapted from Erde *et al.* (2014) [104].

Microproteomic Fractionation and Separation

Loss of peptides due to lower peptide extraction efficiency can discourage the use of gel electrophoresis in microproteomic preparations, despite the raw power of the strategy. However, with stringent sample preparation strategies, gel electrophoresis can be used in microproteomics. Cha *et al.* describe the use of SDS-PAGE for the analysis of 60,000 breast cancer cells per specimen, obtained by laser capture microdissection [149]. After lysis with lithium dodecyl sulfate, proteins were separated by 1D SDS-PAGE. The gel lanes were divided into three sections based on molecular weight and subjected to in-gel digestion. This procedure provided three fractions per sample. Eighteen different samples, nine breast tumors and nine normal tissues, were analyzed. Overall, Cha *et al.* identified 2588 unique protein groups from the 18 samples, with only 3% of unique protein identifications, and about 78 protein groups, found in more than one sample. This study illustrates that, while loss is a real concern for in-gel digestion, useful information can be obtained from small sample amounts using 1D gel fractionation. Gel electrophoresis can be an excellent choice for microgram quantities of sample.

Other electrophoresis techniques exist that are compatible with microproteomics. Jorgenson and Lukacs introduced the concept of zone electrophoresis in silica capillaries [150]. Capillary zone electrophoresis (CZE) is perhaps best known for its use in the Human Genome Project [151]; the technique has grown on the fringe of proteomics [152] but is now gaining wider appeal and has been applied to proteomic separations with excellent results. Li *et al.* demonstrated the use of CZE with a sheath-flow interface for analysis of a moderately complex bacterial proteome [153]. This work has been further advanced by Sun *et al.* for more complex proteomes [154] and quantitation by multiple reaction monitoring on a human cell line, the first analysis of its kind [155]. Zhu *et al.* automated the process using a PrinCE autosampler for *Escherichia coli* digests [156]. Most impressively, Zhu *et al.* demonstrated the use of CZE for the analysis of single nanograms of material for proteomic analysis, identifying more peptides than an UPLC–ESI-MS/MS system [157]. The advantages of CZE lie in its speed and sensitivity, as compared to UPLC; a single separation consisted of 50 min of mass spectrometry time for a 60 centimeter capillary. Similar amounts of time were used for UPLC in this case, but most of the peptides for CZE had eluted by ~30 min for each loading amount. CZE showed much better sensitivity, resolving peptides at the 1 ng amount, where the UPLC showed little to no signal above background. The weakness of CZE is its low loading amount, due in part to the small volume of the 50 μm inner diameter (ID) capillary and zero peptide retention on the stationary phase. This analysis shows that CZE is a viable micro- to nano-proteomic separation technique, lowering sample requirements while retaining sensitivity and providing numerous peptide and protein identifications, albeit complementary to UPLC. CZE has also been applied to top-down proteomics as demonstrated by Li *et al.* [158] and Zhao *et al.* [159].

One Application of Microproteomics: Exploring Cancer Samples

Cancer is one of the leading causes of death in the United States, and the growth and developmental mechanisms of tumors are poorly understood [160]. Tumors have substantial cellular heterogeneity, making tumor biology a critical area of study. Microproteomic techniques are valuable tools to explore the complex proteomic differences within a single tumor [161]. Many of the approaches used in this review are techniques that can be applied to tumor analysis. Indeed, tumor biology and biomarker discovery are major driving forces for proteomic analysis, based on the number of publications on proteomic cancer analysis in recent years.

Proteomics is useful for tumor biology due to the breadth of protein information obtainable. Often, diagnosis and evaluation of tumors are done with histology and immunohistochemical analysis [162]. Biopsies are sliced, stained, and analyzed with microscopy. While accurate, precise, and mature, the throughput of this technique is relatively low. Immunostaining and histochemical staining are commonly used for cancer diagnosis, but they are limited techniques. Only one or two proteins can be visualized with traditional microscopy techniques. Tumors can also be biopsied using needle core or aspiration biopsies. The typical diameter of a needle core biopsy is about 1 mm across [163] and about the length of a grain of rice [164], limiting sample amount and thus the breadth of analyses that may be possible. However, microproteomics can bridge this gap, providing more methods for tumor analysis to complement mature techniques. Thanks to laser capture microdissection, both histology and mass spectrometry can be performed on the same sample. Diagnosis, proteome analysis, and network analysis can be consolidated, and microproteomics can help ensure maximum data from minimal material. Meanwhile, thousands of protein groups can be identified in a single mass spectrometry run. That information is then uploaded into network analysis databases, providing an in-depth look at a tumor's molecular equilibria. Mass spectrometry-based strategies can identify targets for cancer therapy. Analysis of pathways in cancer can not only give scientists and medical personnel insight into the workings of cancer; it can also give more immediate treatment options, possibly ruling out ineffective therapies or encouraging more productive, less deleterious chemotherapies.

Model systems for cancer can be used in place of primary tumor samples, which are precious biological samples. Two-dimensional (2D) cell culture and murine model are both useful model systems with unique advantages and disadvantages. 2D cell culture is least likely to be used for microproteomic analysis; great quantities of cells can be cultured in a single flask, resulting in milligram amounts of protein available. Furthermore, clonal lines can be grown in parallel and combined. For cancer researchers, this is an advantage. 2D cell culture is a high-throughput, relatively low-cost technique, at the expense of the model's accuracy. Murine models, while expensive, have more accurate representations of human tumors. Tumors obtained from mouse models are reproducible, yet unique to the individual organism. The overall size of tumors are smaller, providing less protein; coupled with other analyses, microproteomics is a likely candidate with the sample available. Further, even without direct tumor analysis, inferences and data can be obtained using tests similar to human techniques, such as blood and plasma [165,166]. 3D cell cultures recapitulate the tumor microenvironment to a high degree, but also retain many of the advantages of 2D cell culture. Deriving from clonal cell lines, 3D cultures ("spheroids") are reproducible in their growth patterns, but display intraculture chemical and cellular heterogeneity. Most often, 3D cultures are analyzed in bulk. We have found that, in certain cell lines, these cell cultures consistently provide around 40 μg of protein in a 1 mm HCT116 spheroid [167]. Analysis of single cultures under various conditions using microproteomic techniques is a viable next step in multicellular spheroid characterization.

CONCLUSIONS

The field of proteomics continues to present solutions to unique challenges of small sample analysis. With advances in technology and methodological breakthroughs, microproteomics pushes the boundaries of analysis down to the biological dynamic range. Progress into the lower reaches of biological activity requires greater reproducibility and specificity of techniques, to analyze specific subsets of tissue with greater confidence. Increases in instrument sensitivity and more compatible preparation techniques are necessary advancements for robust analysis of micro- and nano-scale samples, especially in the analysis of membrane proteins.

The future holds many options for the field of microproteomics. The development of methods for efficient removal of contaminants, such as FASP, provides paths to analyze single, fixed mammalian cells with minimal loss. While obligate losses exist, the analysis of single-cell proteomes remains a major goal. Microdissection using lasers has already miniaturized the dimensions of sample collection; separation techniques for UPLC or CZE need to follow this miniaturized approach to provide new workhorses with minimal sample dilution. Robotics may replace hand-held instruments to provide precise handling of specimens and samples, increasing speed of analysis and reproducibility of results. The field of microproteomics can build on the solid foundation already laid, and with some inventiveness and visionary techniques, the limitations of current techniques can be surmounted.

ACKNOWLEDGMENTS

We would like to thank the Mass Spectrometry and Proteomics Facility at the University of Notre Dame, and in particular, the advice of Dr. William Boggess. This report was supported by the National Science Foundation (CAREER Award CHE-1351595) and the National Institutes of Health (1R01GM110406-01) for Amanda B. Hummon. Peter Feist was supported by an Arthur J. Schmidt Presidential Fellowship. The Walther Cancer Foundation provided salary support for Amanda B. Hummon.

Author Contributions

Peter Feist and Amanda B. Hummon composed the manuscript.

Conflicts of Interest

The authors declare no conflict of interest.

REFERENCES

1. Wilson, R. Sensitivity and specificity: Twin goals of proteomics assays. Can they be combined? *Expert Rev. Proteomics* 2013, *10*, 135–149.
2. Tonge, R.; Shaw, J.; Middleton, B.; Rowlinson, R.; Rayner, S.; Young, J.; Pognan, F.; Hawkins, E.; Currie, I.; Davison, M. Validation and development of fluorescence two-dimensional differential gel electrophoresis proteomics technology. *Proteomics* 2001, *1*, 377–396.
3. Lilley, K.S.; Friedman, D.B. All about DIGE: Quantification technology for differential-display 2D-gel proteomics. *Expert Rev. Proteomics* 2004, *1*, 401–409.
4. Karas, M.; Hillenkamp, F. Laser desorption ionization of proteins with molecular masses exceeding 10,000 daltons. *Anal. Chem.* 1988, *60*, 2299–2301.
5. Tanaka, K.; Waki, H.; Ido, Y.; Akita, S.; Yoshida, Y.; Yoshida, T.; Matsuo, T. Protein and polymer analyses up to *m/z* 100,000 by laser ionization time-of-flight mass spectrometry. *Rapid Commun. Mass Spectrom.* 1988, *2*, 151–153.
6. Fenn, J.B.; Mann, M.; Meng, C.K.; Wong, S.F.; Whitehouse, C.M. Electrospray ionization for mass spectrometry of large biomolecules. *Science* 1989, *246*, 64–71.
7. Weaver, E.M.; Hummon, A.B. Imaging mass spectrometry: From tissue sections to cell cultures. *Adv. Drug Deliv. Rev.* 2013, *65*, 1039–1055.
8. Gutstein, H.B.; Morris, J.S.; Annangudi, S.P.; Sweedler,

J.V. Microproteomics: Analysis of protein diversity in small samples. *Mass Spectrom. Rev.* 2008, *27*, 316–330.

9. Aebersold, R.; Mann, M. Mass spectrometry-based proteomics. *Nature* 2003, *422*, 198–207.
10. Scigelova, M.; Makarov, A. Orbitrap mass analyzer—Overview and applications in proteomics. *Proteomics* 2006, *6*, 16–21.
11. Ahlf, D.R.; Compton, P.D.; Tran, J.C.; Early, B.P.; Thomas, P.M.; Kelleher, N.L. Evaluation of the compact high-field orbitrap for top-down proteomics of human cells. *J. Proteome Res.* 2012, *11*, 4308–4314.
12. Hebert, A.S.; Richards, A.L.; Bailey, D.J.; Ulbrich, A.; Coughlin, E.E.; Westphall, M.S.; Coon, J.J. The one hour yeast proteome. *Mol. Cell. Proteomics* 2014, *13*, 339–347.
13. Mann, M.; Kelleher, N.L. Precision proteomics: The case for high resolution and high mass accuracy. *Proc. Natl. Acad. Sci. USA* 2008, *105*, 18132–18138.
14. Stahl-Zeng, J.; Lange, V.; Ossola, R.; Eckhardt, K.; Krek, W.; Aebersold, R.; Domon, B. High sensitivity detection of plasma proteins by multiple reaction monitoring of *N*-glycosites. *Mol. Cell. Proteomics* 2007, *6*, 1809–1817.
15. Lin, S.; Shaler, T.A.; Becker, C.H. Quantification of intermediate-abundance proteins in serum by multiple reaction monitoring mass spectrometry in a single-quadrupole ion trap. *Anal. Chem.* 2006, *78*, 5762–5767.
16. March, R.E.; Todd, J.F.J. *Practical Aspects of Trapped Ion Mass Spectrometry, Volume IV: Theory and Instrumentation*; CRC Press: Boca Raton, FL, USA, 2010; pp. 61–64.
17. Dancík, V.; Addona, T.A.; Clauser, K.R.; Vath, J.E.; Pevzner, P.A. *De novo* peptide sequencing via tandem mass spectrometry. *J. Comput. Biol.* 1999, *6*, 327–342.
18. Medzihradszky, K.F.; Chalkley, R.J. Lessons in *de novo* peptide sequencing by tandem mass spectrometry. *Mass Spectrom. Rev.* 2015, *34*, 43–63.
19. Thiede, B.; Höhenwarter, W.; Krah, A.; Mattow, J.; Schmid, M.; Schmidt, F.; Jungblut, P.R. Peptide mass fingerprinting. *Methods*

(San Diego Calif.) 2005, *35*, 237–247.

20. Webster, J.; Oxley, D. Peptide mass fingerprinting: Protein identification using MALDI-TOF mass spectrometry. *Methods Mol. Biol. (Clifton N.J.)* 2005, *310*, 227–240.
21. Pappin, D.J.; Hojrup, P.; Bleasby, A.J. Rapid identification of proteins by peptide-mass fingerprinting. *Curr. Biol.* 1993, *3*, 327–332.
22. Sommerer, N.; Centeno, D.; Rossignol, M. Peptide mass fingerprinting. In *Plant Proteomics*; Thiellement, H., Zivy, M., Damerval, C., Méchin, V., Eds.; Humana Press: New York, NY, USA, 2007; pp. 219–234.
23. Mao, Y.; Valeja, S.G.; Rouse, J.C.; Hendrickson, C.L.; Marshall, A.G. Top-down structural analysis of an intact monoclonal antibody by electron capture dissociation-Fourier transform ion cyclotron resonance-mass spectrometry. *Anal. Chem.* 2013, *85*, 4239–4246.
24. Syka, J.E.P.; Coon, J.J.; Schroeder, M.J.; Shabanowitz, J.; Hunt, D.F. Peptide and protein sequence analysis by electron transfer dissociation mass spectrometry. *Proc. Natl. Acad. Sci. USA* 2004, *101*, 9528–9533.
25. Zhurov, K.O.; Fornelli, L.; Wodrich, M.D.; Laskay, Ü.A.; Tsybin, Y.O. Principles of electron capture and transfer dissociation mass spectrometry applied to peptide and protein structure analysis. *Chem. Soc. Rev.* 2013, *42*, 5014–5030.
26. Jedrychowski, M.P.; Huttlin, E.L.; Haas, W.; Sowa, M.E.; Rad, R.; Gygi, S.P. Evaluation of HCD- and CID-type fragmentation within their respective detection platforms for murine phosphoproteomics. *Mol. Cell. Proteomics*2011, *10*.
27. Michalski, A.; Neuhauser, N.; Cox, J.; Mann, M. A systematic investigation into the nature of tryptic HCD spectra. *J. Proteome Res.* 2012, *11*, 5479–5491.
28. Papayannopoulos, I.A. The interpretation of collision-induced dissociation tandem mass spectra of peptides. *Mass Spectrom. Rev.* 1995, *14*, 49–73.
29. Medzihradszky, K.F.; Campbell, J.M.; Baldwin, M.A.; Falick, A.M.; Juhasz, P.; Vestal, M.L.; Burlingame, A.L. The characteristics of peptide collision-induced dissociation using a high-

performance MALDI-TOF/TOF tandem mass spectrometer. *Anal. Chem.* 2000, *72*, 552–558.

30. O'Brien, J.P.; Brodbelt, J.S. Structural characterization of gangliosides and glycolipids via ultraviolet photodissociation mass spectrometry. *Anal. Chem.* 2013, *85*, 10399–10407.
31. Brodbelt, J.S. Photodissociation mass spectrometry: New tools for characterization of biological molecules. *Chem. Soc. Rev.* 2014, *43*, 2757–2783.
32. Keough, T.; Youngquist, R.S.; Lacey, M.P. A method for high-sensitivity peptide sequencing using postsource decay matrix-assisted laser desorption ionization mass spectrometry. *Proc. Natl. Acad. Sci. USA* 1999, *96*, 7131–7136.
33. Shen, Y.; Tolić, N.; Xie, F.; Zhao, R.; Purvine, S.O.; Schepmoes, A.A.; Moore, R.J.; Anderson, G.A.; Smith, R.D. Effectiveness of CID, HCD, and ETD with FT MS/MS for degradomic-peptidomic analysis: Comparison of peptide identification methods. *J. Proteome Res.* 2011, *10*, 3929–3943.
34. Scott, C.T.; Caulfield, T.; Borgelt, E.; Illes, J. Personal medicine—The new banking crisis. *Nat. Biotechnol.* 2012, *30*, 141–147.
35. Guidelines for the Use of Human Tissue for Future Unspecified Research Purposes. Available online: http://www.health.govt.nz/publication/guidelines-use-human-tissue-future-unspecified-research-purposes-0 (accessed on 31 October 2014).
36. Salvaterra, E.; Lecchi, L.; Giovanelli, S.; Butti, B.; Bardella, M.T.; Bertazzi, P.A.; Bosari, S.; Coggi, G.; Coviello, D.A.; Lalatta, F.; *et al.* Banking together. A unified model of informed consent for biobanking. *EMBO Rep.* 2008, *9*, 307–313.
37. Raynie, D.E. Modern extraction techniques. *Anal. Chem.* 2010, *82*, 4911–4916.
38. Cañas, B.; Piñeiro, C.; Calvo, E.; López-Ferrer, D.; Gallardo, J.M. Trends in sample preparation for classical and second generation proteomics. *J. Chromatogr.* 2007, *1153*, 235–258.
39. Traditional Methods of Cell Lysis. Available online: http://www.piercenet.com/method/traditional-methods-cell-lysis (accessed on 31 October 2014).

40. Detergents for Cell Lysis and Protein Extraction. Available online: http://www.piercenet.com/method/detergents-cell-lysis-protein-extraction (accessed on 31 October 2014).
41. Swiderek, K.M.; Alpert, A.J.; Heckendorf, A.; Nugent, K.; Patterson, S.D. Structural analysis of proteins and peptides in the presence of detergents: Tricks of the trade. *ABRF News Artic. Methodol.* 1997, *8*, 17–25.
42. Bodzon-Kulakowska, A.; Bierczynska-Krzysik, A.; Dylag, T.; Drabik, A.; Suder, P.; Noga, M.; Jarzebinska, J.; Silberring, J. Methods for samples preparation in proteomic research. *J. Chromatogr.* 2007, *849*, 1–31.
43. Visser, N.F.C.; Lingeman, H.; Irth, H. Sample preparation for peptides and proteins in biological matrices prior to liquid chromatography and capillary zone electrophoresis. *Anal. Bioanal. Chem.* 2005, *382*, 535–558.
44. Hilbrig, F.; Freitag, R. Protein purification by affinity precipitation. *J. Chromatogr.* 2003, *790*, 79–90.
45. Zhou, J.-Y.; Dann, G.P.; Shi, T.; Wang, L.; Gao, X.; Su, D.; Nicora, C.D.; Shukla, A.K.; Moore, R.J.; Liu, T.; *et al.* Simple sodium dodecyl sulfate-assisted sample preparation method for LC–MS-based proteomics applications. *Anal. Chem.* 2012, *84*, 2862–2867.
46. Buxton, T.B.; Crockett, J.K.; Moore, W.L.; Moore, W.L.; Rissing, J.P. Protein precipitation by acetone for the analysis of polyethylene glycol in intestinal perfusion fluid. *Gastroenterology* 1979, *76*, 820–824.
47. Peterson, G.L. A simplification of the protein assay method of Lowry *et al.* which is more generally applicable. *Anal. Biochem.* 1977, *83*, 346–356.
48. Bensadoun, A.; Weinstein, D. Assay of proteins in the presence of interfering materials. *Anal. Biochem.* 1976, *70*, 241–250.
49. Arnold, U.; Ulbrich-Hofmann, R. Quantitative protein precipitation from guanidine hydrochloride-containing solutions by sodium deoxycholate/trichloroacetic acid. *Anal. Biochem.* 1999, *271*, 197–199.
50. Wessel, D.; Flügge, U.-I. A method for the quantitative recovery of protein in dilute solution in the presence of detergents and

lipids. *Anal. Biochem.* 1984, *138*, 141–143.

51. Yeung, Y.-G.; Stanley, E.R. Rapid detergent removal from peptide samples with ethyl acetate for mass spectrometry analysis. *Curr. Protoc. Protein Sci.* 2010.
52. Yeung, Y.-G.; Nieves, E.; Angeletti, R.H.; Stanley, E.R. Removal of detergents from protein digests for mass spectrometry analysis. *Anal. Biochem.* 2008, *382*, 135–137.
53. Levin, B.; Oberholzer, V.G.; Whitehead, T.P. Serum protein fractions: A comparison of precipitation methods with electrophoresis. *J. Clin. Pathol.* 1950, *3*, 260.
54. Fic, E.; Kedracka-Krok, S.; Jankowska, U.; Pirog, A.; Dziedzicka-Wasylewska, M. Comparison of protein precipitation methods for various rat brain structures prior to proteomic analysis. *Electrophoresis* 2010, *31*, 3573–3579.
55. Rabilloud, T. Detergents and chaotropes for protein solubilization before two-dimensional electrophoresis. *Methods Mol. Biol. (Clifton N.J.)* 2009, *528*, 259–267.
56. Chapel, A.; Salvi, D.; Garin, J.; Joyard, J.; de Biologie, D.; et Structurale, M.; VØgØtale, C.; Fourier, U.J. Organic solvent extraction as a versatile procedure to identify hydrophobic chloroplast membrane proteins. *Electrophoresis*2000, *21*, 3517–3526.
57. Englard, S.; Seifter, S. Precipitation techniques. *Methods Enzymol.* 1990, *182*, 285–300.
58. Burgess, R.R. Protein precipitation techniques. *Methods Enzymol.* 2009, *463*, 331–342.
59. Jiang, L.; He, L.; Fountoulakis, M. Comparison of protein precipitation methods for sample preparation prior to proteomic analysis. *J. Chromatogr.* 2004, *1023*, 317–320.
60. Wiśniewski, J.R.; Zielinska, D.F.; Mann, M. Comparison of ultrafiltration units for proteomic and *N*-glycoproteomic analysis by the filter-aided sample preparation method. *Anal. Biochem.* 2011, *410*, 307–309.
61. Gupta, M.N.; Kaul, R.; Guoqiang, D.; Dissing, U.; Mattiasson, B. Affinity precipitation of proteins. *J. Mol. Recognit.*1996, *9*, 356–

359.

62. Holler, C.; Vaughan, D.; Zhang, C. Polyethyleneimine precipitation *versus* anion exchange chromatography in fractionating recombinant β-glucuronidase from transgenic tobacco extract. *J. Chromatogr.* 2007, *1142*, 98–105.
63. Hegg, P.O. Precipitation of egg white proteins below their isoelectric points by sodium dodecyl sulphate and temperature. *Biochim. Biophys. Acta* 1979, *579*, 73–87.
64. Jaffé, W.G. A simple method for the approximate estimation of the isoelectric point of soluble proteins. *J. Biol. Chem.* 1943, *148*, 185–186.
65. Fan, J.; Huang, B.; Wang, X.; Zhang, X.C. Thermal precipitation fluorescence assay for protein stability screening. *J. Struct. Biol.* 2011, *175*, 465–468.
66. Hill, A.R.; Irvine, D.M. Effects of pH on the thermal precipitation of proteins in acid and sweet cheese wheys. *Can. Inst. Food Sci. Technol. J.* 1988, *21*, 386–389.
67. Ingham, K.C. Precipitation of proteins with polyethylene glycol. *Methods Enzymol.* 1990, *182*, 301–306.
68. Ingham, K.C. Protein precipitation with polyethylene glycol. *Methods Enzymol.* 1984, *104*, 351–356.
69. Sim, S.-L.; He, T.; Tscheliessnig, A.; Mueller, M.; Tan, R.B.H.; Jungbauer, A. Protein precipitation by polyethylene glycol: A generalized model based on hydrodynamic radius. *J. Biotechnol.* 2012, *157*, 315–319.
70. Crowell, A.M.J.; Wall, M.J.; Doucette, A.A. Maximizing recovery of water-soluble proteins through acetone precipitation. *Anal. Chim. Acta* 2013, *796*, 48–54.
71. Barritault, D.; Expert-Bezançon, A.; Guérin, M.-F.; Hayes, D. The use of acetone precipitation in the isolation of ribosomal proteins. *Eur. J. Biochem.* 1976, *63*, 131–135.
72. Puchades, M.; Westman, A.; Blennow, K.; Davidsson, P. Analysis of intact proteins from cerebrospinal fluid by matrix-assisted laser desorption/ionization mass spectrometry after two-dimensional liquid-phase electrophoresis. *Rapid Commun. Mass*

Spectrom. 1999, *13*, 2450–2455.

73. Thongboonkerd, V.; Mcleish, K.R.; Arthur, J.M.; Klein, J.B. Proteomic analysis of normal human urinary proteins isolated by acetone precipitation or ultracentrifugation. *Kidney Int.* 2002, *62*, 1461–1469.
74. Srivastava, O.P.; Srivastava, K. Purification of gamma-crystallin from human lenses by acetone precipitation method. *Curr. Eye Res.* 1998, *17*, 1074–1081.
75. Von Hagen, J. *Proteomics Sample Preparation*; John Wiley & Sons Inc.: Hoboken, NJ, USA, 2011.
76. Wu, X.; Xiong, E.; Wang, W.; Scali, M.; Cresti, M. Universal sample preparation method integrating trichloroacetic acid/acetone precipitation with phenol extraction for crop proteomic analysis. *Nat. Protoc.* 2014, *9*, 362–374.
77. Chevallet, M.; Diemer, H.; van Dorssealer, A.; Villiers, C.; Rabilloud, T. Toward a better analysis of secreted proteins: The example of the myeloid cells secretome. *Proteomics* 2007, *7*, 1757–1770.
78. Robinson, P.J.; Liu, J.P.; Chen, W.; Wenzel, T. Activation of protein kinase C *in vitro* and in intact cells or synaptosomes determined by acetic acid extraction of MARCKS. *Anal. Biochem.* 1993, *210*, 172–178.
79. Isaacson, T.; Damasceno, C.M.B.; Saravanan, R.S.; He, Y.; Catalá, C.; Saladié, M.; Rose, J.K.C. Sample extraction techniques for enhanced proteomic analysis of plant tissues. *Nat. Protoc.* 2006, *1*, 769–774.
80. Duan, X.; Young, R.; Straubinger, R.M.; Page, B.; Cao, J.; Wang, H.; Yu, H.; Canty, J.M.; Qu, J. A straightforward and highly efficient precipitation/on-pellet digestion procedure coupled with a long gradient nano-LC separation and orbitrap mass spectrometry for label-free expression profiling of the swine heart mitochondrial proteome. *J. Proteome Res.* 2009, *8*, 2838–2850.
81. De Godoy, L.M.F.; Olsen, J.V.; Cox, J.; Nielsen, M.L.; Hubner, N.C.; Fröhlich, F.; Walther, T.C.; Mann, M. Comprehensive mass-spectrometry-based proteome quantification of haploid *versus* diploid yeast. *Nature* 2008, *455*, 1251–1254.
82. Li, N.; Shaw, A.R.E.; Zhang, N.; Mak, A.; Li, L. Lipid raft

proteomics: Analysis of in-solution digest of sodium dodecyl sulfate-solubilized lipid raft proteins by liquid chromatography-matrix-assisted laser desorption/ionization tandem mass spectrometry. *Proteomics* 2004, *4*, 3156–3166.

83. De Souza, G.A.; Godoy, L.M.F.; Mann, M. Identification of 491 proteins in the tear fluid proteome reveals a large number of proteases and protease inhibitors. *Genome Biol.* 2006, *7*, R72.
84. Go, E.P.; Rebecchi, K.R.; Desaire, H. In-solution digestion of glycoproteins for glycopeptide-based mass analysis. *Methods Mol. Biol. (Clifton N.J.)* 2013, *951*, 103–111.
85. Lasonder, E.; Ishihama, Y.; Andersen, J.S.; Vermunt, A.M.W.; Pain, A.; Sauerwein, R.W.; Eling, W.M.C.; Hall, N.; Waters, A.P.; Stunnenberg, H.G.; *et al.* Analysis of the *Plasmodium falciparum* proteome by high-accuracy mass spectrometry. *Nature* 2002, *419*, 537–542.
86. Pomastowski, P.; Buszewski, B. Two-dimensional gel electrophoresis in the light of new developments. *TrAC Trends Anal. Chem.* 2014, *53*, 167–177.
87. Shevchenko, A.; Wilm, M.; Vorm, O.; Mann, M. Mass spectrometric sequencing of proteins from silver-stained polyacrylamide gels. *Anal. Chem.* 1996, *68*, 850–858.
88. Herbert, B. Advances in protein solubilisation for two-dimensional electrophoresis. *Electrophoresis* 1999, *20*, 660–663.
89. Görg, A.; Weiss, W.; Dunn, M.J. Current two-dimensional electrophoresis technology for proteomics. *Proteomics*2004, *4*, 3665–3685.
90. Rabilloud, T. Two-dimensional gel electrophoresis in proteomics: Old, old fashioned, but it still climbs up the mountains. *Proteomics* 2002, *2*, 3–10.
91. Rabilloud, T.; Chevallet, M.; Luche, S.; Lelong, C. Two-dimensional gel electrophoresis in proteomics: Past, present and future. *J. Proteomics* 2010, *73*, 2064–2077.
92. Granvogl, B.; Plöscher, M.; Eichacker, L.A. Sample preparation by in-gel digestion for mass spectrometry-based proteomics. *Anal. Bioanal. Chem.* 2007, *389*, 991–1002.
93. Shevchenko, A.; Tomas, H.; Havlis, J.; Olsen, J.V.; Mann, M. In-gel

digestion for mass spectrometric characterization of proteins and proteomes. *Nat. Protoc.* 2006, *1*, 2856–2860.

94. Thakur, D.; Rejtar, T.; Wang, D.; Bones, J.; Cha, S.; Clodfelder-Miller, B.; Richardson, E.; Binns, S.; Dahiya, S.; Sgroi, D.; *et al.* Microproteomic analysis of 10,000 laser captured microdissected breast tumor cells using short-range sodium dodecyl sulfate-polyacrylamide gel electrophoresis and porous layer open tubular liquid chromatography tandem mass spectrometry. *J. Chromatogr.* 2011, *1218*, 8168–8174.
95. Rabilloud, T.; Lelong, C. Two-dimensional gel electrophoresis in proteomics: A tutorial. *J. Proteomics* 2011, *74*, 1829–1841.
96. Craven, R.A.; Totty, N.; Harnden, P.; Selby, P.J.; Banks, R.E. Laser capture microdissection and two-dimensional polyacrylamide gel electrophoresis. *Am. J. Pathol.* 2002, *160*, 815–822.
97. Weiner, A.M.; Platt, T.; Weber, K. Amino-terminal sequence analysis of proteins purified on a nanomole scale by gel electrophoresis. *J. Biol. Chem.* 1972, *247*, 3242–3251.
98. Speicher, K.; Kolbas, O.; Harper, S.; Speicher, D. Systematic analysis of peptide recoveries from in-gel digestions for protein identifications in proteome studies. *J. Biomol. Tech.* 2000, *11*, 74–86.
99. Gundry, R.L.; White, M.Y.; Murray, C.I.; Kane, L.A.; Fu, Q.; Stanley, B.A.; van Eyk, J.E. Preparation of proteins and peptides for mass spectrometry analysis in a bottom-up proteomics workflow. *Curr. Protoc. Mol. Biol.* 2009.
100. Zhou, S.; Bailey, M.J.; Dunn, M.J.; Preedy, V.R.; Emery, P.W. A quantitative investigation into the losses of proteins at different stages of a two-dimensional gel electrophoresis procedure. *Proteomics* 2005, *5*, 2739–2747.
101. Manza, L.L.; Stamer, S.L.; Ham, A.-J.L.; Codreanu, S.G.; Liebler, D.C. Sample preparation and digestion for proteomic analyses using spin filters. *Proteomics* 2005, *5*, 1742–1745.
102. Wang, X. *Bioinformatics of Human Proteomics*; Springer Science + Business Media B.V.: Dordrecht, The Netherlands, 2013.
103. Wiśniewski, J.R.; Zougman, A.; Nagaraj, N.; Mann, M. Universal sample preparation method for proteome analysis. *Nat.*

Methods 2009, *6*, 359–362.

104. Erde, J.; Loo, R.R.O.; Loo, J.A. Enhanced FASP (eFASP) to increase proteome coverage and sample recovery for quantitative proteomic experiments. *J. Proteome Res.* 2014, *13*, 1885–1895.

105. Zielinska, D.F.; Gnad, F.; Wiśniewski, J.R.; Mann, M. Precision mapping of an *in vivo* *N*-glycoproteome reveals rigid topological and sequence constraints. *Cell* 2010, *141*, 897–907.

106. Kanshin, E.; Michnick, S.; Thibault, P. Sample preparation and analytical strategies for large-scale phosphoproteomics experiments. *Semin. Cell Dev. Biol.* 2012, *23*, 843–853.

107. Wiśniewski, J.R.; Mann, M. Reply to "Spin filter-based sample preparation for shotgun proteomics". *Nat. Methods*2009, *6*, 785–786.

108. Eggler, A.L.; Luo, Y.; van Breemen, R.B.; Mesecar, A.D. Identification of the highly reactive cysteine 151 in the chemopreventive agent-sensor Keap1 protein is method-dependent. *Chem. Res. Toxicol.* 2007, *20*, 1878–1884.

109. Liebler, D.C.; Ham, A.-J.L. Spin filter-based sample preparation for shotgun proteomics. *Nat. Methods* 2009, *6*, 785.

110. Wiśniewski, J.R.; Ostasiewicz, P.; Mann, M. High recovery FASP applied to the proteomic analysis of microdissected formalin fixed paraffin embedded cancer tissues retrieves known colon cancer markers. *J. Proteome Res.* 2011, *10*, 3040–3049.

111. Weston, L.A.; Bauer, K.M.; Hummon, A.B. Comparison of bottom-up proteomic approaches for LC–MS analysis of complex proteomes. *Anal. Methods* 2013, *5*, 4615–4621.

112. Washburn, M.P.; Wolters, D.; Yates, J.R. Large-scale analysis of the yeast proteome by multidimensional protein identification technology. *Nat. Biotechnol.* 2001, *19*, 242–247.

113. Lipton, M.S.; Pasa-Tolic, L.; Anderson, G.A.; Anderson, D.J.; Auberry, D.L.; Battista, J.R.; Daly, M.J.; Fredrickson, J.; Hixson, K.K.; Kostandarithes, H.; *et al.* Global analysis of the *Deinococcus radiodurans* proteome by using accurate mass tags. *Proc. Natl. Acad. Sci. USA* 2002, *99*, 11049–11054.

114. Xie, F.; Liu, T.; Qian, W.-J.; Petyuk, V.A.; Smith, R.D. Liquid

chromatography–mass spectrometry-based quantitative proteomics. *J. Biol. Chem.* 2011, *286*, 25443–25449.

115. Peng, J.; Elias, J.E.; Thoreen, C.C.; Licklider, L.J.; Gygi, S.P. Evaluation of multidimensional chromatography coupled with tandem mass spectrometry (LC/LC–MS/MS) for large-scale protein analysis: The yeast proteome. *J. Proteome Res.* 2003, *2*, 43–50.

116. Jungblut, P.R. *Protein and Peptide Mass Spectrometry in Drug Discovery*; Gross, M.L., Chen, G., Pramanik, B., Eds.; ChemMedChem: Weinheim, Germany, 2012; Volume 7, pp. 2241–2242.

117. You, J.; Wang, L.; Saji, M.; Olesik, S.V.; Ringel, M.D.; Lucas, D.M.; Byrd, J.C.; Freitas, M.A. High-sensitivity TFA-free LC–MS for profiling histones. *Proteomics* 2011, *11*, 3326–3334.

118. Li, H.; James, C.A. LC–MS bioanalysis of peptides and polypeptides. In *Handbook of LC–MS Bioanalysis*; Li, W., Zhang, J., Tse, F.L.S., Eds.; John Wiley & Sons Inc.: Hoboken, NJ, USA, 2013; pp. 535–549.

119. De Jong, E.P.; Griffin, T.J. Online nanoscale ERLIC–MS outperforms RPLC–MS for shotgun proteomics in complex mixtures. *J. Proteome Res.* 2012, *11*, 5059–5064.

120. Hao, P.; Qian, J.; Dutta, B.; Cheow, E.S.H.; Sim, K.H.; Meng, W.; Adav, S.S.; Alpert, A.; Sze, S.K. Enhanced separation and characterization of deamidated peptides with RP-ERLIC-based multidimensional chromatography coupled with tandem mass spectrometry. *J. Proteome Res.* 2012, *11*, 1804–1811.

121. Zarei, M.; Sprenger, A.; Metzger, F.; Gretzmeier, C.; Dengjel, J. Comparison of ERLIC-TiO_2, HILIC-TiO_2, and SCX-TiO_2 for global phosphoproteomics approaches. *J. Proteome Res.* 2011, *10*, 3474–3483.

122. Hao, P.; Guo, T.; Li, X.; Adav, S.S.; Yang, J.; Wei, M.; Sze, S.K. Novel application of electrostatic repulsion-hydrophilic interaction chromatography (ERLIC) in shotgun proteomics: Comprehensive profiling of rat kidney proteome. *J. Proteome Res.* 2010, *9*, 3520–3526.

123. Wang, H.; Qian, W.-J.; Mottaz, H.M.; Clauss, T.R.W.; Anderson, D.J.; Moore, R.J.; Camp, D.G.; Khan, A.H.; Sforza, D.M.; Pallavicini,

M.; *et al.* Development and evaluation of a micro- and nanoscale proteomic sample preparation method. *J. Proteome Res.* 2005, *4*, 2397–2403.

124. Emmert-Buck, M.R.; Gillespie, J.W.; Paweletz, C.P.; Ornstein, D.K.; Basrur, V.; Appella, E.; Wang, Q.H.; Huang, J.; Hu, N.; Taylor, P.; *et al.* An approach to proteomic analysis of human tumors. *Mol. Carcinog.* 2000, *27*, 158–165.
125. Gozal, Y.M.; Cheng, D.; Duong, D.M.; Lah, J.J.; Levey, A.I.; Peng, J. Merger of laser capture microdissection and mass spectrometry: A window into the amyloid plaque proteome. *Methods Enzymol.* 2006, *412*, 77–93.
126. Picotti, P.; Bodenmiller, B.; Mueller, L.N.; Domon, B.; Aebersold, R. Full dynamic range proteome analysis of S. cerevisiae by targeted proteomics. *Cell* 2009, *138*, 795–806.
127. Anderson, N.L.; Anderson, N.G. The human plasma proteome: History, character, and diagnostic prospects. *Mol. Cell. Proteomics* 2002, *1*, 845–867.
128. Mitchell, P. Proteomics retrenches. *Nat. Biotechnol.* 2010, *28*, 665–670.
129. Wang, N.; Xu, M.; Wang, P.; Li, L. Development of mass spectrometry-based shotgun method for proteome analysis of 500 to 5000 cancer cells. *Anal. Chem.* 2010, *82*, 2262–2271.
130. Sun, L.; Bertke, M.M.; Champion, M.M.; Zhu, G.; Huber, P.W.; Dovichi, N.J. Quantitative proteomics of *Xenopus laevis* embryos: Expression kinetics of nearly 4000 proteins during early development. *Sci. Rep.* 2014, *4*, 4365.
131. Sleno, L. The use of mass defect in modern mass spectrometry. *JMS* 2012, *47*, 226–236.
132. Sun, L.; Zhu, G.; Dovichi, N.J. Comparison of the LTQ-Orbitrap Velos and the Q-Exactive for proteomic analysis of 1–1000 ng RAW 264.7 cell lysate digests. *Rapid Commun. Mass Spectrom.* 2013, *27*, 157–162.
133. Meyer, J.G.; Komives, E.A. Charge state coalescence during electrospray ionization improves peptide identification by tandem mass spectrometry. *J. Am. Soc. Mass Spectrom.* 2012, *23*, 1390–1399.

134. Hahne, H.; Pachl, F.; Ruprecht, B.; Maier, S.K.; Klaeger, S.; Helm, D.; Médard, G.; Wilm, M.; Lemeer, S.; Kuster, B. DMSO enhances electrospray response, boosting sensitivity of proteomic experiments. *Nat. Methods* 2013, *10*, 989–991.

135. Tjernberg, A.; Markova, N.; Griffiths, W.J.; Hallén, D. DMSO-related effects in protein characterization. *J. Biomol. Screen.* 2006, *11*, 131–137.

136. Korfmacher, W.A. *Using Mass Spectrometry for Drug Metabolism Studies*, 2nd ed.; CRC Press: Boca Raton, FL, USA, 2009.

137. Brash, J.L.; ten Hove, P. Effect of plasma dilution on adsorption of fibrinogen to solid surfaces. *Thromb. Haemost.*1984, *51*, 326–330.

138. Horbett, T.A. Chapter 13 Principles underlying the role of adsorbed plasma proteins in blood interactions with foreign materials. *Cardiovasc. Pathol.* 1993, *2*, 137–148.

139. Hervey, W.J.; Strader, M.B.; Hurst, G.B. Comparison of digestion protocols for microgram quantities of enriched protein samples. *J. Proteome Res.* 2007, *6*, 3054–3061.

140. Strader, M.B.; Tabb, D.L.; Hervey, W.J.; Pan, C.; Hurst, G.B. Efficient and specific trypsin digestion of microgram to nanogram quantities of proteins in organic-aqueous solvent systems. *Anal. Chem.* 2006, *78*, 125–134.

141. Kim, S.C.; Chen, Y.; Mirza, S.; Xu, Y.; Lee, J.; Liu, P.; Zhao, Y. A clean, more efficient method for in-solution digestion of protein mixtures without detergent or urea. *J. Proteome Res.* 2006, *5*, 3446–3452.

142. Walsh, G. *Proteins: Biochemistry and Biotechnology*; John Wiley & Sons Inc.: Hoboken, NJ, USA, 2002.

143. Puchades, M.; Westman, A.; Blennow, K.; Davidsson, P. Removal of sodium dodecyl sulfate from protein samples prior to matrix-assisted laser desorption/ionization mass spectrometry. *Rapid Commun. Mass Spectrom.* 1999, *13*, 344–349.

144. Stone, K.L.; Williams, K.R. 2-Enzymatic digestion of proteins and HPLC peptide isolation. In *A Practical Guide to Protein and Peptide Purification for Microsequencing,*, 2nd ed.; Matsudaira, P., Ed.; Academic Press: San Diego, CA, USA, 1993; pp. 43–69.

145. Walker, J.M. *The Proteomics Protocols Handbook*; Humana Press: New York, NY, USA, 2005.

146. Zhou, J.; Zhou, T.; Cao, R.; Liu, Z.; Shen, J.; Chen, P.; Wang, X.; Liang, S. Evaluation of the application of sodium deoxycholate to proteomic analysis of rat hippocampal plasma membrane. *J. Proteome Res.* 2006, *5*, 2547–2553.

147. McDowell, G.S.; Gaun, A.; Steen, H. iFASP: Combining isobaric mass tagging with filter-aided sample preparation. *J. Proteome Res.* 2013, *12*, 3809–3812.

148. Huber, M.L.; Sacco, R.; Parapatics, K.; Skucha, A.; Khamina, K.; Muller, A.C.; Rudashevskaya, E.L.; Bennett, K.L. abFASP-MS: Affinity-based filter-aided sample preparation mass spectrometry for quantitative analysis of chemically labeled protein complexes. *J. Proteome Res.* 2014, *13*, 1147–1155.

149. Cha, S.; Imielinski, M.B.; Rejtar, T.; Richardson, E.A.; Thakur, D.; Sgroi, D.C.; Karger, B.L. *In Situ* proteomic analysis of human breast cancer epithelial cells using laser capture microdissection: Annotation by protein set enrichment analysis and gene ontology. *Mol. Cell. Proteomics* 2010, *9*, 2529–2544.

150. Jorgenson, J.W.; Lukacs, K.D. Zone electrophoresis in open-tubular glass capillaries. *Anal. Chem.* 1981, *53*, 1298–1302.

151. Karger, B.L.; Guttman, A. DNA sequencing by capillary electrophoresis. *Electrophoresis* 2009, *30*, S196–S202.

152. Heemskerk, A.A.M.; Deelder, A.M.; Mayboroda, O.A. CE–ESI-MS for bottom-up proteomics: Advances in separation, interfacing and applications. *Mass Spectrom. Rev.* 2014.

153. Li, Y.; Champion, M.M.; Sun, L.; Champion, P.A.D.; Wojcik, R.; Dovichi, N.J. Capillary zone electrophoresis-electrospray ionization-tandem mass spectrometry as an alternative proteomics platform to ultraperformance liquid chromatography-electrospray ionization-tandem mass spectrometry for samples of intermediate complexity. *Anal. Chem.* 2012, *84*, 1617–1622.

154. Sun, L.; Zhu, G.; Yan, X.; Champion, M.M.; Dovichi, N.J. Capillary zone electrophoresis for analysis of complex proteomes using an electrokinetically pumped sheath flow nanospray interface. *Proteomics* 2014, *14*, 622–628.

155. Sun, L.; Zhu, G.; Mou, S.; Zhao, Y.; Champion, M.M.; Dovichi, N.J. Capillary zone electrophoresis-electrospray ionization-tandem mass spectrometry for quantitative parallel reaction monitoring of peptide abundance and single-shot proteomic analysis of a human cell line. *J. Chromatogr.* 2014, *1359*, 303–308.

156. Zhu, G.; Sun, L.; Yan, X.; Dovichi, N.J. Stable, reproducible, and automated capillary zone electrophoresis–tandem mass spectrometry system with an electrokinetically pumped sheath–flow nanospray interface. *Anal. Chim. Acta* 2014, *810*, 94–98.

157. Zhu, G.; Sun, L.; Yan, X.; Dovichi, N.J. Single-shot proteomics using capillary zone electrophoresis–electrospray ionization-tandem mass spectrometry with production of more than 1250 *Escherichia coli* peptide identifications in a 50 min separation. *Anal. Chem.* 2013, *85*, 2569–2573.

158. Li, Y.; Compton, P.D.; Tran, J.C.; Ntai, I.; Kelleher, N.L. Optimizing capillary electrophoresis for top-down proteomics of 30–80 kDa proteins. *Proteomics* 2014, *14*, 1158–1164.

159. Zhao, Y.; Sun, L.; Champion, M.M.; Knierman, M.D.; Dovichi, N.J. Capillary zone electrophoresis-electrospray ionization-tandem mass spectrometry for top-down characterization of the *Mycobacterium marinum* secretome. *Anal. Chem.* 2014, *86*, 4873–4878.

160. Diamond, D.L.; Jacobs, J.M.; Paeper, B.; Proll, S.C.; Gritsenko, M.A.; Carithers, R.L.; Larson, A.M.; Yeh, M.M.; Camp, D.G.; Smith, R.D.; *et al.* Proteomic profiling of human liver biopsies: Hepatitis C virus-induced fibrosis and mitochondrial dysfunction. *Hepatol. (Baltim. Md.)* 2007, *46*, 649–657.

161. Gerlinger, M.; Rowan, A.J.; Horswell, S.; Larkin, J.; Endesfelder, D.; Gronroos, E.; Martinez, P.; Matthews, N.; Stewart, A.; Tarpey, P.; *et al.* Intratumor heterogeneity and branched evolution revealed by multiregion sequencing. *N. Engl. J. Med.* 2012, *366*, 883–892.

162. Cancer Symptoms, Causes, Treatment—How is cancer diagnosed? Available online: http://www.medicinenet.com/cancer/article.htm (accessed on 3 November 2014).

163. Bruening, W.; Schoelles, K.; Treadwell, J.; Launders, J.; Fontanarosa, J.; Tipton, K. *Comparative Effectiveness of Core-Needle and Open Surgical Biopsy for the Diagnosis of Breast*

Lesions; AHRQ Comparative Effectiveness Reviews; Agency for Healthcare Research and Quality (US): Rockville, MD, USA, 2009.

164. Breast Biopsy—Cancer & Biopsies Susan G. Komen®. Available online: http://ww5.komen.org/BreastCancer/Biopsies.html (accessed on 3 November 2014).

165. Rodenburg, W.; Pennings, J.L.A.; van Oostrom, C.T.M.; Roodbergen, M.; Kuiper, R.V.; Luijten, M.; de Vries, A. Identification of breast cancer biomarkers in transgenic mouse models: A proteomics approach. *Proteomics Clin. Appl.* 2010, *4*, 603–612.

166. Taguchi, A.; Politi, K.; Pitteri, S.J.; Lockwood, W.W.; Faça, V.M.; Kelly-Spratt, K.; Wong, C.-H.; Zhang, Q.; Chin, A.; Park, K.-S.; *et al*. Lung cancer signatures in plasma based on proteome profiling of mouse tumor models. *Cancer Cell* 2011, *20*, 289–299.

167. Feist, P; Sun, L; Liu, X; Dovichi, N.J.; Hummon, A.B. Bottom-up proteomic analysis of single HCT 116 colon carcinoma multicellular spheroids. *Rapid Commun. Mass Spectrom.* 2015. in press.

CHAPTER

2

MAKING PROTEOMICS DATA ACCESSIBLE AND REUSABLE: CURRENT STATE OF PROTEOMICS DATABASES AND REPOSITORIES

Yasset Perez-Riverol, Emanuele Alpi, Rui Wang, Henning Hermjakob, and Juan Antonio Vizcaíno

European Molecular Biology Laboratory, European Bioinformatics Institute (EMBL-EBI), Wellcome Trust Genome Campus, Hinxton, Cambridge, UK

ABSTRACT

Compared to other data-intensive disciplines such as genomics, public deposition and storage of MS-based proteomics, data are still less developed due to, among other reasons, the inherent complexity of the data and the variety of data types and experimental workflows. In order to address this need, several public repositories for MS proteomics experiments have been developed, each with different purposes in mind. The most established resources are the Global Proteome Machine Database (GPMDB), PeptideAtlas, and the PRIDE database. Additionally, there are other useful (in many cases recently developed) resources such as ProteomicsDB, Mass Spectrometry Interactive Virtual Environment (MassIVE), Chorus, MaxQB, PeptideAtlas SRM Experiment Library (PASSEL), Model Organism Protein Expression Database (MOPED), and the Human Proteinpedia. In addition, the ProteomeXchange consortium has been recently developed to enable better integration of public repositories and the coordinated sharing of proteomics information, maximizing its benefit to the scientific community. Here, we will review each of the major proteomics resources independently and some tools that enable the integration, mining and reuse of the data. We will also discuss some of the major challenges and current pitfalls in the integration and sharing of the data

INTRODUCTION

In the age of systems biology and data integration, proteomics data represent a crucial component to understand the "whole picture" of life. Proteomics technologies—particularly MS-based protein identification and quantification approaches—have matured immensely through cumulative advances in high-throughput analytical methodologies [1-5], sample preparation [6], improved instrumentation [7], and the availability of protein sequence databases [8] and computational analysis tools [1, 9]. Therefore, with the development of more powerful and sensitive analytical methods and instrumentation, the identification and quantification of a high proportion of the expressed proteins in a given condition is now achievable in an average experiment [10, 11]. In parallel, as a result, the size of data produced in proteomics laboratories has increased by several orders of magnitude [12].

Compared to other data-intensive fields such as genomics, deposition and storage of original proteomics, data in public resources have been less common [13]. This is regrettable since proteome studies are usually more complex than its counterpart genomics ones. In fact, data interpretation in proteomics can be considerably more complex than in genomics due to the wide variety of analytical approaches [14, 15], bioinformatics tools and pipelines [16, 17], and the related statistical analysis [18, 19].

However, thanks to the guidelines promoted by several scientific journals and funding agencies [20], there is a growing consensus in the community about the need for the public dissemination of proteomics data, which is already facilitating the assessment, reuse, comparative analyses, and extraction of new findings from published data [13, 21].

The complexity of proteomics data is heightened by alternative splicing, PTMs, and protein degradation events, and is further amplified by the interconnectivity of proteins into complexes and signaling networks that are highly divergent in time and space [1]. In order to address this complexity, new analytical and bioinformatics methodologies are developed every year [22, 23], which complicate the data standardization and deposition. Additionally, the audience interested in proteomics data is very heterogeneous. It includes, biologists elucidating the mechanisms of regulation of specific proteins, MS researchers improving the current analytical methods, or computational biologists developing new software tools for the analysis and interpretation of the data [24].

Data sharing in proteomics requires substantial investment and infrastructure. Several public repositories have been developed, each with different purposes in mind. Well-established databases for proteomics data are the Global Proteome Machine Database (GPMDB) [25], PeptideAtlas [26], and the PRIDE database [27]. Additionally, at present there are other resources (many of them recently developed) such as ProteomicsDB [28], MassIVE (Mass Spectrometry Interactive Virtual Environment), Chorus, MaxQB [29], PASSEL (PeptideAtlas SRM Experiment Library) [30], MOPED (Model Organism Protein Expression Database) [31], PaxDb [32], Human Proteinpedia [33], and the human proteome map (HPM) [34]. Furthermore, there are several more specialized resources that will only be cited briefly in this review. It is important to mention here that no single proteomics data resource will be ideally suited to all possible use cases and all potential users. Regrettably, two widely used resources were discontinued due to lack of funding: Peptidome [35] and Tranche [36]. This had a negative impact on the efforts promoting data sharing in the field, as it was perceived by the community that effort invested in data sharing was lost.

Recently, the ProteomeXchange (PX) consortium [37] has been formed to enable a better integration of public repositories, maximizing its benefits to the scientific community through the implementation of standardized submission and dissemination pipelines for proteomics information. By August 2014, PRIDE, PeptideAtlas, PASSEL, and MassIVE are the active members of the consortium.

The aim of this review is to provide an up-to-date overview of the current state of proteomics data repositories and databases, providing a solid starting point for those who want to perform data submission and/or data mining. There are a few comparable reviews available in the literature [24, 38-41], but there is a need for an update since this has been quite a dynamic field over the past few years. In this manuscript, we will not include a thorough review about protein knowledge bases, such as the Universal Protein resource (UniProt) [42] and neXtProt [43], but we will explain how MS proteomics information is made available in these resources.

ORGANIZATION OF PROTEOMICS REPOSITORIES AND DATABASES

The information generated in a typical proteomics experiment can be organized in three different levels [44]: (i) raw data; (ii) processed results, including peptide/protein identification and quantification values; and (iii) the resulting biological conclusions. Technical and/or biological metadata can be provided for each level independently.

These three categories enable the classification of the existing MS proteomics repositories according to their level of specialization (Fig. 1). In our view, these three levels of information should be captured and properly annotated in public databases and repositories, ideally using data standards, when available. In fact, the development of proteomics resources is more feasible due the maturity of some data standard formats [45-47] and open source tools [9], which facilitate public data deposition.

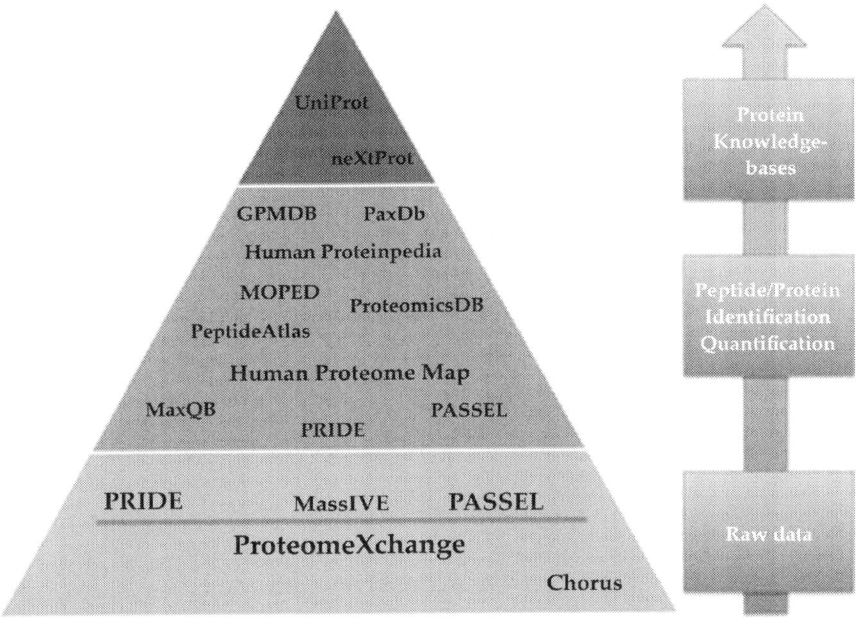

Figure 1. Hierarchy of proteomics data repositories and databases according to the different data types stored: raw MS data repositories, resources that store peptide/protein identification and quantification results, and protein knowledge bases. Some resources are duplicated in different levels because they can be included in more than one category.

Some of the first open formats developed included mzXML (for MS data) [48], pepXML, and protXML (for peptide/protein identifications) that were developed as part of the trans-proteomic pipeline (TPP) [49]. In the context of the Proteomics Standards Initiative (PSI), several standard data formats have been developed over the last few years, which reflect the variety in data types within the field, and therefore those that can be supported by proteomics resources. The main formats developed have been mzML (for MS data) [45], mzIdentML (for peptide and protein identification results) [46], mzQuantML (for capturing a detailed trace of each stage of quantitative analysis) [47], TraML (for representing input transitions in SRM approaches) [50], and the recent mzTab (for capturing a simpler summary of the final results) [51]. The aims and functionalities of the existing resources will be explored in detail in the following sections (Table 1).

Table 1: Main characteristics of the major MS-based proteomics repositories and databases

Repositories	Raw data	Support for targeted approaches	Metadata	Human protein expression information	Species	Quantification data	Related stand-alone tools	Web services URL	URL
PRIDE (June 2014)	X	–	High level	41 835 Protein accessions 269 806 Unique peptide sequences Approximately 101 million spectra	Approximately 450 species	X	PRIDE Inspector, PRIDE Converter 2, Peptide-Shaker	http://www.ebi.ac.uk/pride/ws/archive/	http://www.ebi.ac.uk/pride/
PeptideAtlas (Human, August 2013)	X	X	Medium level	14 018 Proteins 338 013 Peptides Approximately 258 million spectra	*Mus musculus, Candida albicans, Candida, Caenorhabditis elegans, Drosophila melanogaster, Halobacterium, Equuscaballus, Rattus norvegicus, Saccharomycescerevisiae, Danio rerio, Susscrofa, Mycobacterium tuberculosis*	–	TIQAM, TIQAM–Digestor, TIQAM–PeptideAtlas, TIQAM–Viewer, ATAQS, PIPE2, PABST		http://www.peptideatlas.org/
GPMDB (May 2014)	–	X	High level	136 373 Protein accessions 1 786 698 Peptides Approximately 1020 million spectra	*Bos taurus, Canis familiaris, Homo sapiens, M. musculus, R. norvegicus, Gallus gallus, D. rerio, Xenopus tropicalis, Anopheles gambiae, Apis mellifera, D. melanogaster, C. elegans, Oryza sativa, Arabidopsis thaliana, Saccharomyces cerevisiae, Bacillus anthracis* Ames, *Escherichia coli* (K12), *Lactococcus lactis*(Il1403), *M. tuberculosis, Shigella dysenteriae, Salmonella typhi, Salmonella typhimurium*(LT2)	–	–	http://rest.thegpm.org/1	http://gpmdb.thegpm.org/
MassIVE (May 2014)	X	–	Low level	*	*	–	–	–	http://massive.ucsd.edu/
Chorus (May 2014)	X	–	Low level	*	*	–	–	–	https://chorusproject.org/
ProteomicsDB (May 2014)	X	–	High level	18 097 Proteins 739 406 Peptides Approximately 70 million spectra	*H. sapiens*	X	–	–	https://www.proteomicsdb.org/
MOPED (May 2014)	–	–	Medium level	17 141 Proteins 250 000 Unique peptides Approximately 15 million spectra	*H. sapiens, M. musculus, C. elegans, S. cerevisiae*	X	–	–	https://www.proteinspire.org/MOPED/

Human Proteinpedia (May 2014)	X	–	Low level	15 231 Proteins 1 960 352 Peptides Approximately 5 million spectra	*H. sapiens*	–	–	–	http://www.humanproteinpedia.org/
MaxQB (May 2014)	–	–	Medium level	14 732 Proteins 370 551 Peptides Approximately 20 million spectra	*M. musculus, H. sapiens S. cerevisiae*	X	–	–	http://maxqb.biochem.mpg.de/mxdb/
PaxDb (May 2014)	–	–	Low level	10 482 Proteins 143 456 Peptides Approximately 24 million spectra	*A. thaliana, M. musculus, H. sapiens, S. cerevisiae, D. melanogaster, E. coli*(K12), *Microcystis aeruginosa, C. elegans, Leptospira interrogans*serovar *Copenhageni, M. tuberculosis* (H37Rv), *Streptococcus pyogenes M1 GAS, Schizosaccharomyces pombe, B. taurus, S. dysenteriae, B. subtilis, G. gallus, Thermococcus gammatolerans* (EJ3), *Mycoplasma pneumoniae, Halobacterium* (NRC-1), *S. typhimurium* LT2, *R. norvegicus, Deinococcus deserti*(VCD115), *Shigella flexneri, Synechocystis, A. mellifera, C. familiaris, Sus scrofa, X. tropicalis, O. sativa*	X	–	http://pax-db.org/api/search?q=human	http://pax-db.org/
HPM (June 2014)	–	–	Low level	10 482 Proteins 293 000 Unique peptides Approximately 25 million spectra	*H. sapiens*	X	–	–	http://humanproteomemap.org

RESOURCES

The PX consortium

The PX consortium [37] (http://www.proteomexchange.org) was created to promote the collaboration and integration of major stakeholders in the domain of MS proteomics repositories comprising, among others, primary (PRIDE [27] and PASSEL [30]) and secondary resources (PeptideAtlas), proteomics researchers, and representatives from journals regularly publishing proteomics data. Recently, in June 2014, MassIVE joined the consortium. The aim of PX is to provide a common framework for the cooperation of proteomics resources by defining and implementing consistent, harmonized, user-friendly data deposition and dissemination procedures. In addition, another important goal is to enable and provide "mutual backup" if one of the resources has funding issues.

The consortium's members have agreed in providing a sufficient set of common experimental and technical metadata. This information is stored using the PX XML format [37]. Finally, all the submitted datasets get a unique and universal identifier (PXD identifier).

Data Submission and Format Support

By August 2014, two major workflows are fully supported in PX: MS/MS and SRM approaches. In the first stable implementation of the data workflow, PRIDE acts as the initial submission point for MS/MS data, whereas PASSEL has the equivalent role for SRM data. Both workflows will be explained in detail below. At the moment of writing, MassIVE has just joined PX aiming to have an equivalent role to PRIDE.

There are two different PX MS/MS submission modes: "Complete" and "Partial." To perform a "complete" submission means that after all the files have been submitted, it is possible for the receiving repository to connect directly the processed identification results with the mass spectra. This can be achieved if the processed identification results are available in a format supported by the receiving repository (e.g., mzIdentML) and if peak list files are included in the submission. "Complete" submissions get a Digital Object Identifier (DOI) to facilitate its traceability.

On the other hand, after performing a "partial" submission the connection between the spectra and the identification results cannot be done in a straightforward way. In this case, the processed results are not available in a supported format by the receiving repository and the corresponding search engine output files (in heterogeneous formats) are made available for download. For both types of submissions, metadata and raw data are always stored for each dataset.

Although "partial" submissions are searchable by their metadata, peptide and protein identifications cannot be captured by the receiving repository, which decreases the ability of reviewers to check the data and can make data reuse by third parties challenging. For instance, "partial" submissions do not qualify for the requirements on spectra annotation from the journal MCP (*Molecular and Cellular Proteomics*—http://www.mcponline.org/site/misc/ParisReport_Final.xhtml).

Finally, it needs to be highlighted that all PX members support private review of the data during the manuscript review process. The submitted data remains private before manuscript publication and login details are provided to facilitate access for reviewers and journal editors during the manuscript review process.

Data mining and visualization

ProteomeCentral (http://proteomecentral.proteomexchange.org) is the centralized portal for accessing all PX datasets, independently from the original resource where data were stored. ProteomeCentral provides the ability to search metadata associated with datasets in the participating repositories (PRIDE and MassIVE for MS/MS data, PASSEL for SRM data, or reprocessed original PX datasets in PeptideAtlas). It is then possible to query the archive and identify datasets of interest using biological and technical metadata, keywords, tags, or publication information. To monitor the release of new public PX datasets, researchers can subscribe to a Rich Site Summary feed (http://groups.google.com/group/proteomexchange/feed/rss_v2_0_msgs.xml). Next, the main characteristics of the individual members of the consortium will be explained.

PRIDE

The PRIDE database [**27**] (http://www.ebi.ac.uk/pride/) was initially developed at the European Bioinformatics Institute (EBI, Cambridge, UK) to store the experimental data included in publications, supporting the manuscript review process. The main data types stored in PRIDE are peptide/protein identifications (including PTMs), peptide/protein expression values, the analyzed mass spectra (both as raw data and peak lists), and the related technical/biological metadata. In PRIDE, data are stored as originally analyzed by the researchers (the author's analysis view on the data), supporting many popular search engines/analysis workflows. Most of the terms, information and metadata supporting the PRIDE data, are based on ontologies or controlled vocabularies [**52**]. In this context, the "Ontology Lookup Service" [**53**] was developed as a spin-off of PRIDE to enable the querying, browsing, and navigation of biomedical ontologies. Since the inception of PX, the number of datasets submitted to PRIDE has grown considerably (Fig. 2). In parallel, the size of the experiments in terms of spectrum numbers has also increased significantly (Fig. 2).

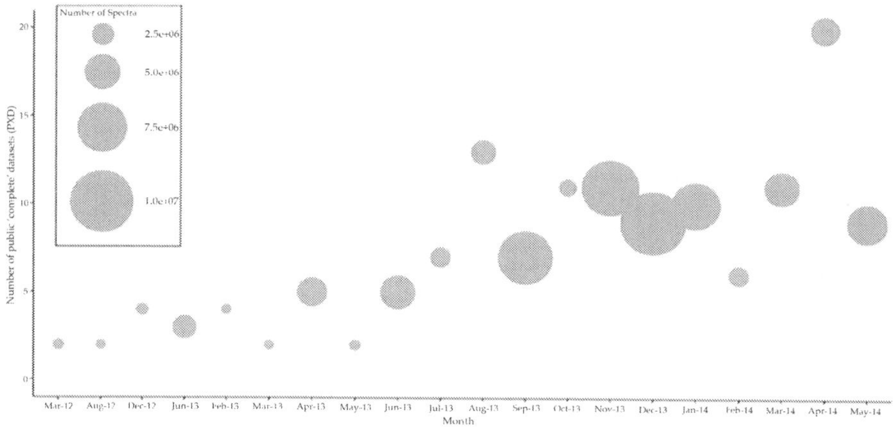

Figure 2: Bubble chart representation of the size of the PX complete submissions to PRIDE (until May 2014). The x-axis includes months with at least one submission, since PX submissions started (from March 2012). The y-axis corresponds to the number of PX "complete" public datasets submitted to PRIDE in each specific month. The size of each bubble represents the total number of mass spectra included in all the datasets in a given month.

Data submission and format support

As a member of the PX consortium, PRIDE supports both "complete" and "partial" submissions. For "complete" submissions, processed identification results need to be provided in PRIDE XML (PRIDE original data format) or the PSI standard mzIdentML (version 1.1) format. If mzIdentML is used, the corresponding peak list files referenced by the mzIdentML files need to be included as well. A "complete" submission ensures that the processed results data can be integrated in PRIDE and visualized using tools, such as PRIDE Inspector [54] (see below). Therefore, for performing a "complete" submission, the output files from the analysis software need to be converted or exported to either mzIdentML (see http://www.psidev.info/tools-implementing-mzidentml and http://www.ebi.ac.uk/pride/help/archive/submission/mzidentml) or PRIDE XML (using PRIDE Converter 2 [55] or other external tools). It is important to highlight that the recently implemented support for mzIdentML makes possible that data from a growing number of tools/ analysis software (that were never supported via PRIDE XML) can now be fully supported by PRIDE.

The PRIDE Converter 2 tool suite (http://pride-converter-2.googlecode.com) [55] is an open source and platform-independent software that allows users to convert several search engine output files to PRIDE XML. The tool suite consists of four different applications: PRIDE Converter 2, PRIDE mzTab generator, PRIDE XML merger, and PRIDE XML filter. All of these tools can be launched using the graphical user interface or from the command line. PRIDE Converter 2 currently supports the output from MASCOT (.dat) [56], X!Tandem (.xml) [57], OMSSA (.csv), among others. It also supports different mass spectra file formats (mzML, dta, mgf, mzData, mzXML, and pkl). As mentioned above, there are other tools not developed by the PRIDE team that can also export to PRIDE XML [37]. It is expected that, as the mzIdentML format becomes more and more popular, the use of PRIDE XML will decrease in time.

The "partial" submission route is aimed at situations where processed identification results cannot be generated in mzIdentML/PRIDE XML if there is no converter/exporter available. However, the "partial" submission mechanism also enables any data from any proteomics workflow to be stored in PRIDE. As a consequence, some datasets are already available coming from workflows, such as top-down proteomics, MS imaging, or SWATH-MS, among others.

The PRIDE/PX submission process has been recently described in detail [58]. The PX submission tool (http://www.proteomexchange.org/submission) [37] is an open-source standalone tool that provides a user-friendly graphical user interface for performing the actual data submission, through a series of steps:

- Select all the files needed for submission.
- Interactively group-related different types of files (e.g., the corresponding raw and processed results files).
- Ensure a minimum level of metadata (according to the PX guidelines).
- Transfer the files via the Aspera (from version 2.1) or FTP file transfer protocols. Aspera (http://asperasoft.com/) can perform

up to 50 times faster than FTP, enabling a convenient way of transferring large datasets.

In addition to the PX submission tool, datasets containing a high number of files can also be submitted using a command line based alternative (http://www.ebi.ac.uk/pride/help/archive/aspera) [58]. Each dataset becomes publicly available on acceptance or publication of the corresponding manuscript, or when the authors tell PRIDE to do so.

Data mining and visualization

PRIDE Inspector (http://pride-toolsuite.googlecode.com) [54] is an open source standalone tool that can be used to efficiently browse and visualize MS proteomics data. PRIDE Inspector can be used by researchers before submission and also by journal editors and reviewers during the manuscript review process. The latest version available at the moment of writing (version 2.1) supports identification results in PRIDE XML and mzIdentML (used in PX "complete" submissions). It also supports spectra files in a variety of formats (mgf, pkl, ms2, mzXML, mzData, and mzML). PRIDE Inspector enables users to visualize and check the data at different levels. It has different panels devoted to experimental metadata, protein, peptide, and spectrum-centric information. Finally, the "summary charts" tab provides eight different charts that can be used to evaluate some aspects of the quality of the dataset. Apart from the visualization functionality, it can access some of the most popular protein databases (UniProt knowledgebase (UniProtKB), Ensembl [59], and the National Center for Biotechnology Information (NCBI) nonredundant database) to retrieve the most up-to-date protein sequences and names for the reported protein identifiers.

The PRIDE Archive website (http://www.ebi.ac.uk/pride/archive/) provides the web interface to query and retrieve the information in PRIDE. It was launched on January 2014 and at the moment of writing, is still under iterative development, so new features are being added constantly. By August 2014, the current version allows querying PRIDE using keywords, publication, species, tissues, diseases, modifications, instruments, peptide sequences, and protein identifiers. When a specific dataset is selected, the users are directed to the dataset summary page, which also lists the assays (equivalent to the old PRIDE experiment numbers) related to the project, in the case of "complete" submissions.

Users can download all the files via FTP or Aspera, and/or visualize the results using PRIDE Inspector. In the coming months, visualization for peptide/protein identifications will be available in the PRIDE web. At present the BioMart interface (http://www.ebi.ac.uk/pride/legacy/prideMart.do) is the easiest way to access this information. However, it is planned that it will be soon replaced by new PRIDE web services.

Additionally, PRIDE data can be accessed using the "PRIDE Cluster" webpage (http://www.ebi.ac.uk/pride/cluster/) [60]. It includes public identified spectra in PRIDE that have been clustered using the "PRIDE Cluster" algorithm (https://code.google.com/p/pride-spectra-clustering/). The PRIDE Cluster resource currently provides two main methods for accessing its data: (i) retrieve all clusters that contain a given peptide identification and (ii) retrieve all clusters with a consensus spectrum similar to a queried spectrum. In addition, spectral libraries for several species are also provided. "PRIDE Cluster" is the first quality control step attempted in a highly heterogeneous MS proteomics repository, such as PRIDE.

PASSEL

PASSEL [30] (http://www.peptideatlas.org/passel/) supports the submission of datasets generated by SRM approaches by storing the experimental results and the corresponding raw data. The submitted raw data are automatically reprocessed in a uniform manner using mQuest, a component of the mProphet software suite [61], and the results are loaded into the database [62]. The original files and the corresponding reprocessed results are made available to the community. Additionally, the measured transitions are incorporated into the SRMAtlas [62] catalog of transitions. Detailed metadata and structured sample information are in accordance with the PX guidelines.

Data submission and format support

PASSEL uses a web interface for the data submission process (http://www.peptideatlas.org/submit) and is now supporting the following data types (http://www.peptideatlas.org/upload/): (i) study metadata, such as dataset title, submitter contact information, sample source, sample preparation, and instrument used; (ii) transition lists describing which transitions were measured for each peptide and targeted ions, along with optional supporting information (collision energy, expected retention time, and expected relative intensities). This information is available in a tab-separated file or in the standard TraML format [50]; and (iii) mass spectrometer output files in mzML [45] or mzXML [48] format. If these are not available, the vendor formats .wiff (AB SCIEX), .raw (Thermo), or .d (Agilent) are also supported. All the files supplied by the users are uploaded to a specially created FTP account and they can be browsed using the "PASSEL Experiment Browser".

Data mining and visualization

The "PASSEL Experiment Browser" enables filtering based on fields, such as research contact information, organism, sample, and instrument type. As mentioned above, it is also possible to download the original files provided (http://db.systemsbiology.net/sbeams/cgi/PeptideAtlas/GetSELExperiments).

The "PASSEL Data Browser" provides a description of each selected transition group and the data collected for it, a link to visualize the trace group using the "Chromavis" chromatogram viewer and further links to the information available in SRMAtlas [62] and PeptideAtlas [26], in order to extract or compare spectral features of the targeted peptides (http://db.systemsbiology.net/sbeams/cgi/PeptideAtlas/GetSELTransitions).

PeptideAtlas

The PeptideAtlas project (http://www.peptideatlas.org) [26, 63, 64] was originally created to serve as the end-point for the TPP processing software [49]. In recent years, PeptideAtlas has grown as a data reprocessing resource and it has served as a research database for the development of spectral libraries [65] and SRM-related tools [66, 67]. For instance, PeptideAtlas provides information about proteotypic peptides using detectability scores [68].

Nowadays, PeptideAtlas is one of the biggest and well-curated protein expression data resources. The initial PeptideAtlas publication reported the identification of 27% of the human genes in Ensembl with a protein false discovery rate (FDR) likely 10% or higher [63]. Over the years, the PeptideAtlas team has developed different tools to control the assignment of incorrect identifications, such as PeptideProphet [69] and ProteinProphet [70], and more recently MAYU [71], to control the protein FDR when different datasets are combined. This platform and the statistically accurate protocol used to curate the protein/peptide identification data have turned PeptideAtlas into a very reliable protein expression database. In 2013, the generated 1% protein-level FDR Human PeptideAtlas had at least one peptide for around 14 000 different UniProtKB/Swiss-Prot entries [26].

In the context of PX, partners have recently started to track the reanalysis of original PX datasets by providing RPXD identifiers to those, and linking them to the original reprocessed datasets. By August 2014, several PX datasets reanalyzed by PeptideAtlas are already publicly available in ProteomeCentral.

Data submission and format support

PeptideAtlas reprocessed data are organized into different builds [72], each includes data from a single proteome or subproteome (see species in Table 1). Each build is generated with raw MS/MS spectra submitted using the "submission form" (http://www.peptideatlas.org/upload/) or the data deposited into another public repository, such as PRIDE. These spectra are searched against a sequence database, a spectral library or both. Peptide and protein identifications are mapped to a comprehensive reference protein database (for the latest human builds, the searched database is a combination of UniProtKB/Swiss-Prot, Ensembl, and sequences from the International Protein Index (IPI)), and postprocessed using the TPP [72]. It also annotates each protein and peptide with supporting data, such as genome mappings, sequence alignments, links to different databases, such as GPMDB or the Human Protein Atlas [73], uniqueness of peptide–protein mappings, observability of peptides, predicted observable peptides, estimated protein abundances and cross-references to other databases, such as RefSeq, UniGene and UniProt. All the processed results are loaded into SBEAMS (systems biology experiment analysis management system) proteomics that is a proteomics analysis database built as a module under the SBEAMS framework.

Data mining and visualization

The PeptideAtlas web search interface can be used to search for proteins by protein accession, peptide sequence, gene name, keyword, or phrase. If a specific protein is requested, the "protein view page" summarizes all the information available for that protein. The top section provides basic information about the protein, including alternative names as well as the total number of corresponding spectra (observations) and distinct peptides. The following two sections, "sequence motifs" and "sequence," summarize the peptide coverage of the protein. Finally, a similar diagram to a genome browser view summarizes all the peptides that map either uniquely or redundantly to the protein, including information on segments unlikely to be observed by MS. Information about signal peptides and transmembrane domains is also provided, where available. One of the best features of PeptideAtlas is the Cytoscape [74] plug-in that allows the user to view the distinct peptides for a particular protein as a network with associated proteins.

Recently, PeptideAtlas implemented the "PeptideAtlas Chromosome Explorer" (http://www.peptideatlas.org/peptideatlasExplorer/) to summarize and classify the identified human proteome using a chromosome-oriented view. The present version shows the number of protein observations and the UniProt entries by chromosome using a circular histogram plot. PeptideAtlas is actively involved in the HUPO chromosome-based Human Proteome Project (C-HPP) [75] and provides a dedicated website to access protein identifications per chromosome (http://www.peptideatlas.org/hupo/c-hpp/).

PeptideAtlas heavily supports targeted proteomics workflows in different ways: (i) SRMAtlas (http://www.srmatlas.org/) is a compendium of targeted proteomics assays to detect and quantify proteins using SRM/MRM-based proteomics workflows; (ii) TIQAM (targeted identification for quantitative analysis by MRM) [76], which is a desktop application to facilitate the selection of peptide and transitions. It consists of three applications: "TIQAM-Digestor," "TIQAM-PeptideAtlas," and "TIQAM-Viewer"; (iii) the automated and targeted analysis with quantitative SRM tool (ATAQS) [77] is a software pipeline tool that contains modules to design, manage, analyze, and validate an MRM assay. ATAQS uses FireGoose [78] to connect to various web services, such as PeptideAtlas (used to select spectra), TIQAM (to generate in silico peptides for a given protein [76]), PIPE2 (to generate a list of proteins, to design an MRM assay, and for other various analysis tasks), and PABST (peptide atlas best SRM transition, used to generate optimal transitions).

MassIVE

The MassIVE data repository (http://massive.ucsd.edu) is a community resource developed by the Center for Computational Mass Spectrometry (University of California, San Diego) to promote the global-free exchange of MS data. MassIVE provides a location for researchers to access public raw datasets and accompanying files, often alongside publication. One of the key features of MassIVE (still under development) compared with other resources is that its functionality is aimed at networking and providing a social platform for researchers. MassIVE users are not only able to browse, download, but also comment on datasets. These comments can be accompanied with new data or new analyses that enrich the original dataset.

Data submission and format support

The submission to MassIVE is based on two main steps: (i) upload the data files to ProteoSAFe (http://massive.ucsd.edu/ProteoSAFe/); and (ii) invoke the MassIVE dataset submission workflow for those files. The MassIVE team strongly recommends that submitters use FTP to upload and organize their dataset files as opposed to the ProteoSAFe file upload web interface.

MassIVE dataset files are organized into the following categories: (i) license files—specifying how and under which conditions the dataset files may be downloaded and used; (ii) spectrum files—any mass spectrum files constituting the main body of a dataset (in mzXML and/or raw binary format); (iii) result files—the output of any search engine; (iv) sequence databases—any protein or other sequence databases that were searched against (.fasta format); (v) spectral libraries—any spectral library files that were searched, if applicable; (vi) methods and protocols, for any open-format files containing explanations or discussions of the experimental procedures used to obtain or analyze a given dataset.

The MassIVE submission web presents an input form including all the files to be uploaded and relevant metadata. The metadata fields include the species, instruments, PTMs, and contact information. When a dataset is submitted, the ProteoSAFe validation is performed.

Data mining and visualization

The "MassIVE datasets" browser (http://massive.ucsd.edu/ProteoSAFe/datasets.jsp) presents a list of all public datasets in a tabular format. Users can sort and filter by any column. Alternatively, users can select specific datasets by checking the boxes next to them. MassIVE tried to rescue as many datasets as possible from the defunct Tranche repository. For this imported data, the Tranche hash is also displayed, enabling the users to search for specific Tranche datasets.

Once data are submitted to the repository, it can be shared with others (e.g., reviewers) using password-protected access to the datasets, or be made publicly available by the submitter. In the near future, it is planned that anyone with access to a given dataset will be able to reanalyze it using the many workflows available in ProteoSAFe. After the reanalysis, the results can then be used to either enrich existing datasets or to create new ones. MassIVE enables on-site reanalysis using a variety of data workflows: standard database searches (MS-GF+) [79], proteogenomics searches against genomics/transcriptomics sequences (ENOSI) [80-82], discovery of unexpected modifications (MODa) [83], identification of mixture spectra using spectral library (MSPLIT) [84] and database (MixDB) search [85], de novo sequencing of peptides (PepNovo) [86] and proteins (Meta-SPS) [87], molecular spectral networks (including both peptides and metabolites), and top-down protein identification (MS-Align+) [88].

Chorus

Chorus (http://chorusproject.org) is a cloud-based application that provides scientists the ability to store, analyze, and share their MS data regardless of the original raw file format generated. The Chorus team's aim is to create a complete catalogue of the mass spectrometric data in a way that can be openly accessed, and make it freely accessible to both the scientific community and the general public. Chorus was originally announced in 2013 and is enabled by Amazon Web Services, Inc. (http://

aws.amazon.com). It is a global cloud-based computing environment that provides customized data analysis tools that convert proprietary data formats to a common "Map Reduce" format for processing large datasets using parallel and distributed algorithms on the cloud. The initial data analysis tools include chromatographic and mass spectral viewers as well as a database search engine for protein sequence identification.

The registered users should first request laboratory membership or create their own laboratory. After that, they can create a project, an instrument, and upload raw files. Once this is done, the users can create experiments and organize them into the created projects.

GPMDB

GPMDB [**25**] (http://gpmdb.thegpm.org) is one of the most well-known protein expression databases. The GPMDB pipeline reprocesses the MS/MS data provided by users or raw data stored in other repositories, such as PX, using the popular open source search engine X!Tandem [**57**]. Peptide and protein identifications are generated and stored in XML files, which are indexed in a MySQL database.

Data submission and format support

The data submission process supports MS data in different formats (.dta, .pkl, .mgf, mzXML, and mzData) and it can take place via the "simple search page." Once the data have been processed using X!Tandem, users can choose whether to submit their data or not to GPMDB. In addition to X!Tandem, users can also use a spectral search engine called X!Hunter (http://xhunter.thegpm.org/) [89] or the proteotypic peptide profiler X!P3 (http://p3.thegpm.org/) [90] to analyze their data. All the identified peptides are matched to the Ensembl [59] genome database.

Data mining and visualization

The GPMDB main entry point is a web-based search interface, where users can search by keywords, data sources, protein identifiers, or gene names. Also, the current GPMDB implementation provides another three alternative pages to quickly access and search the data: (i) dataset search by accession (also called gpm #, http://gpmdb.thegpm.org/gpmnum.html); (ii) sequence search (also called sequence, http://gpmdb.

thegpm.org/seq.html); (iii) access based on GO terms [91], chromosome, Kyoto Encyclopedia of Genes and Genomes (KEGG) pathways [92], BRaunschweig ENzyme DAtabase (BRENDA) [93] tissue ontology (http://gpmdb.thegpm.org/go/index.html); and (iv) the amino acid polymorphisms and PTMs search for proteins.

When a search is performed, the main "protein table" lists all the matched proteins and different features are shown, such as protein accession, number of observations in GPMDB, log(e), and an evidence code that rates the current evidence for the observation of each protein. The "protein view" shows all the peptide identifications for a specific protein and the corresponding sequence coverage to specify the reliability of the identification.

More recently, GPMDB has been actively involved in the C-HPP project [94]. In this context, human protein identification information in GPMDB is now summarized into a collection of spreadsheets called "Guide to the Human Proteome." This guide contains the information organized into separate spreadsheets for each chromosome as well as for the mitochondrial DNA. It also contains protein accession numbers, HGNC (HUGO (Human Genome Organization) Genome Nomenclature Committee) [95] gene names, and chromosomal coordinates taken from Emsembl.

ProteomicsDB

ProteomicsDB (http://www.proteomicsdb.org/) is a human protein expression database that stores protein and peptide identifications and quantification values. The resource was jointly developed by the Technical University of Munich, the company SAP (Walldorf, Germany), and the SAP Innovation Center. It was announced in 2013 and it has been recently highlighted as the main output of a study drafting the human proteome [28]. It contains information of more than 62 projects and more than 300 experiments. The proteins identified in the resource map to over 18 000 human genes, representing around 90% of the human proteome. However, the FDR calculations when different datasets are combined were not taken into account in the analysis [18], as PeptideAtlas does [63, 71]. It currently contains approximately 70 million spectra from human cancer cell lines, tissues, and body fluids. ProteomicsDB enables real-time analysis and is based on the SAP HANA platform (http://www.saphana.com) for rapid data mining and visualization. It has been built to enable public sharing of datasets as well as to enable users to access and review data prior to publication.

Data submission and format support

The submission pipeline is well structured in three different levels: (i) projects that contain the general information, such as publication, title, and an experiment summary; (ii) experiments that include a name, description, and scope (the scope represents a definition of the experiment type, e.g., PTM or full tissue proteome); and (iii) experiment files—when the users upload a set of files, each file is sent to a verification process before it is definitely accepted or rejected. A large number of the experiments in the database were downloaded from repositories such as Tranche, PRIDE/PX, and PeptideAtlas [37], and later reprocessed using two parallel pipelines based on Andromeda/MaxQuant [96] and MASCOT, and quantified using MaxQuant [97].

Data mining and visualization

A user-friendly web interface allows users browse the human proteome, including protein-level information, such as protein function and expression. Protein expression can be visualized across the complete human body. This novel feature is integrated into the "Protein expression tab" and comprises the human body map showing the expression of a given protein in more than 30 tissues, organs and body fluids. Also, the protein expression of cell lines can be projected onto the tissue of origin and experimental details for any sample of interest can be readily obtained. The "chromosome view" shows those proteins identified in each chromosome region, including the description of the proteins, length, unique peptides, unique PSMs (peptide spectrum match), shared PSMs and sequence coverage.

MaxQB

MaxQB (http://maxqb.biochem.mpg.de/mxdb/) [29] was released in 2012 as a database that stores and displays collections of large proteomics projects and allows joint analysis and comparison. MaxQB serves as a generic repository and analysis platform for high-resolution bottom-up experiments. It stores details about protein and peptide identifications together with the corresponding high- or low-resolution fragment spectra and quantitative information, such as protein ratios or label-free derived intensities.

Data submission and format support

Differently to other databases, MaxQB only stores experiment data generated locally by the Mann group. This feature reduces greatly the data heterogeneity and facilitates the data analysis. To enable smooth upload of data, MaxQB is tightly integrated with MaxQuant [**97**]. When a MaxQuant analysis is performed, the user of MaxQuant is asked whether she/he wants to upload the data or not. This new feature will allow the integration of new datasets from others into MaxQB. Alternatively, the data can be manually uploaded through the web interface.

Additional metadata information should be provided in the submission process, such as the project name, experiment name, and workflow parameters. All the data are stored in a relational database. Several human protein sequence databases (UniProt including variants, Ensembl and IPI) were uploaded to MaxQB to build a reference protein database.

Data mining and visualization

The MaxQB web interface allows the users query the database by different fields, such as gene name, organism, and source database. Alternatively, an advance query builder can be used if the user is not familiar with the query syntax. Apart of protein detail information and sequence coverage, MaxQB can also display expression information within any of the proteomes compared with all other quantified proteins in that proteome. The expression of a given protein is estimated by the sum of its peptide signals after normalization of the total proteome signals to each other in MaxQuant. The iBAQ algorithm [98, 99], based on spectral counting, is now implemented in MaxQuant and can also be used to estimate protein quantification values. One of the major conclusions that MaxQB data analysis provides is that the peptide rank order can be used as a component of the protein identification score.

MOPED

MOPED (http://moped.proteinspire.org) [31, 100] is an expanding resource that enables rapid browsing of protein expression information coming from humans and several model organisms. MOPED provides protein-level expression data, meta-analysis capabilities, and quantitative data from standard analyses. In order to address the metadata diversity, the MOPED database developed a multi-"omics" metadata checklist that has been used for collecting metadata, which has been made available to the community through DELSA (www.delsaglobal.org).

Data submission and format support

In order to start a dataset submission, MOPED requires a minimum set of metadata that must be included at the experiment level. Users must then supply a brief experimental description, the source organism, and the journal reference. The MOPED pipeline [101] is based on reanalyzing MS data from other public repositories using SPIRE (Systematic Protein Investigative Research Environment http://proteinspire.org) [102]. SPIRE integrates open source search tools, such X!Tandem, OMSSA, and various statistical models into its pipeline. MOPED estimates protein absolute expression and concentration values using spectral counting. Since the probability of identifying peptides depends on peptide properties, such as the peptide sequence, the APEX quantitative approach weights spectral counts using estimates of these probabilities to improve the accuracy of the provided absolute expression values [101]. Each identified protein links to various protein and pathway databases, including GeneCards [103], UniProt, KEGG and Reactome [104]. The current MOPED database contains data from four of the most studied organisms: human, mouse, *Caenorhabditis elegans* **and** *Saccharomyces cerevisiae.*

Data mining and visualization

The web interface contains three main panels: "protein absolute expression" "protein relative expression" and "gene relative expression." Each of the panels contains the "MOPED search box" that supports queries by keywords, tissues, conditions, and pathways. After a search is performed, the "protein ID and expression summary" section displays expression data. Each protein row in the expression summary table displays the protein accession, description, concentration, organism, localization, sequence coverage, spectral count, and gene name.

The interface also features a "visualization panel" known as the chord diagram. It can break down proteins in experiments by organism, tissue, localization, and condition. The absolute and relative expression matrix in these panels shows the expression of the identified proteins per condition. In the future, MOPED plans to integrate proteomics with transcriptomics and metabolomics data through biological pathways and networks.

PaxDb

The first version of the PaxDb resource (http://pax-db.org/) [**32**] was published and released in 2012. PaxDb is a database dedicated to integrate information on absolute protein abundance levels, based on a deep coverage of the proteome, consistent data postprocessing, and enabling comparability across different organisms. PaxDb is a meta-resource since it takes the information previously published in data repositories, such as PRIDE and PeptideAtlas, and does not accept any direct data submissions.

Data submission and format support

The current PaxDb pipeline analyzes the spectral counting information from PeptideAtlas builds—that is, which peptides have been identified and how often over the whole build. This part of PaxDb's data import is entirely based on the original scoring and quality cutoffs implemented by PeptideAtlas. In the case of identified peptide sequences reported from MS/MS approaches (e.g., PRIDE datasets), the current pipeline remaps each peptide to the corresponding protein, based on sequence matches using reference genomes from the STRING database [105]. Protein abundance values are converted for each dataset into protein abundance estimates, using a consistent value across different techniques. Instead of using "molar concentration" or "molecules per cell," the system expresses all abundances in "parts per million." This means that each protein is listed relatively to all other protein molecules in the same sample. In the case of biochemical, biophysical, or label-free MS experiments, parts per million values are directly computed by rescaling the provided abundance estimates by their total sum. In the case of spectral counting data, the method is based on the likelihood of detectability, as described earlier [106].

Data mining and visualization

The PaxDb website allows an ad hoc query of a protein family of interest, allowing multiple proteins requests simultaneously, as well as browsing and comparing complete datasets. In addition, the user queries are searched against the annotations of all proteins in PaxDb using a full-text search. For each organism, a distinct summary page provides information on the data origin, its coverage and estimated quality, and the distribution of abundance values of each dataset as a histogram. Additionally, the most abundant proteins in the organism are also listed. The PaxDb protein view shows a brief description of the protein functional role as annotated by UniProt and/or by other model organism databases. The species browser allows the navigation through all the proteins for a specific model organism. Furthermore, it shows all the protein abundances in a specific dataset and also for the whole of the PaxDb datasets (addition of all datasets for a given specific species). All PaxDb data are freely available.

Human Proteinpedia

Human Proteinpedia (http://www.humanproteinpedia.org) [33] is a public human proteome repository for sharing protein expression data derived from multiple experimental platforms. It incorporates diverse features of the human proteome, including protein–protein interactions, enzyme–substrate relationships, PTMs, subcellular localization, and expression of proteins in various human tissues and cell lines, in diverse biological conditions (including disease states). The Human Proteinpedia database falls neatly in the annotation-oriented database category [38]. It complements the curated human protein reference database (HPRD) [107] with community-provided annotation. It collects submitted proteomics data and uses the findings to directly annotate the protein entries in HPRD with observed PTMs and subcellular localization [108]. In contrast with the resources previously described, Human Proteinpedia does not only contain MS-based experiments, but also other approaches such as coimmunoprecipitation, Western blotting, fluorescence, immunohistochemistry, and protein and peptide microarrays. All the data should be annotated by the users (http://pdas.hprd.org/) and it is not reprocessed.

Data submission and format support

Users can provide data annotations in four different ways, after registering in the system: (i) data on individual proteins along with experimental evidence through the use of web forms, (ii) upload data via the web in a batch mode, (iii) sending data through FTP/e-mail to the support team, and (iv) distributed annotation service servers setup by the contributing laboratories for the data upload [109]. By June 2014, the current version includes data from more than 249 laboratories, including 2710 distinct experiments, more than 15 000 proteins, almost 2 million peptides, around 5 million spectra, and 2906 annotations related to subcellular localization (Table 1).

The major differences with other repositories are as follows: (i) it does not exclusively contain MS-derived data, as mentioned already; (ii) data from proteomics experiments are viewed in the context of a protein–protein interaction resource (HPRD); (iii) it restricts the data to that derived from human tissues or cell lines; and (iv) data annotation related to various protein features can be done manually.

Data mining and visualization

The query page in Human Proteinpedia (http://www.humanproteinpedia.org/query) enables the search by gene symbol, protein name, accession number, type of protein feature, and type of experimental platform annotated. After a query, the "protein detail view" includes a table where each row shows a "platform" (the type of the experiment, e.g., MS or immunohistochemistry) and related metadata. In the case of MS experiments, each platform contains a table containing all the peptides identified for each protein. All the data are freely available (http://www.humanproteinpedia.org/download).

HPM

The HPM (http://www.humanproteomemap.org) [34] has just been developed as an output of a recent draft study of the human proteome. The original proteomics study [34] was carried out on 30 histologically normal human tissues and primary cells using high-resolution MS. The generated tandem mass spectra correspond to proteins encoded by 17 294 genes, accounting for approximately 84% of the annotated protein-coding genes in the human genome. However, FDR calculations when different datasets are combined are not discussed in enough detail in the original manuscript [18]. The aim of the HPM is to make possible to review, navigate and visualize the protein expression information evidences of gene families, protein complexes, signaling pathways and biomarkers.

Data submission and format support

MS/MS data obtained from all different experiments were searched against the Human RefSeq database using SEQUEST [**110**] and MASCOT [**56**] through the Proteome Discoverer platform (Thermo Fisher Scientific). Then, q **values were estimated using the Percolator algorithm within the Proteome Discoverer suite. Protein and peptide identifications obtained from SEQUEST and MASCOT were converted into MySQL tables. NCBI RefSeq annotations were used as additional information about the genes from various public resources. Normalized spectral counts [34]** were used to represent expression of proteins and peptides. The resource was developed as a protein expression database and do not support the submission of new data from external users.

Data mining and visualization

The web portal was developed using a three-tier web architecture with presentation, application, and persistence layers. Users are able to search using gene or protein identifiers and the information is presented in a tabular format. For each peptide, a high-resolution MS/MS spectrum from the best scoring identification is shown on the spectrum viewer page using the "Lorikeet" JQuery plugin (https://code.google.com/p/lorikeet).

Other proteomics resources

There are other less widely used resources that will not be explained here in detail. First of all, the Cardiac Organellar Protein Atlas Knowledgebase (COPaKB; http://www.heartproteome.org/copa/) [111] is a centralized platform of high-quality cardiac proteomics data, bioinformatics tools, and relevant cardiovascular phenotypes. Currently, COPaKB features eight organellar modules, comprising 4203 MS/MS experiments from human, mouse, Drosophila and C. elegans as well as expression data of 10 924 proteins in the human myocardium. COPaKB has an attractive web interface which provides a number of effective workflows to guide cardiovascular investigators from the actual proteomics data to the systematic biomedical interpretation. The COPaKB team continues to develop innovative workflows to help providing a better understanding of protein functions in cardiovascular diseases. COPaKB will cover additional modules on organelles and cells from cardiovascular-relevant model systems. The content of each module will expand with the available public data.

Pep2pro (http://fgcz-pep2pro.uzh.ch/) [112, 113] is a comprehensive analysis database specifically suitable for performing flexible data analysis. Pep2pro is a further development of the "AtProteome" resource and provides data from Arabidopsis thaliana. The current pipeline employs PepSplice [114] and TPP tools for peptide/protein identification, the characterization of whole genome hits and the PTMs. The database is organized in assemblies, similarly to the PeptideAtlas builds. The Arabidopsis assembly from 2011 contained more than 14 522 protein identifications, 141 235 identified peptides, and around 2 million spectra.

iProX (Integrated proteome resources, http://www.iprox.org/) is a repository jointly developed by Beijing Proteome Research Center and other institutions in China. It has been recently developed reusing part of the PRIDE source code. At the moment of writing, it is not fully operational but contains already some stored datasets.

Proteomics information available through UniProt and neXtProt

UniProt (http://www.uniprot.org) [115] is among the most used of the protein sequence and functional annotation providers. Among the UniProt databases is the UniProtKB, which provides a broad range of protein sequence datasets for a large number of species, specifically tailored for an effective coverage of the sequence space while maintaining a high-quality level of sequence annotations and mappings to the genomics and proteomics information.

MS proteomics data deposited in the main public repositories is flowing into UniProtKB to enrich protein sequence annotations at the level of the evidence, supporting the existence of a protein (isoforms and variant-containing sequences included). This information is thus provided to users mainly in two different ways: (i) indirectly via the UniProtKB protein existence values (http://www.uniprot.org/manual/protein_existence) that are starting to be assigned also on the PSM-level content publicly provided by proteomics repositories (together with the accompanying statistical assessment for each PSM) and (ii) directly through explicit links to relevant cross-referenced resources (http://www.uniprot.org/database), which cover PRIDE, PeptideAtlas, MaxQB and PaxDb, but also others such as PhosphoSitePlus [116].

Organism-specific mappings of the peptides reported in the repositories to the UniProtKB sequences are also shared with groups producing genome builds to improve the corresponding gene annotations. In the future, it is planned that PTMs [117] from proteomics repositories will also be integrated in UniProtKB.

neXtProt (http://www.nextprot.org) [43] is a web-based protein knowledge platform to support research uniquely on human proteins. The set of manually curated annotations extracted from UniProtKB/Swiss-Prot for human, which constitutes the heart of the resource, is constantly complemented with quality-filtered carefully selected high-throughput experiments from different scientific research areas concerning abundance, distribution, subcellular localization, interactions, and cellular functions. neXtProt is part of the consortium, which is driving the C-HPP project. Thus, neXtProt labels as "missing proteins" those ones, which so far lack experimental evidence [43].

All the relevant proteomics MS-related information (like for instance PTM-related MS-based papers on *N*-glycosylation, phosphorylation, S-nitrosylation, ubiquitination, and sumoylation) has been integrated into neXtProt and is available via the web interface in the "Proteomics view" of the "Protein perspective" layout.

DATA REUSE FROM PUBLIC RESOURCES

Since the current volume of proteomics data deposition is rapidly increasing, new approaches based on the reanalysis of the data and/or new uses of the stored data are being developed. The same public dataset can be analyzed using different pipelines to discover, confirm, or highlight new biological evidences. Resources such as GPMDB and PeptideAtlas have been doing this for many years already, emphasizing control of the number of false-positives at both peptide and protein level.

Data availability per se in repositories has enabled data reanalysis by third parties triggering a discussion in the field about controversial datasets [118-122]. Furthermore, targeted reanalysis of public data by individual groups is now starting to flourish. Remarkably, the draft of the human proteome reported recently and available in ProteomicsDB includes a big proportion of reanalyzed public datasets (roughly 40% of the MS runs) [28]. In addition, new PTMs have also been described as a result of reanalysis of public datasets with a different purpose in mind. For instance, two recent studies described new PTMs after reanalyzing available phosphoproteomics studies [123, 124]. Some proteogenomics studies have also made use of public datasets [125].

Data from repositories is often used in the design of SRM/MRM transitions for targeted proteomics approaches [126]. A mentioned already, GPMDB and PeptideAtlas provide already this functionality. Data from PRIDE is being used by the MRMAID resource, with the same purpose in mind [127]. Another popular data reuse is the building of spectral libraries. Several repositories build their own libraries (e.g., PeptideAtlas, PRIDE) that can be used in spectral searches. In addition, meta-analysis of data in PRIDE (without reprocessing the data) has already happened, enabling the extraction of new knowledge [128-132]. Finally, data in proteomics repositories can also be used to improve the content of the protein sequence databases [133].

In this context, PeptideShaker (http://peptide-shaker.googlecode.com/) is a recently developed Java application [134, 135] that allows the postprocessing and visualization of protein identification experiments and it greatly facilitates the reprocessing of PRIDE datasets by using the

"PRIDE Reshake" functionality. One of the key features of PeptideShaker is the validation of the results using different values, such as FDR and false-negative rate [136]. These values are displayed in the "FDR/false-negative rate plot" and the availability of these metrics make possible the generation of a "cost/benefit" curve, also known as receiver operating characteristic curve, which enables the users to optimize the quality thresholds.

Finally, it is also important to highlight that not only the publicly available raw data files are used for reanalysis purposes. The availability of output files from pipelines/software can also help third parties to develop new tools and demonstrate its utility. This was the case for the recently developed SpliceVista visualization tool [137].

We strongly encourage proteomics researchers to upload raw data and processed results to public repositories. Before starting a new project, public data can be used to guide new research. This is already common practice in the case of targeted proteomics workflows.

PITFALLS AND FUTURE CHALLENGES

Data sharing and dissemination is a nontrivial task since it requires substantial investment in infrastructure and software development [138, 139]. As mentioned before, one of the aims of the PX consortium is to provide backup when one of the resources has funding problems. As a proof of concept, PRIDE and the defunct Peptidome joined forces to transfer all data from Peptidome to PRIDE [140]. In parallel, as mentioned already, MassIVE has tried to rescue as many datasets originally stored in Tranche as possible, but it has been quite a challenging task.

In June 2012, PRIDE started to accept and handle raw data as part of the PX data workflow. PX and other resources have proven that public data dissemination in a decentralized and well-structured mode is possible and actually important for the proteomics community. In our opinion, the next step would be to foster a higher integration for these resources. At present, the main layer of integration among PX resources happens at the level of the metadata.

The alternative is to access knowledge bases, such as UniProt or neXtProt, but the amount of data coming from MS proteomics repositories in these resources is still limited. Ideally it should be possible for users to look for all available data for a given protein or peptide (including PTMs) in a more straightforward way. At present, scientists need to access the different proteomics resources when they want to get all existing information about a given protein.

To illustrate this idea, we studied the protein expression evidences for the UniProtKB/Swiss-Prot human proteome, including only canonical sequences (release 2014_05, 20,265 entries) in seven resources that have a uniform data processing pipeline: GPMDB, PeptideAtlas, MaxQB, Human Proteinpedia, PaxDb, ProteomicsDB and the HPM. This proved to be a far from trivial task (see all the retrieved data in Supporting Information, including the protein lists taken from each resource). A total of 7880 proteins represent the "core" proteome since they are stored in all the resources (Fig. 3A). In addition, 4244, 3074, 2008, 1135, 815 and 824 proteins were identified in six, five, four, three, two and only one resource, respectively (Fig. 3A). Only 285 proteins were not found in any of these seven resources. In fact, most of the described resources shared more than 60% of the protein identifications among them (Supporting Information). In our opinion, this redundancy can be used as a "reliability" measure of the evidence of protein expression. For instance, the 13 016 proteins identified in GPMDB, PeptideAtlas, and ProteomicsDB are more likely to be true-positives than the 1171 proteins identified only by ProteomicsDB (Fig. 3B). At the same time, the "exclusive" identifications in ProteomicsDB are new valuable evidences, but probably need further investigation.

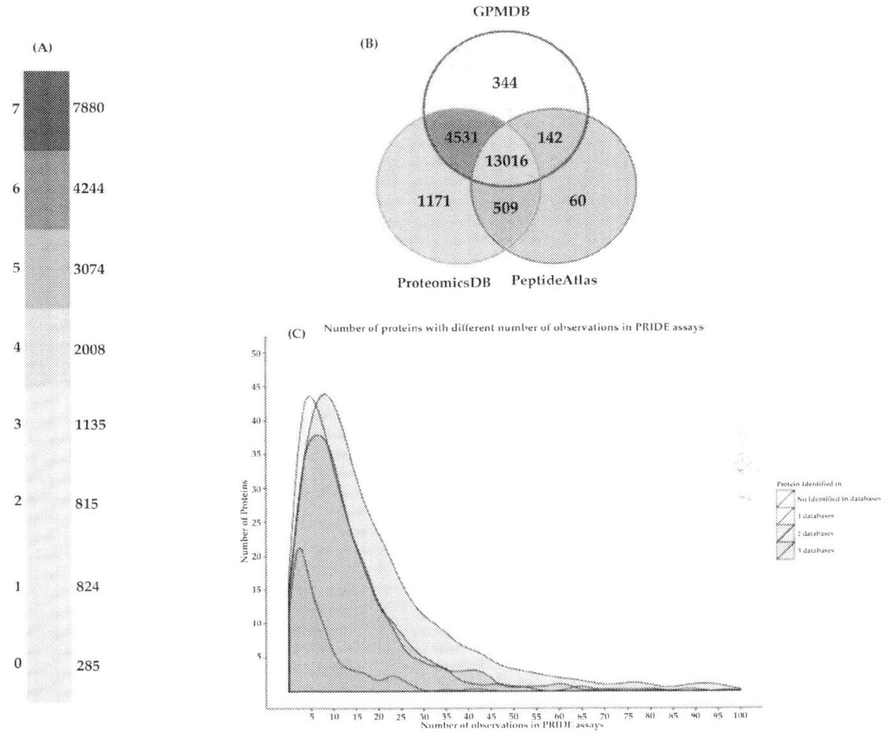

Figure 3: (A) Number of UniProtKB/Swiss-Prot human proteins (release 2014_05, 20,265 entries) observed in different proteomics resources that have a uniform data processing pipeline (GPMDB, ProteomicsDB, PeptideAtlas, HPM, PaxDb, MaxQB, Human Proteinpedia; PRIDE is not included); (B) Venn diagram representing the human protein identifications observed in GPMDB, PeptideAtlas, and ProteomicsDB; (C) Area chart showing the distribution of the number of PRIDE assays for those proteins present in three, two, and one proteomics resources, or for those proteins not identified at all.

As mentioned already, the identification of false-positives in big resources and/or when different datasets are combined is still a problem [18, 26] and protein expression resources should be as thorough as possible in the statistical analysis. With the identification of a high proportion of the proteins in the human proteome [28, 34, 63], new analysis tools should be implemented to retrieve protein expression evidences from different resources and to detect/highlight those identifications considered as the most and/or least reliable.

In our opinion it is also needed to increase the reuse and reanalysis of the available data, since they can provide new valuable scientific knowledge. To illustrate this idea, Fig. 3C shows the distribution of the number of PRIDE assays for those proteins present in three, two, and one proteomics resources (Fig. 3A), or for those proteins not identified at all. It can be observed that, when a protein is identified in three resources, those proteins are likely to be identified, in average, in 13 different PRIDE assays. More interesting is the case of the least observed proteins. For example, if a protein is observed in only one resource, it is observed in nine different PRIDE assays in average. In addition, those 285 proteins without any evidence can be observed in an average of five different PRIDE assays. Therefore, the number of times one given protein is reported in PRIDE (especially if not present in other resources) could be used to select datasets to be reanalyzed.

Finally, another challenge is that at present, studies integrating different "omics" technologies are becoming more and more popular. This type of studies poses a challenge for traditional repositories (which are usually field-specific) and researchers since it is not straightforward to link data from different approaches, for instance MS proteomics and RNAseq data obtained in the same study. Some big institutes such as the EBI and NCBI have implemented specific databases for BioSamples [141] to enable the linking between different studies performed using the same sample. The use of sample identifiers is now starting to facilitate the connection between different "omics" datasets.

CONCLUSIONS

In recent years, there has been a big progress in the development of proteomics repositories and protein expression databases. However, more integration between those databases is required. More tools, such as PRIDE Inspector and PeptideShaker, should be developed to make easier the visualization, reuse, and reanalysis of the data in repositories. Simultaneously, in order to engage users, it seems advisable to monitor download activity of public datasets and measure the community interest on them. Since proteomics has become such a fundamental part of biological research, it is expected that the amount of information available in proteomics resources will keep growing in the coming years. We encourage researchers to make use of this plethora of data.

ACKNOWLEDGMENTS

Y.P.R is supported by the BBSRC "PROCESS" grant (reference BB/K01997X/1). R.W. is supported by the BBSRC "Quantitative Proteomics" grant (reference BB/I00095X/1). J.A.V. is supported by the Wellcome Trust (grant number WT101477MA) and the EU FP7 grants "ProteomeXchange" (grant number 260558) and PRIME-XS (grant number 262067).

The authors thank Michael MacCoss (Chorus), Christoph Schaab (MaxQB), Eugene Kolker and Roger Higdon (MOPED), Advani Jayshree (Human Proteinpedia), Ronald Beavis (GPMDB), Mingcong Wang (PaxDb), and Nuno Bandeira (MassIVE) for providing the original data used in Fig. 3 and in the Supporting Information, and for the general feedback about their resources.

Glossary

C-HPP	chromosome-based Human Proteome Project
COPaKB	Cardiac Organellar Protein Atlas Knowledgebase
FDR	false discovery rate
HPM	human proteome map
HPRD	human protein reference database
MassIVE	Mass Spectrometry Interactive Virtual Environment
MOPED	Model Organism Protein Expression Database
NCBI	National Center for Biotechnology Information
PASSEL	PeptideAtlas SRM Experiment Library
PSM	peptide spectrum match
PX	ProteomeXchange
TIQAM	targeted identification for quantitative analysis by MRM
TPP	trans-proteomic pipeline

REFERENCES

1. Altelaar AF, Munoz J, Heck AJ. Next-generation proteomics: towards an integrative view of proteome dynamics. Nat. Rev. Genet. 2013;14:35–48.
2. Betancourt LH, De Bock PJ, Staes A, Timmerman E. SCX charge

state selective separation of tryptic peptides combined with 2D-RP-HPLC allows for detailed proteome mapping. J. Proteomics. 2013;91:164–171. et al.

3. Branca RM, Orre LM, Johansson HJ, Granholm V. HiRIEF LC-MS enables deep proteome coverage and unbiased proteogenomics. Nat. Methods. 2014;11:59–62. et al.
4. Ramos Y, Gutierrez E, Machado Y, Sanchez A. Proteomics based on peptide fractionation by SDS-free PAGE. J. Proteome Res. 2008;7:2427–2434. et al.
5. Ramos Y, Garcia Y, Perez-Riverol Y, Leyva A. Peptide fractionation by acid pH SDS-free electrophoresis. Electrophoresis. 2011;32:1323–1326. et al.
6. Wisniewski JR, Zougman A, Nagaraj N, Mann M. Universal sample preparation method for proteome analysis. Nat. Methods. 2009;6:359–362.
7. Michalski A, Damoc E, Hauschild JP, Lange O. Mass spectrometry-based proteomics using Q Exactive, a high-performance benchtop quadrupole Orbitrap mass spectrometer. Mol. Cell. Proteomics. 2011;10:M111 011015. et al.
8. UniProt C. Update on activities at the Universal Protein Resource (UniProt) in 2013. Nucl. Acids Res. 2013;41:D43–D47.
9. Perez-Riverol Y, Wang R, Hermjakob H, Muller M. Open source libraries and frameworks for mass spectrometry based proteomics: a developer›s perspective. Biochim. Biophys. Acta. 2014;1844:63–76. et al.
10. Beck M, Schmidt A, Malmstroem J, Claassen M. The quantitative proteome of a human cell line. Mol. Syst. Biol. 2011;7:549. et al.
11. Nagaraj N, Wisniewski JR, Geiger T, Cox J. Deep proteome and transcriptome mapping of a human cancer cell line. Mol. Syst. Biol. 2011;7:548. et al.
12. Csordas A, Ovelleiro D, Wang R, Foster JM. PRIDE: quality control in a proteomics data repository. Database. 2012;2012:bas004. et al.
13. Credit where credit is overdue. Nat. Biotechnol. 2009;27:579.
14. DeSouza LV, Siu KW. Mass spectrometry-based quantification. Clin. Biochem. 2013;46:421–431.

15. Neilson KA, Ali NA, Muralidharan S, Mirzaei M. Less label, more free: approaches in label-free quantitative mass spectrometry. Proteomics. 2011;11:535–553. et al.
16. Allmer J. Algorithms for the de novo sequencing of peptides from tandem mass spectra. Expert Rev. Proteomics. 2011;8:645–657.
17. Hoopmann MR, Moritz RL. Current algorithmic solutions for peptide-based proteomics data generation and identification. Curr. Opin. Biotechnol. 2013;24:31–38.
18. Ezkurdia I, Vazquez J, Valencia A, Tress M. Analyzing the first drafts of the human proteome. J. Proteome Res. 2014;13:3854–3855.
19. Gupta N, Pevzner PA. False discovery rates of protein identifications: a strike against the two-peptide rule. J. Proteome Res. 2009;8:4173–4181.
20. Kinsinger CR, Apffel J, Baker M, Bian X. Recommendations for mass spectrometry data quality metrics for open access data (corollary to the Amsterdam principles) Proteomics. 2012;12:11–20. et al.
21. Anonymous. A home for raw proteomics data. Nat. Methods. 2012;9:419.
22. Distler U, Kuharev J, Navarro P, Levin Y. Drift time-specific collision energies enable deep-coverage data-independent acquisition proteomics. Nat. Methods. 2014;11:167–170. et al.
23. Lanucara F, Holman SW, Gray CJ, Eyers CE. The power of ion mobility-mass spectrometry for structural characterization and the study of conformational dynamics. Nat. Chem. 2014;6:281–294.
24. Riffle M, Eng JK. Proteomics data repositories. Proteomics. 2009;9:4653–4663.
25. Craig R, Cortens JP, Beavis RC. Open source system for analyzing, validating, and storing protein identification data. J. Proteome Res. 2004;3:1234–1242.
26. Farrah T, Deutsch EW, Omenn GS, Sun Z. State of the human proteome in 2013 as viewed through PeptideAtlas: comparing the kidney, urine, and plasma proteomes for the biology- and disease-driven Human Proteome Project. J. Proteome Res. 2014;13:60–75. et al.

27. Vizcaino JA, Cote RG, Csordas A, Dianes JA. The PRoteomics IDEntifications (PRIDE) database and associated tools: status in 2013. Nucl. Acids Res. 2013;41:D1063–D1069. et al.
28. Wilhelm M, Schlegl J, Hahne H, Moghaddas Gholami A. Mass-spectrometry-based draft of the human proteome. Nature. 2014;509:582–587. et al.
29. Schaab C, Geiger T, Stoehr G, Cox J, Mann M. Analysis of high accuracy, quantitative proteomics data in the MaxQB database. Mol. Cell. Proteomics. 2012;11:M111.014068.
30. Farrah T, Deutsch EW, Kreisberg R, Sun Z. PASSEL: the PeptideAtlas SRM experiment library. Proteomics. 2012;12:1170–1175. et al.
31. Montague E, Stanberry L, Higdon R, Janko I. MOPED 2.5—an integrated multi-omics resource: multi-omics profiling expression database now includes transcriptomics data. Omics J. Integr. Biol. 2014;18:335–343. et al.
32. Wang M, Weiss M, Simonovic M, Haertinger G. PaxDb, a database of protein abundance averages across all three domains of life. Mol. Cell. Proteomics. 2012;11:492–500. et al.
33. Kandasamy K, Keerthikumar S, Goel R, Mathivanan S. Human Proteinpedia: a unified discovery resource for proteomics research. Nucl. Acids Res. 2009;37:D773–D781. et al.
34. Kim MS, Pinto SM, Getnet D, Nirujogi RS. A draft map of the human proteome. Nature. 2014;509:575–581. et al.
35. Slotta DJ, Barrett T, Edgar R. NCBI peptidome: a new public repository for mass spectrometry peptide identifications. Nat. Biotechnol. 2009;27:600–601.
36. Smith BE, Hill JA, Gjukich MA, Andrews PC. Tranche distributed repository and ProteomeCommons.org. Methods Mol. Biol. 2011;696:123–145.
37. Vizcaino JA, Deutsch EW, Wang R, Csordas A. ProteomeXchange provides globally coordinated proteomics data submission and dissemination. Nat. Biotechnol. 2014;32:223–226.et al.
38. Martens L. Proteomics databases and repositories. Methods Mol. Biol. 2011;694:213–227.
39. Vizcaino JA, Foster JM, Martens L. Proteomics data repositories:

providing a safe haven for your data and acting as a springboard for further research. J. Proteomics. 2010;73:2136–2146.

40. Mead JA, Bianco L, Bessant C. Recent developments in public proteomic MS repositories and pipelines. Proteomics. 2009;9:861–881.
41. Mead JA, Shadforth IP, Bessant C. Public proteomic MS repositories and pipelines: available tools and biological applications. Proteomics. 2007;7:2769–2786.
42. The UniProtConsortium. Activities at the Universal Protein Resource (UniProt) Nucl. Acids Res. 2014;42:D191–D198.
43. Gaudet P, Argoud-Puy G, Cusin I, Duek P. neXtProt: organizing protein knowledge in the context of Human Proteome Projects. J. Proteome Res. 2013;12:293–298. et al.
44. Olsen JV, Mann M. Effective representation and storage of mass spectrometry-based proteomic data sets for the scientific community. Sci. Signal. 2011;4:pe7.
45. Martens L, Chambers M, Sturm M, Kessner D. mzML—a community standard for mass spectrometry data. Mol. Cell. Proteomics. 2011;10:R110.000133. et al.
46. Jones AR, Eisenacher M, Mayer G, Kohlbacher O. The mzIdentML data standard for mass spectrometry-based proteomics results. Mol. Cell. Proteomics. 2012;11:M111.014381. et al.
47. Walzer M, Qi D, Mayer G, Uszkoreit J. The mzQuantML data standard for mass spectrometry-based quantitative studies in proteomics. Mol. Cell. Proteomics. 2013;12:2332–2340. et al.
48. Pedrioli PG, Eng JK, Hubley R, Vogelzang M. A common open representation of mass spectrometry data and its application to proteomics research. Nat. Biotechnol. 2004;22:1459–1466. et al.
49. Deutsch EW, Mendoza L, Shteynberg D, Farrah T. A guided tour of the trans-proteomic pipeline. Proteomics. 2010;10:1150–1159. et al.
50. Deutsch EW, Chambers M, Neumann S, Levander F. TraML—a standard format for exchange of selected reaction monitoring transition lists. Mol. Cell. Proteomics. 2012;11:R111.015040. et al.
51. Griss J, Jones AR, Sachsenberg T, Walzer M. The mzTab data exchange format: communicating MS-based proteomics and

metabolomics experimental results to a wider audience. Mol. Cell. Proteomics. 2014;pii:mcp.O113.036681. et al.

52. Mayer G, Jones AR, Binz PA, Deutsch EW. Controlled vocabularies and ontologies in proteomics: overview, principles and practice. Biochim. Biophys. Acta. 2014;1844:98–107.et al.
53. Cote R, Reisinger F, Martens L, Barsnes H. The ontology lookup service: bigger and better. Nucl. Acids Res. 2010;38:W155–W160. et al.
54. Wang R, Fabregat A, Rios D, Ovelleiro D. PRIDE inspector: a tool to visualize and validate MS proteomics data. Nat. Biotechnol. 2012;30:135–137. et al.
55. [55] Cote RG, Griss J, Dianes JA, Wang R. The PRoteomics IDEntification (PRIDE) converter 2 framework: an improved suite of tools to facilitate data submission to the PRIDE database and the ProteomeXchange consortium. Mol. Cell. Proteomics. 2012;11:1682–1689. et al.
56. Perkins DN, Pappin DJ, Creasy DM, Cottrell JS. Probability-based protein identification by searching sequence databases using mass spectrometry data. Electrophoresis. 1999;20:3551–3567.
57. Craig R, Beavis RC. TANDEM: matching proteins with tandem mass spectra. Bioinformatics. 2004;20:1466–1467.
58. Ternent T, Csordas A, Qi D, Gomez-Baena G. Standardization and guidelines: how to submit MS proteomics data to ProteomeXchange via the PRIDE database. Proteomics. 2014et al.,, in press.
59. Flicek P, Amode MR, Barrell D, Beal K. Ensembl 2014. Nucl. Acids Res. 2014;42:D749–D755. et al.
60. Griss J, Foster JM, Hermjakob H, Vizcaino JA. PRIDE cluster: building a consensus of proteomics data. Nat. Methods. 2013;10:95–96.
61. Reiter L, Rinner O, Picotti P, Huttenhain R. mProphet: automated data processing and statistical validation for large-scale SRM experiments. Nat. Methods. 2011;8:430–435. et al.
62. Picotti P, Lam H, Campbell D, Deutsch EW. A database of mass spectrometric assays for the yeast proteome. Nat. Methods. 2008;5:913–914. et al.

63. Farrah T, Deutsch EW, Hoopmann MR, Hallows JL. The state of the human proteome in 2012 as viewed through PeptideAtlas. J. Proteome Res. 2013;12:162–171. et al.
64. Desiere F, Deutsch EW, Nesvizhskii AI, Mallick P. Integration with the human genome of peptide sequences obtained by high-throughput mass spectrometry. Genome Biol. 2005;6:R9.et al.
65. Lam H, Aebersold R. Building and searching tandem mass (MS/MS) spectral libraries for peptide identification in proteomics. Methods. 2011;54:424–431.
66. Picotti P, Rinner O, Stallmach R, Dautel F. High-throughput generation of selected reaction-monitoring assays for proteins and proteomes. Nat. Methods. 2010;7:43–46. et al.
67. Deutsch EW, Lam H, Aebersold R. PeptideAtlas: a resource for target selection for emerging targeted proteomics workflows. EMBO Rep. 2008;9:429–434.
68. Mallick P, Schirle M, Chen SS, Flory MR. Computational prediction of proteotypic peptides for quantitative proteomics. Nat. Biotechnol. 2007;25:125–131. et al.
69. Keller A, Nesvizhskii AI, Kolker E, Aebersold R. Empirical statistical model to estimate the accuracy of peptide identifications made by MS/MS and database search. Anal. Chem. 2002;74:5383–5392.
70. Nesvizhskii AI, Keller A, Kolker E, Aebersold R. A statistical model for identifying proteins by tandem mass spectrometry. Anal. Chem. 2003;75:4646–4658.
71. Reiter L, Claassen M, Schrimpf SP, Jovanovic M. Protein identification false discovery rates for very large proteomics data sets generated by tandem mass spectrometry. Mol. Cell. Proteomics. 2009;8:2405–2417. et al.
72. Farrah T, Deutsch EW, Aebersold R. Using the Human Plasma PeptideAtlas to study human plasma proteins. Methods Mol. Biol. 2011;728:349–374.
73. Uhlen M, Bjorling E, Agaton C, Szigyarto CA. A human protein atlas for normal and cancer tissues based on antibody proteomics. Mol. Cell. Proteomics. 2005;4:1920–1932. et al.

74. Saito R, Smoot ME, Ono K, Ruscheinski J. A travel guide to cytoscape plugins. Nat. Methods. 2012;9:1069–1076. et al.

75. Paik YK, Jeong SK, Omenn GS, Uhlen M. The chromosome-centric Human Proteome Project for cataloging proteins encoded in the genome. Nat. Biotechnol. 2012;30:221–223. et al.

76. Lange V, Malmstrom JA, Didion J, King NL. Targeted quantitative analysis of Streptococcus pyogenes virulence factors by multiple reaction monitoring. Mol. Cell. Proteomics. 2008;7:1489–1500. et al.

77. Brusniak MY, Kwok ST, Christiansen M, Campbell D. ATAQS: a computational software tool for high throughput transition optimization and validation for selected reaction monitoring mass spectrometry. BMC Bioinform. 2011;12:78. et al.

78. Shannon PT, Reiss DJ, Bonneau R, Baliga NS. The Gaggle: an open-source software system for integrating bioinformatics software and data sources. BMC Bioinform. 2006;7:176.

79. Kim S, Gupta N, Pevzner PA. Spectral probabilities and generating functions of tandem mass spectra: a strike against decoy databases. J. Proteome Res. 2008;7:3354–3363.

80. Castellana NE, Payne SH, Shen Z, Stanke M. Discovery and revision of *Arabidopsis* genes by proteogenomics. Proc. Natl. Acad. Sci. USA. 2008;105:21034–21038. et al.

81. Woo S, Cha SW, Merrihew G, He Y. Proteogenomic database construction driven from large scale RNA-seq data. J. Proteome Res. 2014;13:21–28. et al.

82. Castellana NE, Shen Z, He Y, Walley JW. An automated proteogenomic method uses mass spectrometry to reveal novel genes in *Zea mays*. Mol. Cell. Proteomics. 2014;13:157–167. et al.

83. Na S, Bandeira N, Paek E. Fast multi-blind modification search through tandem mass spectrometry. Mol. Cell. Proteomics. 2012;11:M111.010199.

84. Wang J, Perez-Santiago J, Katz JE, Mallick P, Bandeira N. Peptide identification from mixture tandem mass spectra. Mol. Cell. Proteomics. 2010;9:1476–1485.

85. Wang J, Bourne PE, Bandeira N. Peptide identification by database

search of mixture tandem mass spectra. Mol. Cell. Proteomics. 2011;10:M111.010017.

86. Frank A, Pevzner P. PepNovo: de novo peptide sequencing via probabilistic network modeling. Anal. Chem. 2005;77:964–973.

87. Guthals A, Clauser KR, Bandeira N. Shotgun protein sequencing with meta-contig assembly. Mol. Cell. Proteomics. 2012;11:1084–1096.

88. Liu X, Sirotkin Y, Shen Y, Anderson G. Protein identification using top-down. Mol. Cell. Proteomics. 2012;11:M111.008524. et al.

89. Craig R, Cortens JC, Fenyo D, Beavis RC. Using annotated peptide mass spectrum libraries for protein identification. J. Proteome Res. 2006;5:1843–1849.

90. Craig R, Cortens JP, Beavis RC. The use of proteotypic peptide libraries for protein identification. Rapid Commun. Mass Spectrom. 2005;19:1844–1850.

91. Ashburner M, Ball CA, Blake JA, Botstein D. Gene ontology: tool for the unification of biology. Nat. Genet. 2000;25:25–29. et al.,. The Gene Ontology Consortium.

92. Kanehisa M, Goto S. KEGG: Kyoto encyclopedia of genes and genomes. Nucl. Acids Res. 2000;28:27–30.

93. Gremse M, Chang A, Schomburg I, Grote A. The BRENDA tissue ontology (BTO): the first all-integrating ontology of all organisms for enzyme sources. Nucl. Acids Res. 2011;39:D507–D513. et al.

94. Lane L, Bairoch A, Beavis RC, Deutsch EW. Metrics for the Human Proteome Project 2013–2014 and strategies for finding missing proteins. J. Proteome Res. 2014;13:15–20. et al.

95. Wain HM, Bruford EA, Lovering RC, Lush MJ. Guidelines for human gene nomenclature. Genomics. 2002;79:464–470. et al.

96. Cox J, Neuhauser N, Michalski A, Scheltema RA. Andromeda: a peptide search engine integrated into the MaxQuant environment. J. Proteome Res. 2011;10:1794–1805. et al.

97. Cox J, Mann M. MaxQuant enables high peptide identification rates, individualized p.p.b.-range mass accuracies and proteome-wide protein quantification. Nat. Biotechnol. 2008;26:1367–1372.

98. Schwanhausser B, Busse D, Li N, Dittmar G. Global quantification

of mammalian gene expression control. Nature. 2011;473:337–342. et al.

99. Schwanhausser B, Busse D, Li N, Dittmar G. Corrigendum: global quantification of mammalian gene expression control. Nature. 2013;495:126–127. et al.

100. Kolker E, Higdon R, Haynes W, Welch D. MOPED: model organism protein expression database. Nucl. Acids Res. 2012;40:D1093–D1099. et al.

101. Higdon R, Stewart E, Stanberry L, Haynes W. MOPED enables discoveries through consistently processed proteomics data. J. Proteome Res. 2014;13:107–113. et al.

102. Kolker E, Higdon R, Morgan P, Sedensky M. SPIRE: systematic protein investigative research environment. J. Proteomics. 2011;75:122–126. et al.

103. Safran M, Dalah I, Alexander J, Rosen N. GeneCards Version 3: the human gene integrator. Database. 2010;2010:baq020. et al.

104. D›Eustachio P. Reactome knowledgebase of human biological pathways and processes. Methods Mol. Biol. 2011;694:49–61.

105. Szklarczyk D, Franceschini A, Kuhn M, Simonovic M. The STRING database in 2011: functional interaction networks of proteins, globally integrated and scored. Nucl. Acids Res. 2011;39:D561–D568. et al.

106. Weiss M, Schrimpf S, Hengartner MO, Lercher MJ, von Mering C. Shotgun proteomics data from multiple organisms reveals remarkable quantitative conservation of the eukaryotic core proteome. Proteomics. 2010;10:1297–1306.

107. Keshava Prasad TS, Goel R, Kandasamy K, Keerthikumar S. Human protein reference database—2009 update. Nucl. Acids Res. 2009;37:D767–D772. et al.

108. Goel R, Harsha HC, Pandey A, Prasad TS. Human protein reference database and Human Proteinpedia as resources for phosphoproteome analysis. Mol. BioSyst. 2012;8:453–463.

109. Muthusamy B, Thomas JK, Prasad TSK, Pandey A. Access guide to human proteinpedia. Curr. Protoc. Bioinformatics. 2013 *Chapter 1* Unit 1.21.

110. Eng JK, McCormack AL, Yates JR. An approach to correlate tandem mass spectral data of peptides with amino acid sequences in a protein database. J. Am. Soc. Mass Spectrom. 1994;5:976–989.

111. Zong NC, Li H, Li H, Lam MP. Integration of cardiac proteome biology and medicine by a specialized knowledgebase. Circ. Res. 2013;113:1043–1053. et al.

112. Baerenfaller K, Hirsch-Hoffmann M, Svozil J, Hull R. pep2pro: a new tool for comprehensive proteome data analysis to reveal information about organ-specific proteomes in *Arabidopsis thaliana*. Integr. Biol. 2011;3:225–237. et al.

113. Hirsch-Hoffmann M, Gruissem W, Baerenfaller K. pep2pro: the high-throughput proteomics data processing, analysis, and visualization tool. Front. Plant Sci. 2012;3:123.

114. Roos FF, Jacob R, Grossmann J, Fischer B. PepSplice: cache-efficient search algorithms for comprehensive identification of tandem mass spectra. Bioinformatics. 2007;23:3016–3023.et al.

115. UniProt C. Activities at the Universal Protein Resource (UniProt) Nucl. Acids Res. 2014;42:D191–D198.

116. Hornbeck PV, Kornhauser JM, Tkachev S, Zhang B. PhosphoSitePlus: a comprehensive resource for investigating the structure and function of experimentally determined post-translational modifications in man and mouse. Nucl. Acids Res. 2012;40:D261–D270. et al.

117. Olsen JV, Mann M. Status of large-scale analysis of post-translational modifications by mass spectrometry. Mol. Cell. Proteomics. 2013;12:3444–3452.

118. Asara JM, Schweitzer MH, Freimark LM, Phillips M, Cantley LC. Protein sequences from mastodon and *Tyrannosaurus rex* revealed by mass spectrometry. Science. 2007;316:280–285.

119. Asara JM, Garavelli JS, Slatter DA, Schweitzer MH. Interpreting sequences from mastodon and T. rex. Science. 2007;317:1324–1325. et al.

120. Bern M, Phinney BS, Goldberg D. Reanalysis of *Tyrannosaurus rex* mass spectra. J. Proteome Res. 2009;8:4328–4332.

121. Bromenshenk JJ, Henderson CB, Wick CH, Stanford MF. Iridovirus and microsporidian linked to honey bee colony decline. PloSOne.

2010;5:e13181. et al.

122. Foster LJ. Bromenshenk et al. (PLoS One, 2011, 5(10):e13181) have claimed to have found peptides from an invertebrate iridovirus in bees. Mol. Cell. Proteomics. 2012;11:A110.0063871.

123. Hahne H, Kuster B. Discovery of O-GlcNAc-6-phosphate modified proteins in large-scale phosphoproteomics data. Mol. Cell. Proteomics. 2012;11:1063–1069.

124. Matic I, Ahel I, Hay RT. Reanalysis of phosphoproteomics data uncovers ADP-ribosylation sites. Nat. Methods. 2012;9:771–772.

125. Brosch M, Saunders GI, Frankish A, Collins MO. Shotgun proteomics aids discovery of novel protein-coding genes, alternative splicing, and "resurrected" pseudogenes in the mouse genome. Genome Res. 2011;21:756–767. et al.

126. Mohammed Y, Domanski D, Jackson AM, Smith DS. PeptidePicker: a scientific workflow with web interface for selecting appropriate peptides for targeted proteomics experiments. J. Proteomics. 2014;106C:151–161. et al.

127. Fan J, Mohareb F, Bond NJ, Lilley KS, Bessant C. MRMaid 2.0: mining PRIDE for evidence-based SRM transitions. OMICS. 2012;16:483–488.

128. Griss J, Cote RG, Gerner C, Hermjakob H, Vizcaino JA. Published and perished? The influence of the searched protein database on the long-term storage of proteomics data. Mol. Cell. Proteomics. 2011;10:M111.008490.

129. Klie S, Martens L, Vizcaino JA, Cote R. Analyzing large-scale proteomics projects with latent semantic indexing. J. Proteome Res. 2008;7:182–191. et al.

130. Mueller M, Vizcaino JA, Jones P, Cote R. Analysis of the experimental detection of central nervous system-related genes in human brain and cerebrospinal fluid datasets. Proteomics. 2008;8:1138–1148. et al.

131. Gonnelli G, Hulstaert N, Degroeve S, Martens L. Towards a human proteomics atlas. Anal. Bioanal. Chem. 2012;404:1069–1077.

132. Foster JM, Degroeve S, Gatto L, Visser M. A posteriori quality control for the curation and reuse of public proteomics data.

Proteomics. 2011;11:2182–2194. et al.

133. Griss J, Martin M, O›Donovan C, Apweiler R. Consequences of the discontinuation of the International Protein Index (IPI) database and its substitution by the UniProtKB "complete proteome" sets. Proteomics. 2011;11:4434–4438. et al.
134. Barsnes H, Vaudel M, Colaert N, Helsens K. Compomics utilities: an open-source Java library for computational proteomics. BMC Bioinformatics. 2011;12:70. et al.
135. Vaudel M, Barsnes H, Berven FS, Sickmann A, Martens L. SearchGUI: an open-source graphical user interface for simultaneous OMSSA and X!Tandem searches. Proteomics. 2011;11:996–999.
136. Vaudel M, Burkhart JM, Sickmann A, Martens L, Zahedi RP. Peptide identification quality control. Proteomics. 2011;11:2105–2114.
137. Zhu Y, Hultin-Rosenberg L, Forshed J, Branca RM. SpliceVista, a tool for splice variant identification and visualization in shotgun proteomics data. Mol. Cell. Proteomics. 2014;13:1552–1562. et al.
138. Martens L. Resilience in the proteomics data ecosystem: how the field cares for its data. Proteomics. 2013;13:1548–1550.
139. Perez-Riverol Y, Hermjakob H, Kohlbacher O, Martens L. Computational proteomics pitfalls and challenges: HavanaBioinfo 2012 workshop report. J. Proteomics. 2013;87:134–138.et al.
140. Csordas A, Wang R, Rios D, Reisinger F. From Peptidome to PRIDE: public proteomics data migration at a large scale. Proteomics. 2013;13:1692–1695. et al.
141. Gostev M, Faulconbridge A, Brandizi M, Fernandez-Banet J. The BioSample database (BioSD) at the European Bioinformatics Institute. Nucl. Acids Res. 2012;40:D64–D70. et al.

CHAPTER

3

CONTROLLED VOCABULARIES AND ONTOLOGIES IN PROTEOMICS: OVERVIEW, PRINCIPLES AND PRACTICE

Gerhard Mayer [a], Andrew R. Jones [b], Pierre-Alain Binz [c], Eric W. Deutsch [d], Sandra Orchard [e], Luisa Montecchi-Palazzi [e], Juan Antonio Vizcaíno [e], Henning Hermjakob [e], David Oveillero [e], Randall Julian [f], Christian Stephan [a,g], Helmut E. Meyer [a], Martin Eisenacher [a,]

[a] Medizinisches Proteom Center (MPC), Ruhr-Universität Bochum, D-44801 Bochum, Germany

[b] Institute of Integrative Biology, University of Liverpool, Liverpool L69 7ZB, UK

[c] SIB Swiss Institute of Bioinformatics, Swiss-Prot group, Rue Michel-Servet 1, CH-1211 Geneva 4, Switzerland

[d] Institute for Systems Biology, 401 Terry Avenue North, Seattle, WA 98109, USA

[e] EMBL-EBI, Wellcome Trust Genome Campus, Hinxton, Cambridge, CB10 1SD, UK

[f] Indigo BioSystems, Indianapolis, IN 46240, USA

[g] Kairos GmbH, Universitätsstraße 136, D-44799 Bochum, Germany

ABSTRACT

This paper focuses on the use of controlled vocabularies (CVs) and ontologies especially in the area of proteomics, primarily related to the work of the Proteomics Standards Initiative (PSI). It describes the relevant proteomics standard formats and the ontologies used within them. Software and tools for working with these ontology files are also discussed. The article also examines the "mapping files" used to ensure correct controlled vocabulary terms that are placed within PSI standards and the fulfillment of the MIAPE (Minimum Information about a Proteomics Experiment) requirements. This article is part of a Special Issue entitled: Computational Proteomics in the Post-Identification Era. Guest Editors: Martin Eisenacher and Christian Stephan.

INTRODUCTION

In science the unique definition of the terms used for describing the subject under inquiry is of prime importance to ensure the reproducibility of the analysis and interpretation of the empirically obtained data. A collection of terms for describing a certain modeling domain is called a controlled vocabulary (CV). Around 1735 Carl von Linné [1] introduced the concept of taxonomies into biology for the unique naming of the taxa of animals and plants. These taxonomies complement the controlled vocabularies by adding a hierarchical ordering for the used terms. Later librarians developed the concept of thesauri, which supplements such a hierarchy of terms by relations for similarity and synonyms between the terms. This means that they added other orthogonal dimensions to the mere subordination relation of a hierarchy, which helped them to improve the indexing of literature. Whereas in taxonomies we have only a tree-like structuring of the used terms, thesauri can be used also to represent the collection of terms in a more network- or graph-like structure [2]. Well-known large thesauri in the biomedical area are for instance MeSH (Medical Subject Headings) [3]and ICD (International Classification of Diseases) [4], which are used in medicine for documentation purposes. It has been announced that the next release of the ICD-11 will also be released in a formal ontology format [5].

Ontologies can be seen as a further step in the attempt to structure the terms used in describing a certain domain of interest. Ontologies are used as a means for knowledge representation by defining the objects and concepts as well as their properties and relations used in a modeling domain. Historically ontologies have a long tradition in philosophy, where they were first introduced by Aristotle (384–322 BC) [6] to describe the study of being. Another root of ontologies goes back to computational linguistics, where they are used to avoid interpretation problems due to synonyms, homonyms, acronyms, case ambiguities and misspellings. With the increasing reliance on computing and software in sciences, the need arose to represent the knowledge contained in thesauri in a formal way so that it can be easily processed and interpreted by a computer. Nowadays ontologies are widely used in the modeling of nearly every scientific field to allow easier computational processing of free text, and for defining a unique vocabulary for use in standard data formats. Therefore formal ontologies, which can be seen as the representation of the information contained in a thesaurus, were developed in a variety of formal ontology representation languages that differ by the degree of their expressiveness. In the ideal case the formal ontology has such a rich and formal logic-based expressiveness that it even enables automated reasoning and logic inference processes to take place on the represented data, which lead to the vision of the semantic web [7,8].

In bioinformatics, ontologies are available for many domain areas. An overview about the different ontologies used in biomedicine and bioinformatics, e.g. to ease data integration, is given in [2,9–11] and by the websites of the OBO (Open Biological and Biomedical Ontologies) Foundry [12] (http://obofoundry.org), NCBO (National Center for Biomedical Ontology) BioPortal [13] website (http://bioportal.bioontology.org) or the OLS (Ontology Lookup Service) at the European Bioinformatics Institute (EBI) [14].

In this article we confine ourselves to ontologies in the area of proteomics and show how they are used in the modern XML-based proteomics standard formats defined by the HUPO-PSI (Human Proteome Organization-Proteomics Standards Initiative) consortium. Then using the example of the mass spectrometry ontology PSI-MS [Mayer et al., in submission] we will describe the maintenance of these ontologies and mention important software, editors and tools for use in ontology engineering.

STANDARDIZED FORMATS AND ONTOLOGIES USED IN PROTEOMICS

Standardized formats are important for several reasons. First, more and more journals require that the data underlying a proteomics study should be made public [15–18] either on the journal website or in a public and free repository for mass spectrometry (MS)-based proteomics data like PRIDE [19] (PRoteomics IDEntifications database) or PeptideAtlas [20], which provide long-term storage of the data. In order to ease the task of data submission the EU-funded consortium project 'ProteomeXchange' (http://www.proteomexchange.org) was founded. Its goal is to provide a single point of data submission using the community data standard formats and to promote the data exchange between the main MS proteomics data repositories. Furthermore, the use of a standardized format makes it much easier to develop sophisticated software (converters, viewers and other tools) for analyzing the data, because one has to implement readers and writers only for the standard formats and not for the plethora of available proprietary formats. The use of standard formats also makes it easier to compare data from different sources or reproduce the results of analysis. Collaborative projects and fraud detection are made easier. And, in addition, the use of standard formats makes the reuse of data for analysis with improved methods or for answering new research questions more feasible.

JCAMP-DX [21] (Joint Committee on Atomic and Molecular Physical data-Data eXchange format), an IUPAC (International Union for Pure and Applied Chemistry) ASCII-based format, and ANDI-MS/netCDF [22] (Analytical Data Interchange format for Mass Spectrometry/Network Common Data Format), a format originally developed for chromatography-MS data, are older standardized mass spectrometry formats which were developed before the rise of the proteomics era. They are today mainly used in metabolomics for storing and exchanging MS information of small molecules, although it is in principle possible to store proteomics results in them. These two formats make no use of ontologies. The same is true for AniML (Analytical Information Markup Language) [23], an ASTM (American Society for Testing and Materials) standard for representing analytical data, but it is planned that AniML will incorporate parts of the PSI-MS ontology in the future [Mark Bean, personal communication, 2012].

In contrast, the modern XML-based data formats developed by the HUPO-PSI(likemzML[24–26],mzIdentML[27,28],mzQuantML[29,30], TraML [31], GelML [32], spML [33]), PEFF (PSI Extended Fasta Format [34]) and associated standards such as imzML [35,36] are well suited for storing the large data sets encountered in proteomics and allow the referencing of terms from controlled vocabularies defined in ontology files. Other HUPO-PSI formats are PSI-MI [37] for storing molecular interaction data and PSI-PAR [38], a format for describing Protein Affinity Reagents. mzML [24–26] is designed to store data generated by a mass spectrometry experiment; mzIdentML [27,28] captures the process and results of a protein a peptide identification experiment based on mass spectrometry data; mzQuantML [29,30] represents the results of a mass spectrometry quantitative experiment. TraML [31] is an exchange format for defining the transitions used in selected reaction monitoring (SRM), a technique also for quantitative proteomics analysis [39]. GelML [32] and spML [33] are standard formats for describing protein separation techniques. PEFF [34] is a proposed extension for the protein and nucleotide sequence format FASTA [40].

YAFMS [41] (Yet Another Format for Mass Spectrometry) and mz5 [42] are recently proposed non-XML based standards for the storage and exchange of proteomics data sets, which need less space than the unzipped XML-based standard formats. YAFMS stores the data as 'Blobs' (Binary Large Objects) in a relational database whereas mz5 uses HDF5 [43] (Hierarchical Data Format) for storing the data, a format especially developed for the storage of very large data sets in high performance computing. Both formats, YAFMS and mz5, allow the referencing of controlled vocabulary terms.

The imzML [35,36] format for MALDI (Matrix Assisted Laser Desorption Ionization) imaging data uses a compromise between data descriptiveness and memory efficiency by storing the metadata part in an XML (.imzML) file, whereas the spectral data are stored in a separate binary format (.ibd) file. Also mzML makes use of the base64 encoding [44] to store the spectra and chromatograms inside the mzML files. This base64 encoding is a method for representing and compressing data as text by encoding them using a subset of 64 characters from the ASCII character set. mzTab [45] is a proposal for a simplified tab-separated-value standard format which allows the use of spreadsheet programs for easily accessing and reporting proteomics identification and quantification results. It is currently in the HUPO-PSI document process [46], which ensures a critical review of proposed standards before their official release. Another tab-based format is MITAB [47], an extension of the PSI-MI format [37].

There are several possible strategies for accessing data in these standard formats. One is the utilization of a common API (Application Programming Interface) [48]. Another possibility is to use standard-specific APIs, as realized for the XML-based formats developed by the HUPO-PSI working group, which developed several Java libraries for the memory-efficient reading and writing of the information contained in the respective standard formats: jmzML [49], jTraML [50], jmzIdentML [51], jmzReader [52] and jmzQuantML [53]. The mzML format is the successor of the merged formats mzData [54] and mzXML [55]. In addition, the alternative de facto standard formats pepXML [56] and protXML [57], which are used by the TPP (Trans-Proteomic Pipeline) [58] for reporting peptide and protein assignments, are still in use. Since the XML-based files have the disadvantage that they can be very large in size, several format reader implementations make use of a sophisticated XPath [59] based XML indexer implemented in the xxindex Java library developed at the EBI (European Bioinformatics Institute) in order to make the processing of these files possible even on standard PCs [49].

An overview about the mass spectrometry standard formats used in proteomics, their usage of CV terms, and their associated web pages is given in Table 1. A more detailed description of some of the standard formats in proteomics is given by the articles of [60] and [Gonzalez-Galarza et al., this issue].

Table 1. Important standard formats for use in proteomics.

Standard format	**Use of CV/ ontology**	**Website (accessed 11/2012)**
JCAMP-DX [21]	None	http://www.jcamp-dx.org
ANDI-MS / netCDF [22]	None	http://enterprise.astm.org/filtrexx40.cgi?+REDLINE_PAGES/E1947.htm
mz5 / HDF5 [42,43]	Possible	http://software.steenlab.org/mz5
YAFMS [41]	PSI-MS	http://omics.pnl.gov/software/YAFMS.php

pepXML [56]	None	http://tools.proteomecenter.org/wiki/index.php?title=Formats:pepXML
protXML [57]	None	http://tools.proteomecenter.org/wiki/index.php?title=Formats:protXML
PSI-MI [37]	PSI-MI	http://www.psidev.info/mif
PSI-PAR [38]	PAR-CV	http://www.psidev.info/psi-par
mzML [24–26]	PSI-MS	http://www.psidev.info/mzml
TraML [31]	PSI-MS	http://www.psidev.info/traml
mzIdentML [27,28]	PSI-MS	http://www.psidev.info/mzidentml
mzQuantML [29,30]	PSI-MS	http://www.psidev.info/mzquantml
mzTab [45]	PSI-MS	https://code.google.com/p/mztab
imzML [35,36]	Imaging MS	http://www.maldi-msi.org
GelML [32]	sepCV	http://www.psidev.info/gelml
spML [33]	sepCV	http://www.psidev.info/search/node/spML

Whereas these standard formats define only the syntax of representing mass spectrometry data, ontologies support flexible definitions of semantics of the represented data. This additional semantic dimension makes the data not only computer readable, but also interpretable by computers, and is a prerequisite for more sophisticated software tools for analyzing and mining the data. The semantic information is defined independently of the standard formats by using ontologies. This means on the one hand that the semantic information can be easily reused by the various standards and on the other hand that it is in principle possible to change the representation format of the semantics without the need for redefining the standard format itself. Furthermore the controlled vocabulary can be extended independently, i.e. without the need to change the structure of the released standard format.

The most important ontologies that can be used to report proteomics experiments are listed in Table 2. They are used by the XML-based proteomics standards defined by the HUPO PSI working groups [61] and some of them can of course be used in other biological disciplines.

Table 2. Important ontologies, which are used in the proteomics field.

Ontology/CV	Prefix	Ontology file name	Website (accessed 11/2012)
Brenda tissue [62]	BTO	BrendaTissueOBO.obo	http://www.brenda-enzymes.info/ontology/tissue/tree/update/update_files/BrendaTissueOBO

Chemical entities of biological interest [63]	CHE-BI	chebi.obo	http://obo.cvs.sourceforge.net/obo/obo/ontology/chemical/chebi.obo
Gene ontology [64]	GO	gene_ontology.obo	http://obo.cvs.sourceforge.net/obo/obo/ontology/genomic-proteomic/gene_ontology.obo
MALDI imaging ontology [65]	IMS	imagingMS.obo	http://www.maldi-msi.org/download/imzml/imagingMS.obo
PSI-Molecular Interactions [66–68]	MI	psi-mi.obo	http://obo.cvs.sourceforge.net/obo/obo/ontology/genomic-proteomic/protein/psi-mi.obo
PSI-Protein modifications [69]	MOD	PSI-MOD.obo	http://psidev.cvs.sourceforge.net/viewvc/psidev/psi/mod/data/PSI-MOD.obo
PSI-Mass Spectrometry [70]	MS	psi-ms.obo	http://psidev.cvs.sourceforge.net/viewvc/psidev/psi/psi-ms/mzML/controlledVocabulary/psi-ms.obo
Ontology for Biomedical Investigations [71]	OBI	obi.owl	http://www.obofoundry.org/cgi-bin/detail.cgi?id=obi
Phenotype Attribute Trait Ontology [72]	PATO	quality.obo	http://obo.cvs.sourceforge.net/obo/obo/ontology/phenotype/quality.obo
PRIDE [19] CV	PRIDE	pride_cv.obo	http://code.google.com/p/ebi-pride/source/browse/trunk/pride-core/schema/pride_cv.obo
Protein ontology [73]	PRO	pro.obo	http://obo.cvs.sourceforge.net/obo/obo/ontology/genomic-proteomic/pro.obo
OBO Relationship Ontology [74]	OBO_REL	relationship.obo	http://obo.cvs.sourceforge.net/obo/obo/ontology/OBO_REL/relationship.obo
PSI-Sample Processing and Separations [75]	SEP	sep.obo	https://psidev.svn.sourceforge.net/svnroot/psidev/psi/sepcv/trunk/sep.obo
Unimod modifications [76]	UNIMOD	unimod.obo	http://www.unimod.org/obo/unimod.obo
Units of measurement [77]	UO	unit.obo	http://obo.cvs.sourceforge.net/obo/obo/ontology/phenotype/unit.obo

It should be mentioned that Unimod [76] is not an ontology in a strict sense – as no relations are defined and therefore no hierarchy is built – and therefore not supported by the OLS (Open Lookup Service). It contains modifications defined by Mascot [78]and converted by a XSLT (eXtensible Stylesheet Language Transformation) [79] script into the obo format.

ONTOLOGY FORMATS

For the formal representation of ontologies several representation formats exist, which differ in their degree of expressiveness. The most important of these are OWL (Web Ontology Language, version 2) [80], RDF(S) (Resource Description Framework (Schema)) [81], Topic Maps [82], Description Logic (DL) [8,83] and the obo flat file text format.

The obo format is used by the open source editor *OBO-Edit*[84], which replaced the older *DAG-Edit* editor. The obo format [85,86] is the simplest and currently most widespread used ontology format in bioinformatics. Those who are interested in the obo format can subscribe to the dedicated mailing list [87].

The obo format first lists some header tags containing meta-information like for instance the date, the version and other imported ontologies. After the header a list of type definitions, a list of terms and a list of instances follow. The format can contain three types of stanzas: [Typedef], [Term] and [Instance], where each stanza can be described by a collection of allowed tags for the respective stanza type. So the format distinguishes in total between 4 types of tags: header tags, typedef tags, term tags and instance tags. The obo flat file format specification recommends that the [Term], [Typedef], and [Instance] stanzas should be serialized in alphabetical order on the value of their id tag and also for the specification of the tags inside the stanzas a certain order is recommended [86].

As an example within psi-ms.obo, the definition for 'ionization energy' (term *MS:1000219*) is shown below. It defines the term together with an identifier, a short human readable definition of the term's meaning, a synonym and the value type for this term. In addition here two relationships are given: the relationship "is_a" states that the ionization energy is a specialization of an ion attribute and the relationship "has_units" states that the ionization energy has to be given in electron volts. Other relationships used in psi-ms.obo are for instance "part_of" and "has_regexp". The relation "has_regexp" for instance is used to describe the cleavage sites of restriction enzymes. Most terms are by default used as "flat" enumeration types, i.e. with the meaning only given by their name and description. The 'xref: value-type' entry allows stating that terms require a value, in this case of type float. An overview about the possible relationships is given in the OBO Relation Ontology [74,88].

[Term]

id MS:1000219

name: ionization energy

def: "The minimum energy required to remove an electron from an atom or molecule to produce a positive ion." [PSI:MS]

synonym: "IE" EXACT []

xref: value-type:xsd\:float "The allowed value-type for this CV term."

is_a: MS:1000507 ! ion attribute

relationship: has_units UO:0000266 ! electronvolt

The usage of this CV term in a standard format file is shown later in Section 5.

As shown in the next example for a quadrupole ion trap, it is possible to define more than one synonym for a given term, which allows to model cases where many terms are in use for the same meaning, so that redundancy on term level is avoided.

[Term]

id:MS:1000082

name: quadrupole ion trap

def: "Quadrupole Ion Trap mass analyzer captures the ions in a three dimensional ion trap and then selectively ejects them by varying the RF and DC potentials." [PSI:MS]

synonym: "Paul Ion trap" EXACT []

synonym: "QIT" EXACT []

synonym: "Quistor" EXACT []

is_a: MS:1000264 ! ion trap

Sometimes a merging, splitting, replacement or deprecation of an ontology term is necessary, e.g. due to upcoming new technologies or instruments or changes in standard formats. Montecchi-Palazzi et al. [89] demand that the old terms must be obsoleted, but they must stay inside the ontology and any new terms replacing them must get a new identifier. This is important for backward compatibility, so that instance files with old identifiers are still valid and contain reasonable content. This marking as obsolete is only necessary, if the meaning of a term changes. In contrast, changes in wording only can be made without marking a term obsolete. An example for an obsolete term is for instance:

[Term]

id: MS:1001849

name: sum of MatchedFeature values

def: "OBSOLETE Peptide quantification value calculated as sum of MatchedFeature quantification values." [PSI:PI]

comment: This term was made obsolete because the concept Matched-Feature was dropped.

is_a: MS:1001805 ! quantification datatype

is_obsolete: true

Here the relation "is_obsolete" was added and set to true, the 'def:' tag begins with 'OBSOLETE:' and the following definition now contains a hint which term should be used instead. In this example it is mentioned that the concept of a MatchedFeature was dropped, so that there is now no need for using the CV term anymore.

Inside the obo file one can also reference terms defined in other ontologies by using database cross reference ("dbxref") lists. This way, one cannot only refer to other ontologies, but also to databases or web pages. For instance the example term (*MS:1000219*) for the 'ionization energy' shown above contains a "dbxref" list after the "def:" term tag, stating the source where the term was originally defined. In the example it references with *[PSI:MS]* to itself. Analogously the relationship "has_units" refers with the "dbxref" '*UO:0000266*' to the "Unit" ontology [77]. Another example would be the term tag *def: "Enzyme leukocyte elastase*

(EC 3.4.21.37)." [BRENDA:3.4.21.37], which states that the BRENDA ontology is the original source of reference for the enzyme "leukocyte elastase". A list of allowed "dbxref" terms can be found online at the gene ontology website [90].

Other formal languages for ontology representation like OWL [80], RDF(S) [81] and Description Logic (DL) [8,83] allow much more expressive semantics than the relatively simple obo format and can be used for automatic reasoning procedures and are the basis for building up the semantic web [91–93].

Description Logics [8,83] are decidable parts of first-order predicate logics and differ from one another by their degree of expressivity. This means that they have more expressiveness than propositional logic, but decision problems based on them are more efficiently decidable than the general first-order predicate logic. The complexity [94] of the decision problems depends on the different allowed and not allowed language constructs of the used description logic. RDF [81] is based on XML and describes data based on a graph model consisting of triples of subject, predicate and object. Comparable to XML schema for XML, RDFS describes the allowed structures for RDF documents. OWL resp. OWL 2 buildup on the top of RDF(S) and are thus more expressive. OWL 2 defines the three so-called "profiles" OWL 2 EL, QL and RL [95] differing in allowed language constructs determining the level of expressiveness. Ontologies for the OBO Foundry must be either in obo or OWL format and must use the OBO Relation Ontology [74]. From the ontologies mentioned in Table 2 only the OBI ontology is in OWL format, all others are represented in the obo format. It should be mentioned here, that several tools exist to automatically convert obo files into some of these other formats like OWL or RDF [96–99]. Of course, the resulting files cannot contain more information than the simple obo files, but they can be used as a starting point for a semantically more detailed modeling of the ontology information.

SOFTWARE AND TOOLS FOR ACCESSING, BROWSING, CREATING, EDITING AND MANIPULATING ONTOLOGY FILES

Because all the formats OBO, OWL, RDF(S) are text files one can in principle edit them with a normal text editor. However, for working more efficiently with them, some specialized editors exist. In addition to an ASCII editor they have additional useful functions, like for instance visualizing the hierarchy or performing some validity checks before storing a changed version of the ontology file. The most important of these specialized ontology editors are listed in Table 3. A good overview about tools for ontology engineering is given in [100].

Table 3. Software programs for accessing, browsing, creating, editing and manipulating ontology files.

Name	Category	Website (accessed 11/2012)
OBO-Edit [84]	Ontology editor	http://oboedit.org
Protégé [101]	Ontology editor	http://protege.stanford.edu
OLS (Ontology Lookup Service) [14]	Web service interface, Web portal	http://www.ebi.ac.uk/ontology-lookup
OLS dialog [103]	Java plug-in component	https://code.google.com/p/ols-dialog
OLSVis [104]	Visual browser	http://ols.wordvis.com
OBO-Explorer [105]	Ontology editor	http://www.aiai.ed.ac.uk/project/cobra-ct

Name	Category	Website (accessed 11/2012)
OBO-Edit [84]	Ontology editor	http://oboedit.org
Protégé [101]	Ontology editor	http://protege.stanford.edu
OLS (Ontology Lookup Service) [14]	Web service inter-face, Web portal	http://www.ebi.ac.uk/ontology-lookup
OLS dialog [103]	Java plug-in compo-nent	https://code.google.com/p/ols-dialog
OLSVis [104]	Visual browser	http://ols.wordvis.com
OBO-Explorer [105]	Ontology editor	http://www.aiai.ed.ac.uk/project/cobra-ct
NCBI BioPortal [13]	Web portal	http://bioportal.bioontology.org

OBO-Edit[84] for instance contains a configurable verification manager (Fig. 1), where one can specify which checks the editor should perform during loading, saving or changing of an obo ontology file. Whereas *OBO-Edit* and OLS [14] work only with files in the obo format, the *Protégé* editor and the *OBO-Explorer* support also OWL. *Protégé*[101] furthermore supports the RDF(S) ontology format. With OLS one can either browse interactively through the ontologies by using the web interface [102] or access them from within a Java class by using the web service implemented in the available ols-client.jar file of the EBI.

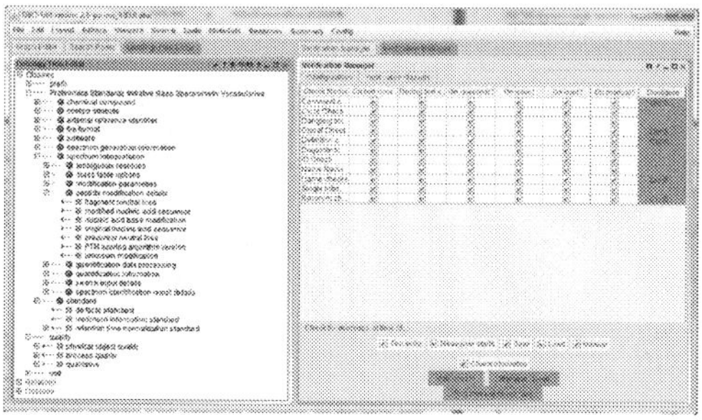

Figure. 1. The OBO-Edit [84] user interface showing the ontology tree editor and the verification manager.

For accessing the ontology files, the Open Lookup Service (OLS) [14] allows the browsing, searching and accessing of the obo file contents either interactively via a web-site interface or automatically by computer programs via a web service interface. Internally, OLS uses an indexing based on Apache Lucene [106], for case-insensitive indexing of all the terms and their synonyms [107]. This allows converter programs like PRIDEConverter 2 [108] or ProCon (PROteomics CONversion tool) [109] to easily access the ontology files during the creation process of proteomics data files.

USE OF CONTROLLED VOCABULARIES IN THE XML-BASED PROTEOMICS STANDARD FORMATS OF THE HUPO-PSI

The HUPO-PSI formats mzML, TraML, mzIdentML, mzQuantML and GelML, as well as the PSI-associated format imzML and the non-XML mzTab [45] and MITAB [47]formats all make intensive use of controlled vocabulary terms defined in ontologies. Therefore these formats allow the usage of < cvParam > elements at various places in an instance data file. All these standard format instance files have at their beginning an element < CvList >, in which the used controlled vocabularies are first defined with their name, their ID, their version and the URI (Uniform Resource Identifier). The latter specifies a name space-like unique identifier and can – if it is a URL – also specify, where to find the actual ontology files:

< CvList >

< Cv fullName = "Proteomics Standards Initiative Protein Modifications" version = "1.010.7" uri = "http://psidev.cvs.sourceforge.net/viewvc/psidev/psi/mod/data/PSI-MOD.obo" id = "MOD"/>

< Cv fullName = "Proteomics Standards Initiative Mass Spectrometry Vocabulary" version = "3.34.0" uri =http://psidev.cvs.sourceforge.net/viewvc/psidev/psi/psi-ms/mzML/controlledVocabulary/psi-ms.oboid = "MS"/>

< Cv fullName = "UNIMOD CV for modifications" version = "1.0" uri =http://www.unimod.org/obo/unimod.obo

id = "UNIMOD"/>

< Cv fullName = "Unit Ontology" version = "1.0"

uri = "http://obo.cvs.sourceforge.net/viewvc/obo/obo/ontology/phenotype/unit.obo" id = "UO"/>

</CvList >

Later in the instance data file these defined controlled vocabularies and their terms can be referenced, as shown in the following mzIdentML example specifying the original result file, the spectra data files, their formats and the used search database using < cvParam > XML elements:

< Inputs >

< SourceFile location = “D:\TestingProteinGrouping\Testing Decoy-Dash.msf” id = “SF_1”>

< FileFormat >

< cvParam accession = “MS:1001107” cvRef = “MS” name = “data stored in database”/>

</FileFormat >

</SourceFile >

< SourceFile location = “C:\Users\Gerhard\AppData\Local\Temp\Testing Decoy-Dash_2.prot.xml” id = “SF_2”>

< FileFormat >

< cvParam accession = “MS:1001422” cvRef = “MS” name = “protXML file”/>

</FileFormat >

</SourceFile >

< SearchDatabase location = “uniprot_sprot_human_target_decoy.dashed.fasta” name = “uniprot_sprot_human_target_decoy.dashed.fasta” id = “SDB”>

< FileFormat >

< cvParam accession = “MS:1001348” cvRef = “MS” name = “FASTA format”/>

</FileFormat >

< DatabaseName >

< userParam value = “uniprot_sprot_human_target_decoy.dashed.fasta” name = “database name”/>

</DatabaseName >

</SearchDatabase >

< SpectraData location = "D:\HPP_VallHebron_MRMvelos_120719_Fr04_04.mgf" id = "HPP_VallHebron_MRMvelos_Test1_120719_Fr04_04.mgf">

< ExternalFormatDocumentation>http://www.psidev.info/files/mzIdentML1.1.0.xsd </ExternalFormatDocumentation >

< FileFormat >

< cvParam accession = "MS:1001062" cvRef = "MS" name = "Mascot MGF file"/>

</FileFormat >

< SpectrumIDFormat >

< cvParam accession = "MS:1000774" cvRef = "MS" name = "multiple peak list nativeID format"/>

</SpectrumIDFormat >

</SpectraData >

</Inputs >

To make sure that the CV terms are used only at correct positions in the files, a mapping file exists for each of the standards, which exactly defines where and in which combination with other CV terms a certain CV term can occur inside the data file. The schema for this CV mapping file is shown in Fig. 2. Such a mapping file contains a < CvReferenceList > element, which contains a list of CVs that are required in an instance data file and a < CvMappingRuleList > element, which contains the mapping rules for the various elements of the data file.

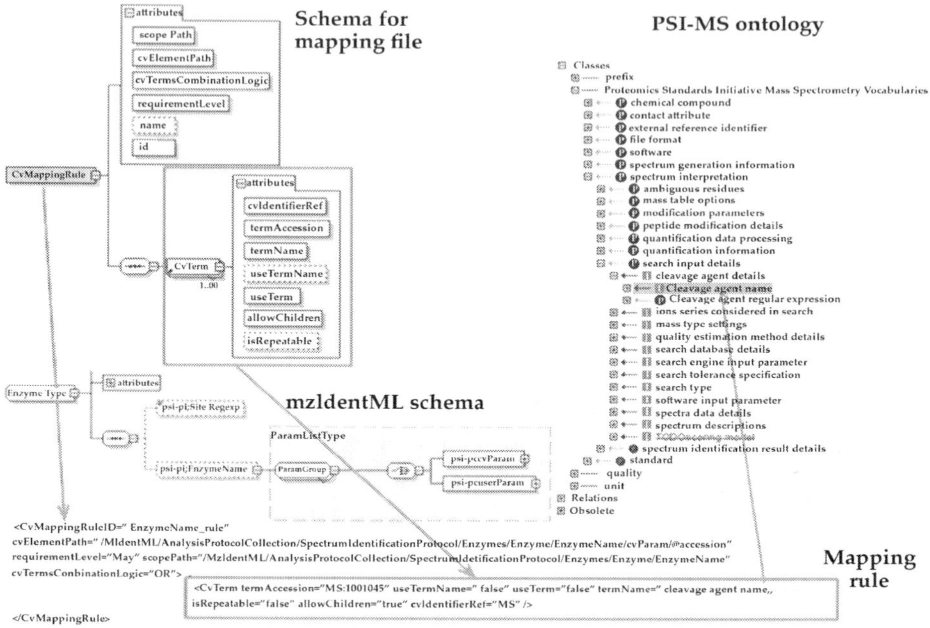

Figure. 2. Mapping rule for using a CV term in the correct position (XPath) of an XML data file.

Each < CvMappingRule > element has an attribute 'cvElementPath', which describes in XPath expression syntax [59] the path to the element in the standard file to which the current CV mapping rule applies. The attribute 'cvTermsCombinationLogic' is a Boolean operator describing how the subordinate < CvTerm > elements of the < CvMappingRule > are logically combined. The 'requirementLevel' attribute can have the values MAY, SHOULD or MUST depending on whether the association with the CV term is optional, recommended or mandatory. The attributes 'useTerm' and 'allowChildren' of the < CvTerm > element state, if the term itself or children of it can be used for data annotation at this place inside a data instance file. The attribute 'isRepeatable' states if the term can be repeated at this position or not and the Boolean value 'useTermName' specifies if the checking of the CV term is done on the 'termName' (if true) or on the termAccession (if false).

An example of such a < CvMappingRule > is given in the following, which states that in a mzIdentML file it is recommended that under the

XPath "/MzIdentML/AuditCollection/Person/" there are < cvParam > elements describing the contact data of a person. The < cvParam > elements allowed here are all logical OR combinations of the three CV terms 'contact address', 'contact URL' and 'contact email':

< CvMappingRule id = "AuditCollectionPerson_rule"

cvElementPath = "/MzIdentML/AuditCollection/person/cvParam/@accession" requirementLevel = "SHOULD"

scopePath = "/MzIdentML/AuditCollection/person" cvTermsCombinationLogic = "OR">

< CvTerm termAccession = "MS:1000587" useTermName = "false" useTerm = "true" termName = "contact address"

isRepeatable = "true" allowChildren = "false" cvIdentifierRef = "MS" />

< CvTerm termAccession = "MS:1000588" useTermName = "false" useTerm = "true" termName = "contact URL"

isRepeatable = "true" allowChildren = "false" cvIdentifierRef = "MS" />

< CvTerm termAccession = "MS:1000589" useTermName = "false" useTerm = "true" termName = "contact email"

isRepeatable = "true" allowChildren = "false" cvIdentifierRef = "MS" />

</CvMappingRule >

In addition to the standard syntactic checks for well-formedness (i.e. if the XML file fulfills the XML syntax rules) and validity (i.e. if the XML file follows the structure defined in the corresponding XML schema), these mapping files thus allow an additional semantic checking of CV term usage in XML files [112–116]. In general, there might exist more than one mapping file per format, which could allow for different levels of stringency checking, e.g. checking MIAPE compliance (see next paragraph) or compliance to specific journal guidelines [15–18].

MIAPE COMPLIANCE

To ensure that published experimental data fulfill basic requirements regarding reproducibility, transparency and secondary usage of the data, the MIBBI (Minimal Information for Biological and Biomedical Investigations) [110] project was founded. It describes minimal information checklists that data and metadata describing an experiment should fulfill. For proteomics, the MIAPE (Minimum Information about a Proteomics Experiment) [111] guidelines describe what information should be reported about an experiment, for example in a text document or a data file. A basic (text-based) mapping table defined together with each standard lists the possible locations of MIAPE requirements within the standard. Additional (computer-readable) mapping files and validators may be developed to allow checks for e.g. all steps between a "minimal sensible file" and a "strictly MIAPE-conform file". A first implementation is [114]. Currently there are the following MIAPE guidelines defined: MIAPE-MS [117], MIAPE-MSI [118], MIAPE-GI [119], MIAPE-GE [120], MIAPE-CC [121], MIAPE-CE [122] and MIAPE-Quant [123]. The validators are either based on the PSI semantic validator framework [124], the underlying Java library used for developing the validators for the various HUPO XML-based proteomics standard formats, or are implemented locally or in web environments. The MIAPE compliance can also be tested by using the ProteoRed MIAPE web toolkit [125]. On the website [126] one can find links to the available validators for the various HUPO-PSI proteomics standards. All these validators check if the rules specified in the mapping file for the respective standards are fulfilled by a given instance data file.

MAINTENANCE OF THE CONTROLLED VOCABULARIES AND ONTOLOGIES

In the PSI community practice document [89] the HUPO-PSI working groups defined some guidelines for the development of controlled vocabularies. Since ongoing technological progress and the upcoming of new instruments and methods, an ontology is never complete, and steadily grows over time. Therefore the ontologies need a continuous maintenance. For the PSI-MS [70] ontology the maintenance procedure is as follows: Everyone in the proteomics community is free to subscribe to the psidev-ms-vocab mailing list [127] and to make proposals for new terms and/or improvements of the already existing psi-ms.obo ontology terms. After receiving a request for a new CV term the PSI ontology coordinator checks if the proposed term and its description, data type, parent terms and relations are sensible. It is also checked if the term is already part of other ontologies, e.g. MALDI imaging obo [65] or ChEBI [63] and if it is better to add them there or if the term isn't necessary because there exists already an attribute in the standard files, which describes the same fact. A term which passes all these checks is then included into the next release candidate of the obo file, which is sent to the three mailing lists psidev-ms-vocab@lists.sourceforge.net, psidev-pi-dev@lists.sourceforge.net and psidev-ms-vocab@lists.sourceforge.net for public discussion. If the proteomics community comes to consensus with the new term, then it is added to the next release version of the obo file, which is then made public at a CVS repository [128] and announced via the three mentioned mailing lists. A more detailed description of the PSI-MS maintenance process can be found at [Mayer et al., 2012, in submission].

SUMMARY

In the last 10 years the proteomics community defined several modern standard formats (most of them XML-based) useful for the representation of the complex and large data sets faced in proteomics today. Because it is necessary to enrich these data with semantic information in order to annotate and make use of them effectively, the data standards refer to controlled vocabularies defined in ontology formats, of which the obo format is the one predominantly used today. In this manuscript, we briefly described the obo format and discussed some software tools for easily working with these files.

The integration of the terms defined in the ontologies into the XML data standards made it necessary to develop semantic validators for checking the correct use of the CV terms. For this, the validators make use of mapping files that complement the standard format defining XML schemas, and contain the rules for the correct usage of the CV terms. Also the conformance to the MIAPE and/or journal guidelines can be assured by additional mapping files governing the use of specific terms. Finally, the current procedure for maintaining the PSI mass spectrometry ontology psi-ms.obo was presented.

ACKNOWLEDGEMENTS

GM, JAV and PAB are funded by the European Union project 'ProteomeXchange'(http://www.proteomexchange.org, EU FP7 grant number 260558). JAV is also supported by the Wellcome Trust [grant number WT085949MA]. PAB is funded also by the Swiss Federal Government through the Federal Office of Education and Science. ME is funded by P.U.R.E. (http://www.pure.rub.de, Protein Unit for Research in Europe), a project of Nordrhein-Westfalen, a federal state of Germany. ARJ gratefully acknowledges funding from the UK BBSRC [BB/I000909/1 and BB/H024654/1]. EWD is funded in part by NIGMS grants R01 GM087221, P50 GM076547/Center for Systems Biology, and from the Luxembourg Centre for Systems Biomedicine and the University of Luxembourg.

REFERENCES

1. Carl von Linné, Systema Naturae, Johan Wilhelm de Groot, Leiden, 1735.
2. L.J. Jensen, P. Bork, Ontologies in quantitative biology: a basis for comparison, integration, and discovery, PLoS Biol. 8 (2010) e1000374.
3. C.E. Lipscomb, Medical Subject Headings (MeSH), Bull. Med. Libr. Assoc. 88 (2000) 265–266.
4. World Health Organisation, International Statistical Classification of Diseases and Related Health Problems 10th Revision, WHO, 2010. (http://apps.who.int/ classifications/icd10/browse/2010/en, accessed 11/2012).
5. http://www.bioontology.org/ICD11-2, (accessed 11/2012).
6. http://plato.stanford.edu/entries/aristotle-metaphysics/, (accessed 10/2012).
7. P. Hitzler, M. Krötzsch, S. Rudolph, Foundations of Semantic Web Technologies, CRC Press, Boca Raton ; London, 2010. (http://semantic-web-book.org, accessed 11/2012).
8. F. Baader, The Description Logic Handbook: Theory, Implementation, and Applications, 2nd ed. Cambridge University Press, Cambridge, 2007.
9. M. Courtot, N. Juty, C. Knupfer, D. Waltemath, A. Zhukova, A. Drager, M. Dumontier, A. Finney, M. Golebiewski, J. Hastings, S. Hoops, S. Keating, D.B. Kell, S. Kerrien, J. Lawson, A. Lister, J. Lu, R. Machne, P. Mendes, M. Pocock, N. Rodriguez, A. Villeger,

D.J. Wilkinson, S. Wimalaratne, C. Laibe, M. Hucka, N. Le Novere, Controlled vocabularies and semantics in systems biology, Mol. Syst. Biol. 7 (2011) 543.

10. P.N. Robinson, F. Bauer, Introduction to Bio-Ontologies, Chapmann & Hall / CRC Press, 2011. (http://bio-ontologies-book.org, accessed 11/2012).
11. S. Eckstein, Informationsmanagement in der Systembiologie, Springer, Berlin, Heidelberg, 2011.
12. B. Smith, M. Ashburner, C. Rosse, J. Bard, W. Bug, W. Ceusters, L.J. Goldberg, K. Eilbeck, A. Ireland, C.J. Mungall, N. Leontis, P. Rocca-Serra, A. Ruttenberg, S.A. Sansone, R.H. Scheuermann, N. Shah, P.L. Whetzel, S. Lewis, The OBO Foundry: coordinated evolution of ontologies to support biomedical data integration, Nat. Biotechnol. 25 (2007) 1251–1255.
13. N.F. Noy, N.H. Shah, P.L. Whetzel, B. Dai, M. Dorf, N. Griffith, C. Jonquet, D.L. Rubin, M.A. Storey, C.G. Chute, M.A. Musen, BioPortal: ontologies and integrated data resources at the click of a mouse, Nucleic Acids Res. 37 (2009) W170–W173.
14. R. Coté, F. Reisinger, L. Martens, H. Barsnes, J.A. Vizcaíno, H. Hermjakob, The Ontology Lookup Service: bigger and better, Nucleic Acids Res. 38 (2010) W155–W160.
15. R.A. Bradshaw, A.L. Burlingame, S. Carr, R. Aebersold, Reporting protein identifi- cation data: the next generation of guidelines, Mol. Cell Proteomics 5 (2006) 787–788.
16. H. Rodriguez, M. Snyder, M. Uhlen, P. Andrews, R. Beavis, C. Borchers, R.J. Chalkley, S.Y. Cho, K. Cottingham, M. Dunn, T. Dylag, R. Edgar, P. Hare, A.J. Heck, R.F. Hirsch, K. Kennedy, P. Kolar, H.J. Kraus, P. Mallick, A. Nesvizhskii, P. Ping, F. Ponten, L. Yang, J.R. Yates, S.E. Stein, H. Hermjakob, C.R. Kinsinger, R. Apweiler, Recommendations from the 2008 International Summit on Proteomics Data Release and Sharing Policy: the Amsterdam principles, J. Proteome Res. 8 (2009) 3689–3692.
17. http://www.mcponline.org/site/misc/ParisReport_Final.xhtml, (accessed 11/2012).
18. http://www.mcponline.org/site/misc/PhialdelphiaGuidelinesFINALDRAFT.pdf, (accessed 11/2012).

19. J.A. Vizcaíno, R. Coté, F. Reisinger, H. Barsnes, J.M. Foster, J. Rameseder, H. Hermjakob, L. Martens, The Proteomics Identifications database: 2010 update, Nucleic Acids Res. 38 (2010) D736–D742.

20. E.W. Deutsch, H. Lam, R. Aebersold, PeptideAtlas: a resource for target selection for emerging targeted proteomics workflows, EMBO Rep. 9 (2008) 429–434.

21. P. Lampen, H. Hillig, A.N. Davies, M. Linscheid, JCAMP-DX for mass-spectrometry, Appl. Spectrosc. 48 (1994) 1545–1552.

22. B. Erickson, ANDI MS standard finalized, Anal. Chem. 72 (2000) 103a-103a.

23. T. Davies, Herding AnIMLs, Chem. Int. 29 (2007) 21–23.

24. L. Martens, M. Chambers, M. Sturm, D. Kessner, F. Levander, J. Shofstahl, W.H. Tang, A. Rompp, S. Neumann, A.D. Pizarro, L. Montecchi-Palazzi, N. Tasman, M. Coleman, F. Reisinger, P. Souda, H. Hermjakob, P.A. Binz, E.W. Deutsch, mzML— a community standard for mass spectrometry data, Mol. Cell Proteomics 10 (2011), (R110 000133).

25. E.W. Deutsch, Mass spectrometer output file format mzML, Methods Mol. Biol. 604 (2010) 319 331.

26. M. Turewicz, E.W. Deutsch, Spectra, chromatograms, metadata: mzML—the standard data format for mass spectrometer output, Methods Mol. Biol. 696 (2011) 179–203.

27. A.R. Jones, M. Eisenacher, G. Mayer, O. Kohlbacher, J. Siepen, S.J. Hubbard, J.N. Selley, B.C. Searle, J. Shofstahl, S.L. Seymour, R. Julian, P.A. Binz, E.W. Deutsch, H. Hermjakob, F. Reisinger, J. Griss, J.A. Vizcaíno, M. Chambers, A. Pizarro, D. Creasy, The mzIdentML data standard for mass spectrometry-based proteomics results, Mol. Cell Proteomics 11 (2012), (M111 014381).

28. M. Eisenacher, mzIdentML: an open community-built standard format for the results of proteomics spectrum identification algorithms, Methods Mol. Biol. (Clifton N. J.) 696 (2011) 161–177.

29. M. Walzer, O. Kohlbacher, F. Reisinger, J.A. Medina-Aunon, J. Uszkoreit, G. Mayer, M. Eisenacher, A.R. Jones, mzQuantML: exchange format for quantitation values associated with peptides, proteins and small molecules from mass spectra, HUPO PSI Draft

Recommendation, 2012.

30. M. Walzer, D. Qi, G. Mayer, J. Uszkoreit, M. Eisenacher, T. Sachsenberg, F.F. Gonzalez-Galarza, J. Fan, C. Bessant, E.W. Deutsch, F. Reisinger, J.A. Vizcaíno, J. A. Medina-Aunon, J.P. Albar, O. Kohlbacher, A.R. Jones, The mzQuantML data standard for mass spectrometry-based quantitative studies in proteomics, Mol. Cell Proteomics, under review

31. E.W. Deutsch, M. Chambers, S. Neumann, F. Levander, P.A. Binz, J. Shofstahl, D.S. Campbell, L. Mendoza, D. Ovelleiro, K. Helsens, L. Martens, R. Aebersold, R.L. Moritz, M.Y. Brusniak, TraML—a standard format for exchange of selected reaction monitoring transition lists, Mol. Cell Proteomics 11 (2012), (R111 015040).

32. F. Gibson, C. Hoogland, S. Martinez-Bartolome, J.A. Medina-Aunon, J.P. Albar, G. Babnigg, A. Wipat, H. Hermjakob, J.S. Almeida, R. Stanislaus, N.W. Paton, A.R. Jones, The gel electrophoresis markup language (GelML) from the Proteomics Standards Initiative, Proteomics 10 (2010) 3073–3081.

33. N.W. Paton, A.R. Jones, C. Taylor, spML: sample processing markup language, HUPO PSI SP Working Group, 2007. (http://www.psidev.info/search/node/ spML%20Milestone%20Documents, accessed 11/2012).

34. http://psidev.info/node/363, (accessed 11/2012).

35. A. Römpp, T. Schramm, A. Hester, I. Klinkert, J.P. Both, R.M. Heeren, M. Stöckli, B. Spengler, imzML: Imaging Mass Spectrometry Markup Language: A common data format for mass spectrometry imaging, Methods Mol. Biol. 696 (2011) 205–224.

36. T. Schramm, A. Hester, I. Klinkert, J.P. Both, R.M. Heeren, A. Brunelle, O. Laprevote, N. Desbenoit, M.F. Robbe, M. Stöckli, B. Spengler, A. Römpp, imzML—a common data format for the flexible exchange and processing of mass spectrometry imaging data, J. Proteome 75 (2012) 5106–5110.

37. S. Kerrien, S. Orchard, L. Montecchi-Palazzi, B. Aranda, A.F. Quinn, N. Vinod, G.D. Bader, I. Xenarios, J. Wojcik, D. Sherman, M. Tyers, J.J. Salama, S. Moore, A. Ceol, A. Chatr-Aryamontri, M. Oesterheld, V. Stumpflen, L. Salwinski, J. Nerothin, E. Cerami,

M.E. Cusick, M. Vidal, M. Gilson, J. Armstrong, P. Woollard, C. Hogue, D. Eisenberg, G. Cesareni, R. Apweiler, H. Hermjakob, Broadening the horizon— level 2.5 of the HUPO-PSI format for molecular interactions, BMC Biol. 5 (2007) 44.

38. D.E. Gloriam, S. Orchard, D. Bertinetti, E. Bjorling, E. Bongcam-Rudloff, C.A. Borrebaeck, J. Bourbeillon, A.R. Bradbury, A. de Daruvar, S. Dubel, R. Frank, T.J. Gibson, L. Gold, N. Haslam, F.W. Herberg, T. Hiltke, J.D. Hoheisel, S. Kerrien, M. Koegl, Z. Konthur, B. Korn, U. Landegren, L. Montecchi-Palazzi, S. Palcy, H. Rodriguez, S. Schweinsberg, V. Sievert, O. Stoevesandt, M.J. Taussig, M. Ueffing, M. Uhlen, S. van der Maarel, C. Wingren, P. Woollard, D.J. Sherman, H. Hermjakob, A community standard format for the representation of protein affinity reagents, Mol. Cell Proteomics 9 (2010) 1–10.

39. S. Gallien, E. Duriez, B. Domon, Selected reaction monitoring applied to proteomics, J. Mass Spectrom. 46 (2011) 298–312.

40. W.R. Pearson, Flexible sequence similarity searching with the FASTA3 program package, Methods Mol. Biol. 132 (2000) 185–219.

41. A.R. Shah, J. Davidson, M.E. Monroe, A.M. Mayampurath, W.F. Danielson, Y. Shi, A.C. Robinson, B.H. Clowers, M.E. Belov, G.A. Anderson, R.D. Smith, An efficient data format for mass spectrometry-based proteomics, J. Am. Soc. Mass Spectrom. 21 (2010) 1784–1788.

42. M. Wilhelm, M. Kirchner, J.A. Steen, H. Steen, mz5: space- and time-efficient storage of mass spectrometry data sets, Mol. Cell Proteomics 11 (2012), (O111 011379).

43. http://www.hdfgroup.org/, (accessed 11/2012).

44. S. Josefsson, RFC 4648: The Base16, Base32, and Base64 Data Encodings, Internet Engineering Task Force, 2006. (http://tools.ietf.org/pdf/rfc4648.pdf, accessed 11/2012).

45. J. Griss, T. Sachsenberg, M. Walzer, O. Kohlbacher, A.R. Jones, H. Hermjakob, J.A. Vizcaíno, mzTab: Exchange Format for Proteomics and Metabolomics Results, HUPO PSI Recommendation, 2012. (https://code.google.com/p/mztab/, accessed 11/2012).

46. R. Julian, N.W. Paton, Proteomics Standards Initiative Document

Process and Requirements, HUPO-PSI, 2010. (http://www.psidev.info/psi-doc-process, accessed 11/2012).

47. http://code.google.com/p/psicquic/wiki/MITAB27Format, (accesed 11/2012).

48. M. Askenazi, J.R. Parikh, J.A. Marto, mzAPI: a new strategy for efficiently sharing mass spectrometry data, Nat. Methods 6 (2009) 240–241.

49. R.G. Coté, F. Reisinger, L. Martens, jmzML, an open-source Java API for mzML, the PSI standard for MS data, Proteomics 10 (2010) 1332–1335.

50. K. Helsens, M.Y. Brusniak, E. Deutsch, R.L. Moritz, L. Martens, jTraML: an open source Java API for TraML, the PSI standard for sharing SRM transitions, J. Proteome Res. 10 (2011) 5260–5263.

51. F. Reisinger, R. Krishna, F. Ghali, D. Ríos, H. Hermjakob, J.A. Vizcaíno, A.R. Jones, jmzIdentML API: a Java interface to the mzIdentML standard for peptide and protein identification data, Proteomics 12 (2012) 790–794.

52. J. Griss, F. Reisinger, H. Hermjakob, J.A. Vizcaíno, jmzReader: a Java parser library to process and visualize multiple text and XML-based mass spectrometry data formats, Proteomics 12 (2012) 795–798.

53. http://code.google.com/p/jmzquantml/, (accessed 11/2012).

54. http://www.psidev.info/mass-spectrometry#mzdata, (accessed 11/2012).

55. http://tools.proteomecenter.org/wiki/index.php?title=Formats:mzXML, (accessed 11/2012).

56. http://tools.proteomecenter.org/wiki/index.php?title=Formats:pepXML, (accessed 11/2012).

57. http://tools.proteomecenter.org/wiki/index.php?title=Formats:protXML, (accessed 11/2012).

58. A. Keller, D. Shteynberg, Software pipeline and data analysis for MS/MS proteomics: the trans proteomic pipeline, Methods Mol. Biol. 694 (2011) 169–189.

59. http://www.w3.org/TR/xpath/, (accessed 10/2012).
60. E.W. Deutsch, File formats commonly used in mass spectrometry proteomics, 2012 (http://dx.doi.org/10.1074/mcp.R112.019695.
61. http://www.psidev.info/groups/proteomics-informatics, (accessed 11/2012).
62. M. Gremse, A. Chang, I. Schomburg, A. Grote, M. Scheer, C. Ebeling, D. Schomburg, The BRENDA Tissue Ontology (BTO): the first all-integrating ontology of all organisms for enzyme sources, Nucleic Acids Res. 39 (2011) D507–D513.
63. K. Degtyarenko, P. de Matos, M. Ennis, J. Hastings, M. Zbinden, A. McNaught, R. Alcantara, M. Darsow, M. Guedj, M. Ashburner, ChEBI: a database and ontology for chemical entities of biological interest, Nucleic Acids Res. 36 (2008) D344–D350.
64. M. Ashburner, C.A. Ball, J.A. Blake, D. Botstein, H. Butler, J.M. Cherry, A.P. Davis, K. Dolinski, S.S. Dwight, J.T. Eppig, M.A. Harris, D.P. Hill, L. Issel-Tarver, A. Kasarskis, S. Lewis, J.C. Matese, J.E. Richardson, M. Ringwald, G.M. Rubin, G. Sherlock, Gene ontology: tool for the unification of biology. The Gene Ontology Consortium, Nat. Genet. 25 (2000) 25–29.
65. http://www.maldi-msi.org/download/imzml/imagingMS.obo, (accessed 11/2012).
66. S. Orchard, Molecular interaction databases, Proteomics 12 (2012) 1656–1662.
67. S. Orchard, S. Kerrien, Molecular interactions and data standardisation, Methods Mol. Biol. 604 (2010) 309–318.
68. S. Orchard, L. Montecchi-Palazzi, H. Hermjakob, R. Apweiler, The use of common ontologies and controlled vocabularies to enable data exchange and deposition for complex proteomic experiments, Pac. Symp. Biocomput. 2005 (2005) 186–196.
69. L. Montecchi-Palazzi, R. Beavis, P.A. Binz, R.J. Chalkley, J. Cottrell, D. Creasy, J. Shofstahl, S.L. Seymour, J.S. Garavelli, The PSI-MOD community standard for representation of protein modification data, Nat. Biotechnol. 26 (2008) 864–866.
70. http://psidev.cvs.sourceforge.net/viewvc/psidev/psi/psi-ms/mzML/controlled Vocabulary/psi-ms.obo?revision=1.201, (accessed

11/2012).

71. R.R. Brinkman, M. Courtot, D. Derom, J.M. Fostel, Y. He, P. Lord, J. Malone, H. Parkinson, B. Peters, P. Rocca-Serra, A. Ruttenberg, S.-A. Sansone, L.N. Soldatova, C.J. Stoeckert Jr., J.A. Turner, J. Zheng, O.B.I. consortium, modeling biomedical experimental processes with OBI, J. Biomed. Semant. 1 (Suppl. 1) (2010) S7.
72. http://obofoundry.org/wiki/index.php/PATO:Main_Page, (accessed 11/2012).
73. D.A. Natale, C.N. Arighi, W.C. Barker, J.A. Blake, C.J. Bult, M. Caudy, H.J. Drabkin, P. D'Eustachio, A.V. Evsikov, H. Huang, J. Nchoutmboube, N.V. Roberts, B. Smith, J. Zhang, C.H. Wu, The Protein Ontology: a structured representation of protein forms and complexes, Nucleic Acids Res. 39 (2011) D539–D545.
74. B. Smith,W. Ceusters, B. Klagges, J. Kohler, A. Kumar, J. Lomax, C. Mungall, F. Neuhaus, A.L. Rector, C. Rosse, Relations in biomedical ontologies, Genome Biol. 6 (2005) R46.
75. http://www.psidev.info/sepcv, (accessed 11/2012).
76. http://www.unimod.org/obo/unimod.obo, (accessed 11/2012).
77. G.V. Gkoutos, P.N. Schofield, R. Hoehndorf, The Units Ontology: a tool for integrating units of measurement in science, Database (Oxford) 2012 (2012) bas033.
78. D.N. Perkins, D.J. Pappin, D.M. Creasy, J.S. Cottrell, Probability-based protein identification by searching sequence databases using mass spectrometry data, Electrophoresis 20 (1999) 3551 3567.
79. http://www.w3.org/TR/xslt, (accessed 11/2012).
80. http://www.w3.org/TR/owl2-overview/, (accessed 11/2012).
81. http://www.w3.org/RDF/, (accessed 11/2012).
82. http://topicmaps.org/xtm/, (accesses 10/2012).
83. M. Krötzsch, F. Simancík, I. Horrocks, Description Logic Primer, http://arxiv.org/ pdf/1201.4089v1.pdf, (accessed 11/2012).
84. J. Day-Richter, M.A. Harris, M. Haendel, S. Lewis, OBO-Edit—an ontology editor for biologists, Bioinformatics 23 (2007) 2198–2200.
85. C. Mungall, A. Ireland, The OBO Flat File Format Guide, version 1.4 (draft), http:// www.geneontology.org/GO.format.obo-1_4.shtml, (accessed 11/2012).

86. J. Day-Richter, The OBO Flat File Format Specification, version 1.2, 2006, http:// www.geneontology.org/GO.format.obo-1_2.shtml, (accessed 11/2012).
87. https://lists.sourceforge.net/lists/listinfo/obo-format, (accessed 11/2012).
88. http://www.geneontology.org/GO.ontology.relations.shtml and http://obofoundry. org/ro/, accessed 11/2012
89. L. Montecchi-Palazzi, F. Gibson, D. Schober, S. Sansone, Guidelines for the development of Controlled Vocabularies, Proteomics Standards Initiative/Metabolomics Standards Initiative, http://www.psidev.info/node/47.
90. http://www.geneontology.org/cgi-bin/xrefs.cgi, (accessed 11/2012).
91. T. Berners-Lee, J. Hendler, O. Lassila, The Semantic Web — a new form of Web content that is meaningful to computers will unleash a revolution of new possibilities, Sci. Am. 284 (2001) 34–43.
92. H. Chen, T. Yu, J.Y. Chen, Semantic Web meets Integrative Biology: a survey, Brief. Bioinform. (2012).
93. G. Mayer, Data Management in Systems Biology II — Outlook Towards the Semantic Web, 1 13, arxiv.org, 2009. (http://arxiv.org/pdf/0912.2822.pdf, accessed 11/2012).
94. http://www.cs.man.ac.uk/~ezolin/d, (accessed 11/2012).
95. http://www.w3.org/TR/owl2-profiles, (accessed 11/2012).
96. S.H. Tirmizi, S. Aitken, D.A. Moreira, C. Mungall, J. Sequeda, N.H. Shah, D.P. Miranker, Mapping between the OBO and OWL ontology languages, J. Biomed. Semant. 2 (Suppl. 1) (2011) S3.
97. R. Hoehndorf, A. Oellrich, M. Dumontier, J. Kelso, D. Rebholz-Schuhmann, H. Herre, Relations as patterns: bridging the gap between OBO and OWL, BMC Bioinforma. 11 (2010) 441.
98. OBO download matrix, http://www.berkeleybop.org/ontologies/, (accessed 11/2012).
99. D.A. Moreira, C.J. Mungall, N.H. Shah, S. Aitken, J.-D. Richter, T. Redmond, M.A. Musen, The NCBO OBO to OWL mapping, Nature

Proceedings, 2009, (hdl:10101/ npre.2009.3938.1, http://precedings.nature.com/documents/3938/version/1, accessed 11/2012).

100. I. Horrocks, Tool support for ontology engineering, in: D. Fensel (Ed.), Foundations for the Web of Information and Services, Springer, 2011, pp. 103–122.

101. J.H. Gennari, M.A. Musen, R.W. Fergerson, W.E. Grosso, M. Crubezy, H. Eriksson, N.F. Noy, S.W. Tu, The evolution of Protégé: an environment for knowledge-based systems development, Int. J. Hum. Comput. Stud. 58 (2003) 89–123.

102. http://www.ebi.ac.uk/ontology-lookup/, (accessed 11/2012).

103. H. Barsnes, R.G. Coté, I. Eidhammer, L. Martens, OLS dialog: an open-source front end to the Ontology Lookup Service, BMC Bioinforma. 11 (2010).

104. S. Vercruysse, A. Venkatesan, M. Kuiper, OLSVis: an animated, interactive visual browser for bio-ontologies, BMC Bioinforma. 13 (2012) 116.

105. S. Aitken, Y. Chen, J. Bard, OBO Explorer: an editor for Open Biomedical Ontologies in OWL, Bioinformatics 24 (2008) 443–444.

106. http://lucene.apache.org/core/, (accessed 02/2013).

107. R.G. Coté, P. Jones, R. Apweiler, H. Hermjakob, The Ontology Lookup Service, a lightweight cross-platform tool for controlled vocabulary queries, BMC Bioinforma. 7 (2006) 97.

108. R.G. Cote, J. Griss, J.A. Dianes, R. Wang, J.C. Wright, H.W. van den Toorn, B. van Breukelen, A.J. Heck, N. Hulstaert, L. Martens, F. Reisinger, A. Csordas, D. Ovelleiro, Y. Perez-Rivevol, H. Barsnes, H. Hermjakob, J.A. Vizcaino, The Proteomics IDEntification (PRIDE) Converter 2 framework: an improved suite of tools to facilitate data submission to the PRIDE database and the ProteomeXchange consortium, Mol. Cell Proteomics 11 (2012) 1682–1689.

109. http://www.medizinisches-proteom-center.de/ProCon, (accessed 02/2013).

110. C.F. Taylor, D. Field, S.A. Sansone, J. Aerts, R. Apweiler, M. Ashburner, C.A. Ball, P.A. Binz, M. Bogue, T. Booth, A. Brazma, R.R. Brinkman, A. Michael Clark, E.W. Deutsch, O. Fiehn, J. Fostel, P. Ghazal, F. Gibson, T. Gray, G. Grimes, J.M. Hancock, N.W. Hardy, H. Hermjakob, R.K. Julian Jr., M. Kane, C. Kettner, C. Kinsinger, E. Kolker, M. Kuiper, N. Le Novere, J. Leebens-Mack, S.E. Lewis, P. Lord, A.M. Mallon, N. Marthandan, H. Masuya, R. McNally, A. Mehrle, N. Morrison, S. Orchard, J. Quackenbush, J.M. Reecy, D.G. Robertson, P. Rocca-Serra, H. Rodriguez, H. Rosenfelder, J. Santoyo-Lopez, R.H. Scheuermann, D. Schober, B. Smith, J. Snape, C.J. Stoeckert Jr., K. Tipton, P. Sterk, A. Untergasser, J. Vandesompele, S. Wiemann, Promoting coherent minimum reporting guidelines for biological and biomedical investigations: the MIBBI project, Nat. Biotechnol. 26 (2008) 889–896.

111. C.F. Taylor, N.W. Paton, K.S. Lilley, P.A. Binz, R.K. Julian Jr., A.R. Jones, W. Zhu, R. Apweiler, R. Aebersold, E.W. Deutsch, M.J. Dunn, A.J. Heck, A. Leitner, M. Macht, M. Mann, L. Martens, T.A. Neubert, S.D. Patterson, P. Ping, S.L. Seymour, P. Souda, A. Tsugita, J. Vandekerckhove, T.M. Vondriska, J.P. Whitelegge, M.R. Wilkins, I. Xenarios, J.R. Yates III, H. Hermjakob, The minimum information about a proteomics experiment (MIAPE), Nat. Biotechnol. 25 (2007) 887–893.

112. http://www-bs2.informatik.uni-tuebingen.de/services/OpenMS/analysisXML/index. php, (accessed 11/2012).

113. http://www-bs2.informatik.uni-tuebingen.de/services/OpenMS/mzML/, (accessed 11/2012).

114. http://www.proteored.org/MIAPE/tutorials.asp, (accessed 11/2012).

115. http://code.google.com/p/mzquantml-validator/, (accessed 11/2012).

116. http://code.google.com/p/psi-pi/downloads/detail?name=mzIdentMLValidator- 1.3-SNAPSHOT.zip&can=2&q=, (accessed 10 / 2012).

117. http://www.psidev.info/sites/default/files/MIAPE_MS_2.98.pdf, (accessed 11/2012).

118. P.A. Binz, R. Barkovich, R.C. Beavis, D. Creasy, D.M. Horn, R.K. Julian, S.L. Seymour, C.F. Taylor, Y. Vandenbrouck, Guidelines for

reporting the use of mass spectrometry informatics in proteomics, Nat. Biotechnol. 26 (2008) 862-862.

119. C. Hoogland, M. O'Gorman, P. Bogard, F. Gibson, M. Berth, S.J. Cockell, A. Ekefjard, O. Forsstrom-Olsson, A. Kapferer, M. Nilsson, S. Martinez-Bartolome, J.P. Albar, S. Echevarria-Zomeno, M. Martinez-Gomariz, J. Joets, P.A. Binz, C.F. Taylor, A. Dowsey, A.R. Jones, Guidelines for reporting the use of gel image informatics in proteomics, Nat. Biotechnol. 28 (2010) 655–656.
120. http://www.psidev.info/sites/default/files/MIAPE_GE_1_4.pdf, 1–11, (accessed 11/2012).
121. A.R. Jones, K. Carroll, D. Knight, K. MacLellan, P.J. Domann, C. Legido-Quigley, L.H. Huang, L. Smallshaw, H. Mirzaei, J. Shofstahl, N.W. Paton, Guidelines for reporting the use of column chromatography in proteomics, Nat. Biotechnol. 28 (2010) 654-654.
122. http://www.psidev.info/sites/default/files/MIAPE_CE_0.9.3.doc, (accessed 11/2012)
123. http://www.psidev.info/miape-quant-in-docproc, (accessed 11/2012).
124. L. Montecchi-Palazzi, S. Kerrien, F. Reisinger, B. Aranda, A.R. Jones, L. Martens, H. Hermjakob, The PSI semantic validator: a framework to check MIAPE compliance of proteomics data, Proteomics 9 (2009) 5112–5119.
125. J.A. Medina-Aunon, S. Martinez-Bartolome, M.A. Lopez-Garcia, E. Salazar, R. Navajas, A.R. Jones, A. Paradela, J.P. Albar, The ProteoRed MIAPE web toolkit: a user-friendly framework to connect and share proteomics standards, Mol. Cell Proteomics 10 (2011).
126. http://www.psidev.info/validator, (accessed 11/2012).
127. https://lists.sourceforge.net/lists/listinfo/psidev-ms-vocab, (accessed 11/2012).
128. http://psidev.cvs.sourceforge.net/viewvc/psidev/psi/psi-ms/mzML/controlledVocabulary/psi-ms.obo?view=log, (accessed 11/2012).

CHAPTER

4

BASICS AND RECENT ADVANCES OF TWO DIMENSIONAL- POLYACRYLAMIDE GEL ELECTROPHORESIS

Sameh Magdeldin[1,2] , Shymaa Enany[1,3] , Yutaka Yoshida[1] , Bo Xu[1] , Ying Zhang[1] , Zam Zureena[4] , Ilambarthi Lokamani[4] , Eishin Yaoita[1] and Tadashi Yamamoto [1]

[1] Department of Structural Pathology, Institute of Nephrology, Graduate School of Medical and Dental Sciences, Niigata University, 1-757 Asahimachi-dori, Niigata, Japan.

[2] Department of Physiology, Faculty of Veterinary Medicine, Suez Canal University, Ismailia, Egypt.

[3] Department of Microbiology and Immunology, Faculty of Pharmacy, Suez Canal University, Ismailia, Egypt.

[4] UKM Medical Centre, Kuala Lumpur, Malaysia.

ABSTRACT

Gel- based proteomics is one of the most versatile methods for fractionating protein complexes. Among these methods, two dimensional- polyacrylamide gel electrophoresis (2-DE) represents a mainstay orthogonal approach, which is popularly used to simultaneously fractionate, identify, and quantify proteins when coupled with mass spectrometric identification or other immunological tests. Although 2-DE was first introduced more than three decades ago, several challenges and limitations to its utility still exist. This review discusses the principles of 2-DE as well as both recent methodological advances and new applications.

Keywords: Two dimensional electrophoresis Advances Basics Review

INTRODUCTION

Two dimensional polyacrylamide gel electrophoresis (2-DE) is considered a powerful tool used for separation and fractionation of complex protein mixtures from tissues, cells, or other biological samples. It allows separation of hundreds to thousands of proteins in one gel. This technique became more popular and comprehensive after the prime advances and high resolution 2-DE modification [1]. Actually, the modification established by O'Farrell allowed 2-DE technique to resolve up to 5000 protein- representing spots in an even 2 dimensional distribution and enabled precise separation of protein spots with high accuracy [2].

2-DE technique is deemed to be one of the leading powers in the expansion of proteomics and protein studies. It provides the first step for further analysis of the differentially regulated protein spots using mass spectrometry and western blotting. The use of 2-DE has been effectively defined in many cases to disclose both physiological mechanisms and proteins associated with clinical pathologies that can aid in the discovery of biomarkers.

BASIC PRINCIPLES OF 2-DE

Concept of 2-DE

2-DE consists mainly of two steps of separation; first dimension and second dimension. In the first dimension, protein molecules are resolved depending on their isoelectric point (pI) [2]. Separation of proteins under a pH gradient allows intense band recovering using various tactics such as immobilized gradient electrophoresis (IPEG), isoelectric focusing (IEF), or non-equilibrium pH gradient electrophoresis (NEPHGE). In the second dimension, protein separation is performed based on molecular weight using SDS Laemmli or Tris-Tricine buffers. Due to the fact that it is improbable that different protein molecules may have the same physicochemical properties (pI and MW), proteins are efficiently separated by 2-DE rather than 1D-SDS PAGE [2]. A prominent merit of 2-DE is that the resolution acquired during the first dimensional separation is not missed in the second electrophoresis when IEF gel strip is connected to the SDS- PAGE gel [2, 3].

Applications and utilities of 2-DE

2-DE is a powerful and widely used method for analysis of complex protein mixtures with exceptional ability to separate thousands of proteins at once. It provides direct visual confirmation of changes in protein/post-translational modifications (PTMs) abundance, thus providing early justification for downstream analytical steps through detecting post- and co-translational modifications, which cannot be predicted from the genomic sequence. Other applications of 2-DE include whole proteome analysis [4], cell differentiation [5], detection of biomarkers and disease markers, drug discovery, cancer research [6], bacterial pathogenesis [7], purity checks, microscale protein purification, and product characterization.

Advantages and strengths of 2-DE

Robustness

During the last few years, several methodological improvements have contributed to increase the robustness of 2-DE workflows. The use of immobilized isoelectric focusing (IEF) strips, ampholytes- based buffers, highly sensitive dyes, and gel imaging software made the variability most likely from upstream process such as protein loss during extraction [8]. In a recent multi-laboratory study on the feasibility of 2-DE, it is reported that 70-93% of spots were detected with coefficient of variation (CVs) less than 20% within same laboratory researchers [9]. On the other hand, 72% of spots showed CVs with less than 20% across laboratories [9]. This finding proves the feasibility and the robustness of 2-DE. Moreover, 2-DE becomes less variable when multiplexing electrophoresis developed. Differentially labeled samples run at the same time minimized the possibility of artifacts resulted from technical errors. Finally, the recent improvements of the gel image analysis minimized the former high percentage of spot identification failure estimated to reach 60% which considered one of the major contributors to variability seen with 2-DE [10].

Visualized mapping analysis

One of the unique features of 2-DE is its ability to resolve intact full-length proteins (up to 5000 protein) in a single gel. This includes visualized detection of the physico-chemical properties such as MW and pI with possible quantification based on the spot intensity [8]. Proteins of interest could be characterized via peptide mass finger printing or when probed with antibodies. Moreover, 2-DE allows separation and identification of PTMs and protein isoforms (Figure 1). In several cases, PTMs could be recognized by horizontal or vertical shifting of a protein spot as these modifications usually change the protein MW and pI [11].

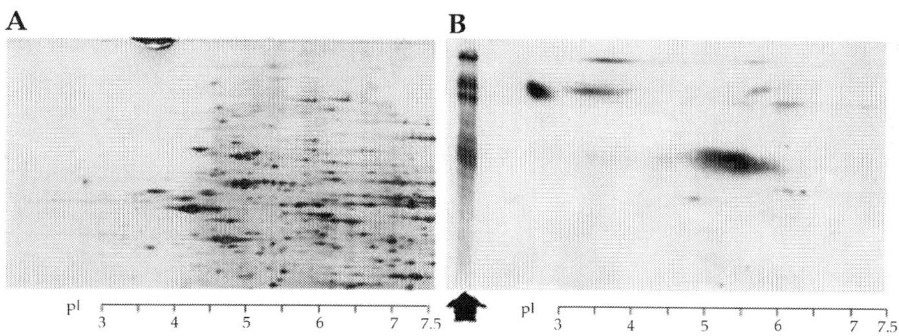

Figure 1: 2-DE showed the resolving of glomerular proteins using 7 cm IPG strip. (A) Gel stained with silver stain (B)western blot of PVDF membrane using P- Tyr- 100 antibody for detecting phosphorylated proteins. Arrow shows 1D- SDS pattern [32].

Compatible platform for further analysis

2-DE gel easily and efficiently couples with many other analysis and biochemical techniques. Thus, it provides a compatible platform for subsequent analysis. For example, stained gels can be followed by spot excision, destaining, protein extraction, digestion, and analysis of the tryptic peptides by mass spectrometry. Although coomassie blue could be reversibly destained and compatible with mass spectrometry, silver staining, it is not compatible because of the usage of formaldehyde or glutaraldehyde during the fixing and sensitization step that results in lysine residue cross-linking within the protein chain interfering MS analysis and thus will hinder trypsin digestion [12, 13]. Various modifications in the silver nitrate stain approach were performed to overcome this drawback. Compatibility of 2-DE includes bottom-up proteomics to identify proteins and characterize their amino acid sequences or alternatively proceeded by the top-down proteomics (shotgun) in which the crude proteins extract is digested directly for analysis. In another powerful combination, antibody- based analysis could be coupled before or after 2-DE. For instance, immune-affinity purification can be used to pre-fractionate a protein of interest prior to running 2-DE such as phosphorylated [2] or ribonucleoproteins [14]. Most commonly, 2-DE fractionated proteins are subjected to either in-gel digestion to prepare tryptic peptides for mass spectrometric analysis or gels are validated for protein of interest using western blotting [15].

Limitations of 2-DE

Low reproducibility

In the traditional method of the first dimension IEF, the carrier ampholytes is utilized to build pH gradient. The carrier ampholytes-based pH gradient made from soft unsupported tube gels (typically 4% acrylamide) is not stable. There is batch to batch variability and prone to cathodic drift (a progressive loss of basic proteins during long running of electro-focusing under electric field), leading to low reproducibility and requiring careful monitoring of electric field [2]. Replacement of carrier ampholyte-based pH gradient in tube gel with the immobilized pH gradients (IPG) was the key development in increasing the reproducibility of 2-DE [16, 17].

Difficulty in separating hydrophobic and extremely acidic or basic proteins

Different types of proteins can always be missing, due to the difficulty in separating membrane- bound (hydrophobic) and extreme proteins [18, 19]. Notably, highly acidic or basic proteins are neither easily extracted nor solubilized. This difficulty in extraction relies mainly on the solubilization power of the buffer used in the IEF step. Many efforts have been performed for better solubilization of membrane proteins using different chaotropes or detergents. For instant, Triton X- 114 and CHAPS showed a powerful recovering power towards hydrophobic proteins on 2-DE when tested immunologically [20]. Another study reported the use of cationic detergent benzyldimethyl-n-hexadecylammonium chloride to improve resolving hydrophobic proteins with GRAVY index as low as 0.56 [21]. Similar improvement was reported when using detergents such as DHPC [22] and 1,4-dithioethanol [23]. Although solubilization of wide range of proteins could be achieved using denaturating solution (Urea and Thiourea) and zwitterionic detergents (SB 3–10) [24], resistance of certain proteins still remains a built-in problem [2, 24].

Narrow dynamic range of 2-DE

Low dynamic range of proteins is one of the challenging problems encountered during 2-DE. Highly abundant peptides mask low abundant ones, which may be reflective of low abundant proteins. Moreover, the visualization of faint protein spots (low abundant) separated on 2-DE gels is also governed by staining sensitivity. For example, classic coomassie has a narrow dynamic range with detection limit of only about 100 ng. Colloidal coomassie is relatively higher in sensitivity with detection limit 10 ng. The dynamic range could be increased to a detection level below 1 ng with the availability of highly sensitive silver-staining method [12] and a diversity of fluorescence dyes such as SYPRO-Ruby and Deep purple flurophore dyes (Figure 2). Therefore, using sensitive stain increases protein sample dynamic range, leads to successful gel imaging, and finally leads to successful mass spectrometric identification and immunological validation [11]. Alternatively, depletion of highly abundant proteins such as albumin and hemoglobin significantly improves the dynamic range of 2-DE by allowing better focusing and mass spectrometric picking (see Prefractionation, enrichment, and depletion prior to 2-DE section). Therefore, the limited detection sensitivity of 2-DE does not cope with the actual dynamic range of protein concentration in cell and tissue extracts, or biological fluids [2]. Other solution to overcome the dynamic range problem is to load more protein sample and using a giant 2-DE gel (24 cm). However, this approach could result in production of overcrowded images with non well-separated spots.

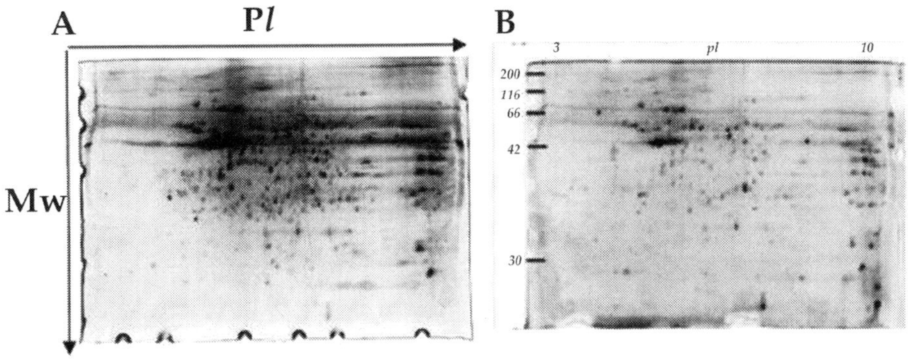

Figure 2: Staining of 2-DE gels. A) 24 cm two dimensional polyacrylamide gel electrophoresis of mouse colon protein stained by silver staining or (B) Deep purple flurophore dye. Visualization of B image was done using a Laser scanner [4].

Low throughput and labor- intensiveness

2-DE is labor-intensive and has a relative low throughput. The throughput of 2-DE is adequate for many small-scale basic research studies, but it may present a serious obstacle for projects that involve screening of a large number of clinical samples. Furthermore, 2-DE requires skills and experience to counteract any possible trouble shooting due to the procedural artifacts. Much care should be considered to minimize the variations during sample processing that lasts up to 3 days in larger gels.

Recent advances and technologies associated with 2-DE

2D- DIGE

The development of image technology has introduced differential imaging gel electrophoresis (DIGE) technique. This method was designed in an attempt to increase sensitivity and reproducibility of 2-DE using multiplexed fluorescent dyes- labeled protein samples. 2D-DIGE is based mainly on running more than one sample (maximum 3 samples) on a single gel at once to address the issue of gel-to gel variability. In this technique, different fluorescent cyanine (Cy) dyes are used for labeling proteins from different samples [25]. After mixing these samples in equal ratio and running them together as one sample, same protein from different samples migrates to the same position on the 2D gel where it could be easily explored and differentiated by the different fluorophore-labeled dye and imaged to calculate its abundance. 2D-DIGE is an important tool, especially for clinical laboratories involved in the determination of protein expression levels and disease biomarker discovery. When absolute biological variation between samples is the main objective, as in biomarker discovery, 2D-DIGE is one of the methods of choice.

Prefractionation, enrichment, and depletion prior to 2-DE

Protein sample prefractionation before 2-DE has been implemented to reduce sample complexity. As a result, the low abundance proteins present in these fractions will be clearly represented in 2-DE. Proteins identified in the prefractionated samples have a higher number of peptides. Moreover, low molecular weight proteins can be clearly detected when sample complexity is reduced [26]. Prefractionation ultimately increases loading capacity of samples onto the 2-DE gel and leads to better resolution, visualization and identification. Sample prefractionation can be performed in several different ways, broadly subdivided into three levels: cellular, subcellular and protein subfractionation [27]. For example, cellular extraction is a generalized approach to screen the whole proteome of a given sample. An extraction buffer containing urea and thiourea with NP-40 usually recovers most proteins [24]. In subcellular prefractionation, a density gradient (isopycnic) centrifugation method [28], or alternatively tissue strainers and buffers approach is used to homogenate a given tissue followed by using selective solvents to dissolve mitochondrial, ER-golgi, or nuclear proteins [29]. Protein subfractionation based on its physico-chemical properties is another level of fractionation. This method could be used to fractionate proteins based on their charge such as strong cation exchange (SCX) method [30] or based on their pI such as Liquid phase IEF Zoom® prefractionator [31]. Enrichment of protein of interest that exists in the sample in low abundance such as phosphorylated proteins (represents 3-4%) might be essential before 2-DE. For example, in one of our experiments to characterize tyrosine phosphorylated glomerular proteins related to slit diaphragm, samples were immunoprecipitated using protein A- Sepharose then subjected to 2-DE [32]. Depending on the nature of the sample, biological fluids such as plasma and urine may require removal of specific proteins (albumin or hemoglobin) to increase the resolution of separation before 2-DE. On the downstream mass spectrometric level, the excess of highly tryptic peptides generated from these abundant proteins bias the identification towards these proteins on the expense of lower abundant ones. In contrast, depletion increases peptides capture possibilities of low abundant proteins [33, 34].

Blue native Gel electrophoresis for membrane proteins studies

Blue-Native polyacrylamide gel electrophoresis (Blue Native PAGE) was originally introduced by Schagger and von Jagow as a technique for separating enzymatically active membrane protein complexes under mild condition [35]. In this approach, the anionic dye Coomassie Brilliant Blue G-250 (5% w/v) is mixed with the protein sample prior to gel loading. This dye has the ability to provide negative charges to the surface of the protein. It induces a charge shift that improves solubilization of hydrophobic proteins especially membrane intrinsic electron/proton transfer complexes in mitochondria providing a global analysis of membrane proteomics. The dye binds to protein complexes and both migrate during electrophoresis. Finally, the gels are stained with coomassie dye again prior to mass analysis. A recent implementation for this technique was fruitful in identifying individual compounds within protein complexities of inflammasomes [36] and integrin and histone complexes in placenta [37]. In addition to its ability to unveil protein complexities, this methodology allows better separation of hydrophobic proteins (membrane proteins) as well [38].

2-DE for post translational modifications (PTMs)

The analysis of protein post-translational modifications (PTMs) has become an important topic for the study of cell biology, disease treatment, and disease prevention. 2-DE provides a direct observation of protein

PTMs in gels as well as its relative abundance. Publicly available web-based tools such as ProMoST [39] and JVirGel [40] could be used for that purpose. In addition, specific staining dyes for PTM monitoring were developed and became widely applied in the proteome studies. In phosphoproteome research, molecular Probes, Pro-Q® Diamond phosphoprotein gel stain is a breakthrough technology that provides a precise method for selectively staining phosphoproteins in polyacrylamide gels. Similarly, for glycoproteomics, commercially available fluorescent dye; Lissamine rhodamine B sulfonyl hydrazine (LRSH) was introduced to specifically stain the glycoproteins. This stain relays on periodate/Schiff base mechanism. Unlike conventional methods used for the characterization of PTMs such as the enrichment strategies either with antibodies or immobilized resin, lectin's binding strategy, or enzymatic based techniques, 2-DE offers an analytical tool with high resolution and high reproducibility by taking the advantage of the well known change in pI and MW induced by many modifications [41].

Outlines on the types and current availability of 2-DE

First dimension electrophoresis

The first dimension electrophoresis can be performed using ionic substances, which reacts as acid or base and is termed as carrier ampholyte pH gradient. It could be blended and optimized for wide or more restricted pH ranges. Several techniques for IEF could be applied in the first dimension electrophoresis as explained below.

Conventional IEF

The conventional method of IEF depends on the carrier ampholyte where proteins migrate in a solution media until reaching the equilibrium state when its net charge equals to zero. Proteins that are in a pH region below its isoelectric point (pI) will be positively charged and will migrate towards the cathode. As it migrates through a gradient of increasing pH, however, the protein's overall charge will decrease until the protein reaches the pH region that corresponds to its pI. At this point, the migration ceases. As a result, the proteins become focused into sharp stationary bands with each protein positioned at a point in the pH gradient corresponding to its pI. [2]. Although this conventional method is easy to prepare and do not require much casting equipments, it has a main disadvantage as the ampholytes have some susceptibility to flow towards cathode and this gradient flow usually causes a reduction in the reproducibility.

Immobilized pH gradient (IPG)

Immobilized pH gradient strip (IPG) is an integrated part of polyacrylamide gel matrix fixed on a plastic strip. Co-polymerization of a set of non-amphoteric buffers with different chemical properties is included [2, 42]. A ready- made IPG strips are available with different lengths and pI. Usually, short length IPG strips are used for fast screening while longer one for maximal and comprehensive analysis. Various models of 2- DE gels are shown in Figure 3. A commercial pre-casted acrylamide gel matrix co-polymerized with a pH gradient on a plastic strip results in a stable pH value over the traditional ampholyte method. It has an ability to avoid cationic accumulation and to produce a better-focused protein with less smearing [2, 42]. There are many other advantages of using IPG strips over ampholytes such as reduced cathodic drift, higher mechanical strength as the strips are casted on a plastic backing that minimizing gel breakage, and higher protein loading capacity due to the sample loading method [17, 43].

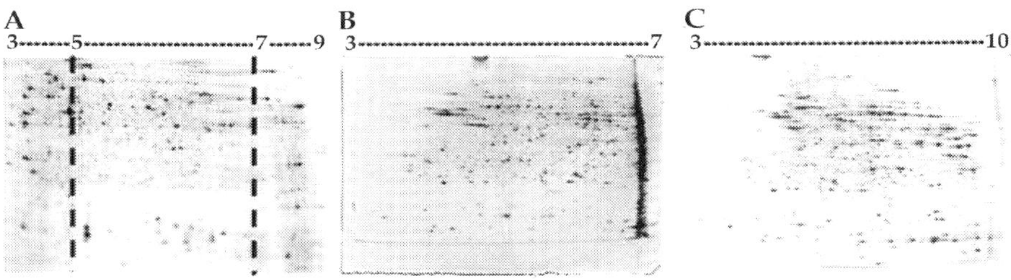

Figure 3: Various IPG strips used in the first dimension step of 2- DE gels. **(A)** Gel with non-liner (NL) IPG strips [pI 5–7]. **(B)**8 cm gel with IPG strips [pI 3–7]. **(C)** Wide- ranged (pI 3–10) IPG strip.

Non-equilibrium pH gel electrophoresis (NEPHGE)

Non-equilibrium pH gel electrophoresis (NEPHGE) technique was developed to resolve proteins with basic to extremely high pI (7.0 to 11.0) [1, 44] that cannot be separated by the traditional method. In contrast, IPG method allows the protein molecules to move at different rate across the gel based on the charge and the volt hours setting that determine speed pattern and reproducibility [2]. A previous study comparing IPG and NEPHGE techniques showed that protein loss was higher in IPG-based method, especially for basic proteins. They found the reproducibility of spots was slightly better in NEPHGE-based method. About half of detected basic protein spots were not reproducible by IPG-based 2-DE, whereas NEPHGE-based method showed excellent reproducibility in the basic gel zone. The reproducibility of acidic proteins was similar in both methods [45].

Second dimension SDS PAGE

This step separates proteins based on their molecular weight using a vertical electrophoretic device with either Laemmli buffer [46] or Tris-Tricine buffer [47]. Instead of loading protein sample within the wells, the first dimension-rehydrated strip is carefully placed on the top of the SDS-PAGE and sealed with agarose. Further technical details are described by Magdeldin *et al.*[4].

The state of art in the analysis of 2-DE images

2- DE software analysis

Most of the software analysis workflow starts with either spot detection or image alignment. Examples of currently available software are listed in Table 1. In general, spot detection software (ex. PDQuest and Proteomweaver) sometimes lead to missing some data because of the mismatching error resulted due to shifted spots between gels or spots overlapping. Furthermore, false positive spots are commonly detected due to staining artifacts especially with poor technical skills. To avoid these problems, an alternative processing workflow approaches relay on aligning gel images before processing were developed (ex. Progenesis samespot and Decodon). This method increased the precision of spot detection, raised the accuracy of spot overlying, enabled alignment of several replicates, and saved time. In a practical example illustrated in Figure 4, same gel images of 2 groups were analyzed using PDquest and progenesis samespot software that utilize initial step of spot detection or image alignment, respectively. The result showed inaccuracy of spot detection in the analysis starting with spot detection by detecting many false positive spots. On contrary, precise spots were recognized when gel images where aligned first. This improvement in gel image processing reduced variability between gels and increased the power to detect differentially expressed spots with less effort of manual editing. The key steps of the workflow of most of these automated methods are as following; image quality control, image alignment, spot detection, automatic analysis, editing of spot detection, review the results, statistical analysis, calibration of spots against either a molecular weight ladder or known proteins, spot picking, and importing the protein ID (Figure 5). These automated methods are helpful for gel identification particularly in the quantitative proteomics. They primarily analyze biomarkers by quantifying individual proteins and showing the separation between one or more protein spots on a scanned image of a 2-DE gel. Additionally, they match spots between gels of similar samples. For example, proteomic differences between early and advanced stages of an illness. However, any change or failure in the spot identification may affect the peptide identification in the downstream proteomics as explained later in the challenges section (Challenges in the analysis of the 2-DE images).

Table 1: Current available commercial software used for 2-DE gel image analysis

Company	Software name	Image analysis approach
Bio-Rad	PDQuest	Spot detection first
	Proteomweaver	Spot detection first
GE Healthcare	Decyder 2D	Spot detection first
GE Healthcare	ImageMaster 2D platinum	Spot detection first
GeneBio	Melanie	Spot detection first
Applied Maths NV	BioNumerics	Spot detection first
Ludesi	Redfin	Spot detection first
Ludesi	Gel IQ*	Spot detection first
Compugen	Z3 (discontinued)	Image alignment first

Decodon	Delta2D	Image alignment first
Nonlinear dynamics	Progenesis SameSpots	Image alignment first
Sourceforge	Flicker*	Image alignment first

*Open source or free of charge.

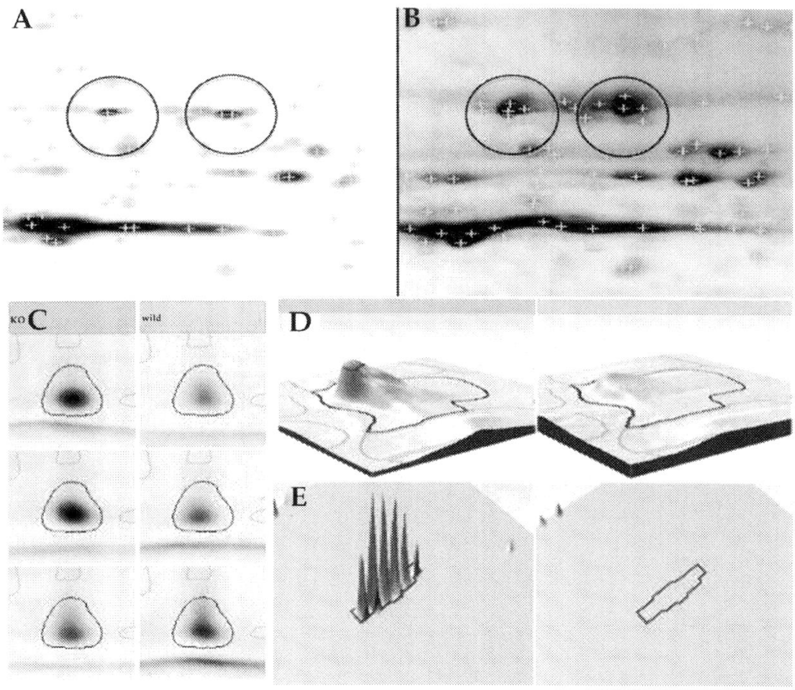

Figure 4: Comparative analysis of gel image using PDQuest and Progenesis SameSpot. Two different experimental groups were analyzed. As shown in A and B, a protein spot of F-cappig protein was recognized as a multiple spots by PDQuest when spot detection approach was used. C shows a precise detection of the same protein spot when images were aligned first (Progenesis SameSpot). The detected spot was confirmed later on a 2D (D) and 3D (E)views to be an actual fragmented peptide ions.

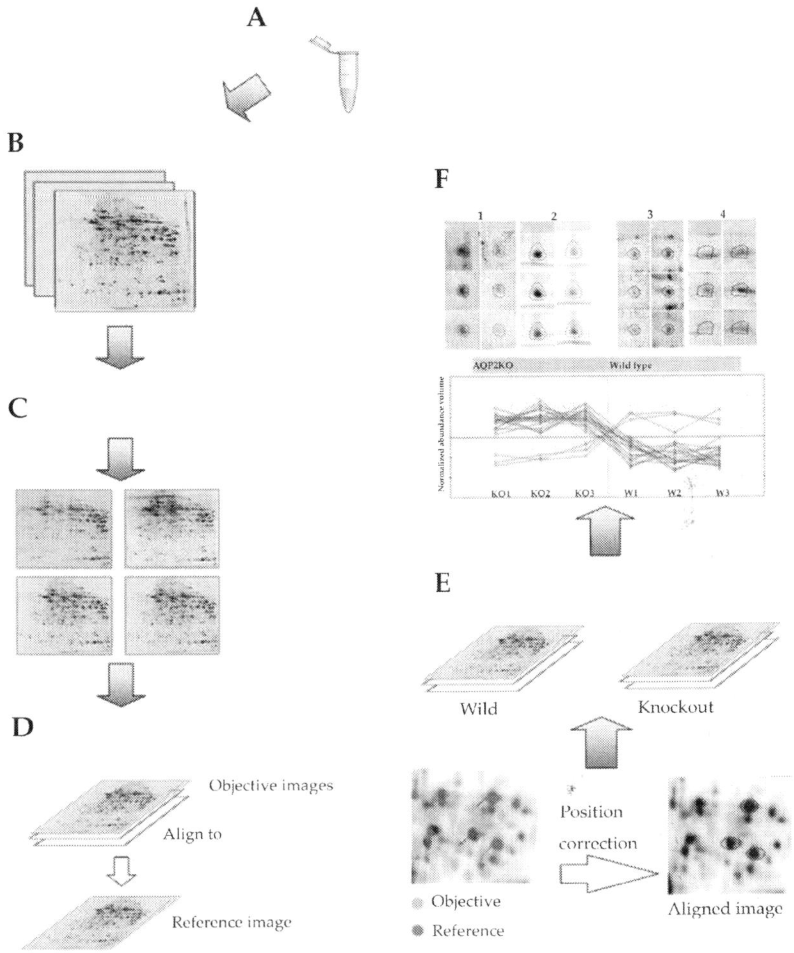

Figure 5: Image analysis workflow of a 2-DE gel electrophoresis experiment using Progenesis SameSpots (Nonlinear Dynamics). A. Sample preparation; B. 2-DE gel electrophoresis and gel staining; C. 2-DE image acquisition and image quality checked by the software automatically; D. Gel image alignment. A 2-DE gel image in the set of images for an experiment is manually set as the reference image (pink) by the user and then other 2-DE gel images (green) are aligned to the reference gel image one by one by a manual and/or automated way. E. After gel image alignment, the aligned images are grouped according to the experimental design. F. Extraction of proteins of interest. Spot volume normalization and calculation are performed by the software automatically.

2-DE proteome databases

2-DE proteome database websites allow experimental data to be uploaded and disseminated for public. They also help to generate reference maps for normal and diseased cases. These databases display annotated and representative reference map for a variety of cell and tissue types. A list of publicly available databases is found at http://www.expasy.org/ch2d-index.html. The Swiss- 2D PAGE website highlights the World- 2D PAGE database server, the World-2 D PAGE Portal, and the World- 2D PAGE repository, which displays the publication links of each reference map and the experimental condition [48]. So far, it is one of the largest databases for 2-DE. It contains more than 36 different 2-DE reference proteome maps from 7 different species including *S. Cerevisiae* and *E. coli* as well as many different human tissues and organs such as kidney and liver [48]. The database annotates protein identification, protein patho-physiological function, physicochemical properties (MW and pl, amino acid composition and peptide masses), and citations. Recently, many other 2-DE reference maps are available online e.g. (http://web.mpiib-berlin.mpg.de/cgi-bin/pdbs/2d-page/extern/index.cgi and http://www.gelbank.anl.gov).

Challenges in the analysis of the 2-DE images

Although bioinformatics software made 2-DE gel analysis much easier, several challenges remain in order to achieve a non-biased, accurate, and reproducible result.

Before software analysis

The key role in a successful computer- based gel imaging relays on the quality of the raw image itself. Therefore, it is necessary to use a high resolution scanner. CCD cameras are time consuming but could be used for several stained gels. laser scanners are more accurate and generate high resolution raw images. In general, image resolution of 100–150 in TIEF format is enough for quantitative analysis. In comparative analysis, it is essential to reproduce replicate gels with minimal noise and background for correct quantification. Certain filters might be applied with care for gel optimization prior to analysis.

During the process of analysis

Differences in the spot positions between gels are major challenging issue in image processing because they impede accurate spot matching. The key step for overcoming spot shifting is to perform a gel image alignment in which certain landmark spots are first pinned between all gels. In a next step, the software tries to align other spots based on these landmarks. However, in some cases, complete alignment could not be obtained if the two patterns are so different. Therefore, in such case, one should avoid excessive manual interventions because this would worsen the reliability of the image control and the reproducibility of the operation between different users. Because of the quantification and the normalization of the spot intensity, one should realize that the relationship between the original protein quantity in the sample and the measured spot intensity is affected by various intervening factors. For example, sample loss occurring during the IEF or while transferring to the second dimension. Given the biochemical diversity of the protein molecules, it is expected that there are some proteins with a nonlinear relation between concentration and intensity. Therefore, one should expect to obtain relatively quantitative results referring to same protein species coming from different samples.

CONCLUDING REMARKS

Protein separation is a core part of proteomics analysis. 2-DE is a basic and fundamental procedure to fractionate and visualize protein complexes. The 2-DE method is superior to visualize each protein as a spot that can be interpreted by its abundance, location, or even its presence or absence. This visual- based result is in most cases confident. Meanwhile, 2-DE procedures need experience and optimization of skills. With the ongoing development and modifications applied to this sophisticated technique, 2-DE is expected to be less labor, more informative, sensitive, rapid, and easily applied.

Notes

Sameh Magdeldin, Shymaa Enany contributed equally to this work.

DECLARATIONS

Acknowledgments

This work was supported by JSPS (Japan Society for Promotion of Science) Grant-in-Aid for scientific research (B) to SM (#23790933) from Ministry of Education, Culture, Sports, Science and Technology of Japan. The funders had no role in the decision to publish, or preparation of the article.

Authors' original submitted files for images

Below are the links to the authors' original submitted files for images.

12014_2013_60_MOESM1_ESM.tif Authors' original file for figure 1

12014_2013_60_MOESM2_ESM.tif Authors' original file for figure 2

12014_2013_60_MOESM3_ESM.tif Authors' original file for figure 3

12014_2013_60_MOESM4_ESM.pdf Authors' original file for figure 4

12014_2013_60_MOESM5_ESM.tiff Authors' original file for figure 5

Competing interest

The authors declare that they have no competing interest.

Authors' contributions

SM, SE, YY, BX, YZ, ZZ, IL, EY drafted the manuscript. TY edited and approved the manuscript. All authors read and approved the final manuscript.

Authors' Affiliations

(1) Department of Structural Pathology, Institute of Nephrology, Graduate School of Medical and Dental Sciences, Niigata University

(2) Department of Physiology, Faculty of Veterinary Medicine, Suez Canal University

(3) Department of Microbiology and Immunology, Faculty of Pharmacy, Suez Canal University

(4) UKM Medical Centre

REFERENCES

1. O'Farrell PH: High resolution two-dimensional electrophoresis of proteins. J Biol Chem. 1975, 250: 4007-4021.
2. Magdeldin S, Zhang Y, Bo X, Yoshida Y, Yamamoto T: Biochemistry, Genetics and Molecular Biology "Gel Electrophoresis - Principles and Basics". 2012, Rijeka, Croatia: INTECH,
3. Anderson NG, Matheson A, Anderson NL: Back to the future: the human protein index (HPI) and the agenda for post-proteomic biology. Proteomics. 2001, 1: 3-12.
4. Magdeldin S, Li H, Yoshida Y, Enany S, Zhang Y, Xu B, Fujinaka H, Yaoita E, Yamamoto T: Comparison of two dimensional electrophoresis mouse colon proteomes before and after knocking out aquaporin 8. J Proteomics. 2010, 73: 2031-2040.
5. Jungblut PR, Seifert R: Analysis by high-resolution two-dimensional electrophoresis of differentiation-dependent alterations in cytosolic protein pattern of HL-60 leukemic cells. J Biochem Biophys Methods. 1990, 21: 47-58.
6. Wu W, Tang X, Hu W, Lotan R, Hong WK, Mao L: Identification and validation of metastasis-associated proteins in head and neck cancer cell lines by two-dimensional electrophoresis and mass spectrometry. Clin Exp Metastasis. 2002, 19: 319-326.
7. Enany S, Yoshida Y, Magdeldin S, Bo X, Zhang Y, Enany M, Yamamoto T: Two dimensional electrophoresis of the exo-proteome produced from community acquired methicillin resistant Staphylococcus aureus belonging to clonal complex 80. Microbiol Res. 2013, 168: 504-511.
8. Rabilloud T, Chevallet M, Luche S, Lelong C: Two-dimensional gel electrophoresis in proteomics: past, present and future. J Proteomics. 2010, 73: 2064-2077.
9. Bech-Serra JJ, Borthwick A, Colome N, ProteoRed C, Albar JP, Wells M, Sanchez del Pino M, Canals F: A multi-laboratory study assessing reproducibility of a 2D-DIGE differential proteomic experiment. J Biomol Tech. 2009, 20: 293-296.

10. Bland AM, Janech MG, Almeida JS, Arthur JM: Sources of variability among replicate samples separated by two-dimensional gel electrophoresis. J Biomol Tech. 2010, 21: 3-8.
11. Chevalier F: Highlights on the capacities of "Gel-based" proteomics. Proteome Sci. 2010, 8: 23-
12. Chevallet M, Luche S, Diemer H, Strub JM, Van Dorsselaer A, Rabilloud T: Sweet silver: a formaldehyde-free silver staining using aldoses as developing agents, with enhanced compatibility with mass spectrometry. Proteomics. 2008, 8: 4853-4861.
13. Shevchenko A, Wilm M, Vorm O, Mann M: Mass spectrometric sequencing of proteins silver-stained polyacrylamide gels. Anal Chem. 1996, 68: 850-858.
14. Pinol-Roma S, Choi YD, Matunis MJ, Dreyfuss G: Immunopurification of heterogeneous nuclear ribonucleoprotein particles reveals an assortment of RNA-binding proteins. Genes Dev. 1988, 2: 215-227.
15. Kjaersgard IV, Jessen F: Two-dimensional gel electrophoresis detection of protein oxidation in fresh and tainted rainbow trout muscle. J Agric Food Chem. 2004, 52: 7101-7107.
16. Gorg A, Postel W, Domscheit A, Gunther S: Two-dimensional electrophoresis with immobilized pH gradients of leaf proteins from barley (Hordeum vulgare): method, reproducibility and genetic aspects. Electrophoresis. 1988, 9: 681-692.
17. Gorg A, Postel W, Gunther S: The current state of two-dimensional electrophoresis with immobilized pH gradients. Electrophoresis. 1988, 9: 531-546.
18. Wilkins MR, Gasteiger E, Sanchez JC, Bairoch A, Hochstrasser DF: Two-dimensional gel electrophoresis for proteome projects: the effects of protein hydrophobicity and copy number. Electrophoresis. 1998, 19: 1501-1505.
19. Corthals GL, Wasinger VC, Hochstrasser DF, Sanchez JC: The dynamic range of protein expression: a challenge for proteomic research. Electrophoresis. 2000, 21: 1104-1115.
20. Santoni V, Rabilloud T, Doumas P, Rouquie D, Mansion M, Kieffer S, Garin J, Rossignol M: Towards the recovery of hydrophobic

proteins on two-dimensional electrophoresis gels. Electrophoresis. 1999, 20: 705-711.

21. Zahedi RP, Meisinger C, Sickmann A: Two-dimensional benzyldimethyl-n-hexadecylammonium chloride/SDS-PAGE for membrane proteomics. Proteomics. 2005, 5: 3581-3588.
22. Babu GJ, Wheeler D, Alzate O, Periasamy M: Solubilization of membrane proteins for two-dimensional gel electrophoresis: identification of sarcoplasmic reticulum membrane proteins. Anal Biochem. 2004, 325: 121-125.
23. Pasquali C, Fialka I, Huber LA: Preparative two-dimensional gel electrophoresis of membrane proteins. Electrophoresis. 1997, 18: 2573-2581.
24. Rabilloud T, Adessi C, Giraudel A, Lunardi J: Improvement of the solubilization of proteins in two-dimensional electrophoresis with immobilized pH gradients. Electrophoresis. 1997, 18: 307-316.
25. Falvo S, Di Carli M, Desiderio A, Benvenuto E, Moglia A, America T, Lanteri S, Acquadro A: 2-D DIGE analysis of UV-C radiation-responsive proteins in globe artichoke leaves. Proteomics. 2011, 12 (3): 448-60.
26. Semaan SM, Sang Q-XA: Prefractionation enhances loading capacity and identification of basic proteins from human breast cancer tissues. Anal Biochem. 2011, 411: 8-
27. Stasyk T, Huber LA: Zooming in: fractionation strategies in proteomics. Proteomics. 2004, 4: 3704-3716.
28. Boisvert FM, Lam YW, Lamont D, Lamond AI: A quantitative proteomics analysis of subcellular proteome localization and changes induced by DNA damage. Mol Cell Proteomics. 2010, 9: 457-470.
29. Cascio S, Zhang L, Finn OJ: MUC1 protein expression in tumor cells regulates transcription of proinflammatory cytokines by forming a complex with nuclear factor-kappaB p65 and binding to cytokine promoters: importance of extracellular domain. J Biol Chem. 2011, 286: 42248-42256.
30. Molloy MP, Herbert BR, Walsh BJ, Tyler MI, Traini M, Sanchez JC, Hochstrasser DF, Williams KL, Gooley AA: Extraction of membrane proteins by differential solubilization for separation

using two-dimensional gel electrophoresis. Electrophoresis. 1998, 19: 837-844.

31. Zhu K, Yan F, O'Neil KA, Hamler R, Lubman DM, Lin L, Barder TJ: Proteomic analysis using 2-D liquid separations of intact proteins from whole-cell lysates. Curr Protoc Protein Sci. 2004, Chapter 23: Unit 23 23-
32. Zhang Y, Yoshida Y, Nameta M, Xu B, Taguchi I, Ikeda T, Fujinaka H, Mohamed SM, Tsukaguchi H, Harita Y, Yaoita E, Yamamoto T: Glomerular proteins related to slit diaphragm and matrix adhesion in the foot processes are highly tyrosine phosphorylated in the normal rat kidney. Nephrol Dial Transplant. 2010, 25: 1785-1795.
33. Karlsson K, Cairns N, Lubec G, Fountoulakis M: Enrichment of human brain proteins by heparin chromatography. Electrophoresis. 1999, 20: 2970-2976.
34. Fountoulakis M, Takacs MF, Berndt P, Langen H, Takacs B: Enrichment of low abundance proteins of Escherichia coli by hydroxyapatite chromatography. Electrophoresis. 1999, 20: 2181-2195.
35. Schagger H, von Jagow G: Blue native electrophoresis for isolation of membrane protein complexes in enzymatically active form. Anal Biochem. 1991, 199: 223-231.
36. Kofoed EM, Vance RE: Blue native polyacrylamide gel electrophoresis to monitor inflammasome assembly and composition. Methods Mol Biol. 2013, 1040: 169-183.
37. Wang F, Wang L, Xu Z, Liang G: Identification and analysis of multi-protein complexes in placenta. PloS one. 2013, 8: e62988-
38. Salzano AM, Novi G, Arioli S, Corona S, Mora D, Scaloni A: Mono-dimensional blue native-PAGE and bi-dimensional blue native/urea-PAGE or/SDS-PAGE combined with nLC-ESI-LIT-MS/MS unveil membrane protein heteromeric and homomeric complexes in Streptococcus thermophilus. J Proteomics. 2013, 94: 240-261.
39. Halligan BD, Ruotti V, Jin W, Laffoon S, Twigger SN, Dratz EA: ProMoST (Protein Modification Screening Tool): a web-based tool for mapping protein modifications on two-dimensional gels. Nucleic Acids Res. 2004, 32: W638-644.

40. Hiller K, Schobert M, Hundertmark C, Jahn D, Munch R: JVirGel: calculation of virtual two-dimensional protein gels. Nucleic Acids Res. 2003, 31: 3862-3865.
41. Farley AR, Link AJ: Identification and quantification of protein posttranslational modifications. Methods Enzymol. 2009, 463: 725-763.
42. Bjellqvist B, Ek K, Righetti PG, Gianazza E, Gorg A, Westermeier R, Postel W: Isoelectric focusing in immobilized pH gradients: principle, methodology and some applications. J Biochem Biophys Methods. 1982, 6: 317-339.
43. Gorg A, Weser J, Westermeier R, Postel W, Weidinger S, Patutschnick W, Cleve H: Isoelectric focusing with immobilized pH gradients for the analysis of transferrin (Tf) subtypes and variants. Hum Genet. 1983, 64: 222-226.
44. Lopez MF: Nonequilibrium pH gel electrophoresis (NEPHGE). Methods Mol Biol. 1999, 112: 129-131.
45. Slibinskas R, Razanskas R, Zinkeviciute R, Ciplys E: Comparison of first dimension IPG and NEPHGE techniques in two-dimensional gel electrophoresis experiment with cytosolic unfolded protein response in Saccharomyces cerevisiae. Proteome Sci. 2013, 11: 36-
46. Laemmli UK: Cleavage of structural proteins during the assembly of the head of bacteriophage T4. Nature. 1970, 227: 680-685.
47. Rozalska M, Szewczyk EM: Staphylococcus cohnii hemolysins - isolation, purification and properties. Folia Microbiol (Praha). 2008, 53: 521-526.
48. Hoogland C, Sanchez JC, Tonella L, Binz PA, Bairoch A, Hochstrasser DF, Appel RD: The 1999 SWISS-2DPAGE database update. Nucleic Acids Res. 1999, 2000 (28): 286-288.

SECTION II

PROTEOMIC APPLICATIONS FOR FOOD AUTHENTICATION

CHAPTER

5

REVIEW ON PROTEOMICS FOR FOOD AUTHENTICATION

Ignacio Ortea, Gavin O'Connor, Alain Maquet

European Commission, DG Joint Research Centre (JRC), Directorate F. Health, Consumers and Reference Materials, Retieseweg 111, 2440 Geel, Belgium

ABSTRACT

Consumers have the right to know what is in the food they are eating. Accordingly, European and global food regulations require that the provenance of the food can be guaranteed from farm to fork. Many different instrumental techniques have been proposed for food authentication. Although traditional methods are still being used, new approaches such as genomics, proteomics, and metabolomics are helping to complement existing methodologies for verifying the claims made about certain food products. During the last decade, proteomics (the large-scale analysis of proteins in a particular biological system at a particular time) has been applied to different research areas within food technology. Since proteins can be used as markers for many properties of a food, even indicating processes to which the food has been subjected, they can provide further evidence of the foods labeling claim. This review is a comprehensive and updated overview of the applications, drawbacks, advantages, and challenges of proteomics for food authentication in the assessment of the foods compliance with labeling regulations and policies.

Significance

This review paper provides a comprehensive and critical overview of the application of proteomics approaches to determine the authenticity of several food products updating the performances and current limitations of the applied techniques in both laboratory and industrial environments.

INTRODUCTION

Consumers have the right to know what is in the food they are eating and producers have a duty to inform and not knowingly mislead consumers over the source or contents of the food product. Nevertheless, due to the increasing demand for food and the increased globalization of trade, there is a higher risk that adulteration can occur throughout the food chain. Food products can be deliberately substituted, partially or entirely, with similar, lower quality, and cheaper counterparts or unintentional errors can cause inadvertent mislabeling of products [1]. In both cases, such commercial fraud affects not only the consumers but also the food industry. Examples of common frauds are replacing key ingredients with cheaper alternatives, mislabeling of the animal species used in a food product, incorrectly labeling of ingredient proportions, selling aquaculture fish as wild-caught, and labeling ordinary foods as organic.

Adulteration of food products is a major concern not only for preventing economic fraud, but also for safety reasons. The non-declared introduction of food ingredients such as toxic or allergenic products may be harmful to consumer's health, thus representing a potential public health risk [2]. Moreover, mislabeling may affect the eating habits of certain groups of consumers, such as abstinence on religious grounds or specific lifestyle choices such as vegetarianism. In order to fight against adulteration and the misbranding of food products, many regulations are currently enforced [3–9], outlining the requirement for complete and truthful information about food products that are being traded.

For all these reasons, the authentication of food labeling claims must be guaranteed, and therefore accurate and reliable analytical methods are needed in order to verify that the components used in a food product are of the nature or quality demanded by the purchaser and compatible with the declaration of the seller. Many different instrumental techniques have been proposed for food authentication, such as HPLC, trace element analysis, stable isotope ratio (SIR) analysis, NMR spectroscopy, and more recently, genomics, metabolomics, and proteomics approaches [10–12].

Proteomics is defined as the large-scale analysis of proteins in a particular biological system at a particular time [13]. The proteome, the total protein content of one particular biological system, is highly dynamic and is constantly changing according to different stimuli. Proteomics includes the structural and functional knowledge of proteins, but also the quantification of their abundance, the study of their modifications, the interactions between them, and the study of their localization. Proteomics, a discipline that up to a few years ago was practically being used only in biomedical research, is currently being revealed as a powerful tool in food science research. During the many stages of food production, proteins can act as ideal indicators for many properties associated with food quality, composition, or origin. Proteome analysis can be applied for the systematic search for new marker proteins/peptides, thus accelerating the development of assays to detect adulteration and deceptive practices. Subsequently, accurate and reliable analytical methodologies can be developed and validated to detect the previously identified marker proteins/peptides.

This review is a comprehensive and updated overview of the application, drawbacks, advantages, and challenges that both a global and specific study of proteins for authentication, by means of proteomics approaches, can offer as a tool to comply with food labeling regulations and policies. Additionally, we describe the updated workflows, technologies, and tools that are being assessed in proteomics-related studies, followed by a review of the specific applications regarding food authenticity and, in some instance, food quality.

PROTEOMICS

General areas

Depending on the general objective, most proteomic studies can be divided in three different areas:

(a) Qualitative proteomics (protein characterization and identification)

The goal for qualitative proteomics is to identify and characterize the complete set of proteins present in a sample, which can include the characterization of the protein post-translational modifications (PTMs). This systematic investigation can be focused on a complete proteome (e.g., all the proteins comprised in a food product) or in a specific subset of the proteins (e.g., sarcoplasmic proteins; gluten proteins; glycosylated proteins). The two most common approaches for protein identification are peptide mass fingerprinting (PMF) [14] and peptide fragmentation fingerprinting (PFF) [15], both of them needing an enzymatic digestion of the studied proteins. Alternatively, MS/MS data from the intact protein can be used in a similar way, in an approach known as top-down proteomics [16]. For all these approaches, the presence in the database of the sequence information of the corresponding protein, or a homologous one, is needed. The vast array of different animal and plant species used by the food industry is likely to result in the absence of many peptides/proteins in the current databases. This combined with the many different ingredients in food makes food proteomics a challenging subject. Great caution and extra scrutiny is often required even when working with simple foods. PTMs play a crucial role since they affect protein activity and stability. PTMs can occur due to a wide range of biological signals, as well as due to food processing methods including cooking and preservation treatments. More than 300 different types of PTMs are known, although only a few of them are being extensively investigated, such as phosphorylation, acetylation, glycosylation, or oxidation [17]. During food processing and storage, non-biological (environmental or process-induced) PTMs, termed non-enzymatic PTMs (nePTMs), such as carbonylation, thiol oxidation, aromatic hydroxylation, Maillard glycation, condensation, elimination of side chains, and peptide backbone breakdown, occur regularly [18,19].

(b) Differential/quantitative proteomics

In food proteomes, the relative amount of proteins can change due mainly to foodstuff composition, technological processing of the food and biological variability of the food components. Quantitative information at the protein level, such as the relative abundance of a specific protein between different samples, or the absolute amount of the protein, can be very helpful when looking for differences between different conditions (e.g., different technological treatments of food products; GM vs non-GM food). Relative quantification can be achieved with different methodologies, which can be classified as gel-based, label-based, and label-free approaches [20]. Gel-based methods consist of the separation of proteins by two-dimensional electrophoresis and the comparison of protein abundance determined as the spot volume between different samples. Each sample being compared can be run on a different gel or, alternatively, up to three samples can be differentially labeled and run on the same gel using the DIGE technology [21], increasing confidence in the detection and quantification of differences in protein abundance. In the label-based methods, proteins or peptides are previously labeled using a mass tag and relative quantification is then obtained from the MS or MS/MS read-outs. In label-free approaches, protein amount is generally calculated based on the MS extracted ion current signal of the peptides/proteins during a liquid chromatography (LC) run. Quantitative proteomic methodologies have been greatly improved with the introduction of selected reaction monitoring experiments (SRM), a highly sensitive LC-MS/MS acquisition mode that is commonly used in biomedicine research to verify and validate candidate biomarker proteins [22].

(c) Functional proteomics

Functional proteomics studies the functional interaction between proteins, or between a protein and other molecules and the consequences of these interactions [23]. The understanding of protein–protein interactions, through network analysis, will be crucial if further improvements in food quality are to be realized. Activity-based proteomics is another area related to functional proteomics, studying the specific activities of the proteins in a sample, such as function and inhibition [24]. MS imaging, a new imaging modes that allows mapping proteins within a tissue or sample section [25,26], has proved to be a relevant tool for functional proteomics, since the location of the different protein isoforms can help to understand their roles.

Proteomic workflows

As stated above, the analysis of a proteome is commonly performed after preparation of the sample, by one or more separation steps followed by mass spectrometric analysis. The great diversity and high dynamic range of proteins in food-related samples present a great challenge for MS analysis. For this reason, sample preparation will usually include enrichment of the proteins of interest. Common protocols when working with complex samples usually include depletion of the most abundant interfering proteins, selective enrichment of the low abundance proteins of interest, or even partial purification of the target proteins [27]. Before the final analysis in the mass spectrometer, further separation is performed at protein and/or peptide level, typically based on gel electrophoresis (gel-based approach) and/or LC (gel-free approach). Depending on whether the MS analysis is carried out on the intact proteins or on the corresponding peptide fragments, two different proteomic workflows can be followed, namely, top-down and bottom-up, respectively. Fig. 1 summarizes the most significant proteomic workflows used in food authentication.

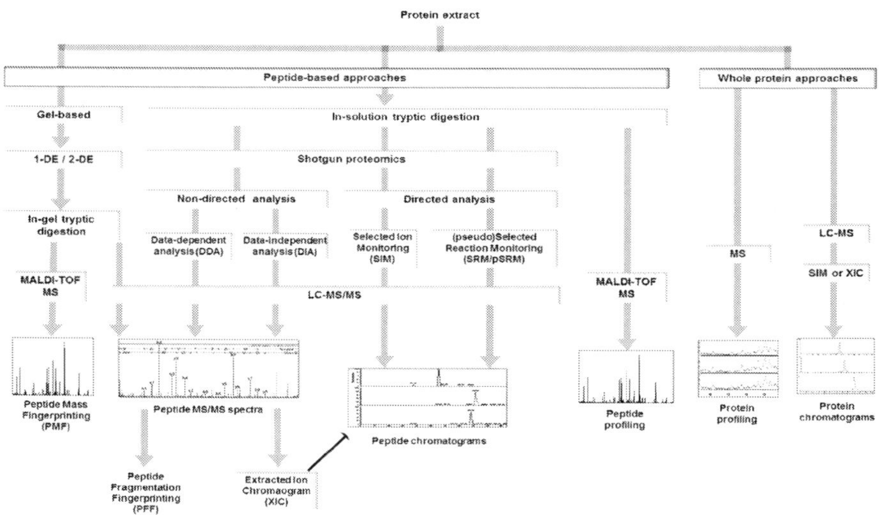

Figure. 1. Common proteomic workflows used in food authentication.

In the bottom-up approach, proteins are enzymatically digested, commonly using trypsin, and the resulting peptides are analyzed by mass spectrometry [13]. Thus, this approach is sometimes known as peptide-based proteomics. Bottom-up approaches can be divided in two groups depending on whether the fractionation step, for reducing sample complexity, is performed at the protein level or at the peptide level, i.e., before or after the enzymatic digestion of the proteins. The first strategy is most commonly used in the two-dimensional electrophoresis (2-DE) gel-based approach, where proteins are isolated based on p*I* and *Mr.* so that they can be individually excised from the gel and digested into peptides. These peptides are then analyzed by MS for protein identification. In the latter approach, referred to as "shotgun" proteomics, the protein mixture is enzymatically digested without previous fractionation and the resulting mixture of peptides are separated by high-throughput LC, typically RP HPLC or ultra-high-performance liquid chromatography (UHPLC). Subsequently, the eluted peptides are analyzed using MS, which is usually coupled online with the LC system. However, depending on the complexity of the peptide mixtures, multidimensional liquid chromatography is sometimes needed to increase the resolving power of the chromatographic separation. SCX columns coupled with RP LC are the most common combination used.

The alternative approach, top-down proteomics [16], characterizes the fragments produced by the dissociation of the intact proteins directly inside the mass spectrometer, avoiding the variable step of protein digestion. Although this kind of intact protein analysis is currently available due to new dissociation mechanisms and the high mass accuracy obtained in the new high-resolution MS equipment, its use is still limited due to instrumental constraints.

Although no common guidelines are established for describing how to perform proteomics-related experiments, the Human Proteome Organization-Proteomics Standards Initiative (HUPO-PSI; http://www.psidev.info) has developed guidance documents specifying the data and metadata that should be collected from several proteomics workflows, known collectively as the "minimum information about a proteomics experiment" (MIAPE) guidelines [28], in order to facilitate data comparison, exchange, and verification. MIAPEs include modules related not only to LC parameters and how MS data is generated but also to study design, sample generation and preparation, and statistical analysis of data [29]. Although not mandatory, there is an increasing trend to follow these guidelines when publishing proteomics experiments in specialized proteomics journal. However, this is not always the case in journals from other areas such as food science and technology.

Sample preparation

Protein extraction is a key step in a proteomic analysis. A specific protocol has to be optimized for each food sample, in order to maximize protein recovery and minimize proteolysis and modifications. Thus, there is no universal extraction buffer composition that can be used. Compounds for stabilizing and solubilizing the proteins, such as pH regulators (e.g., Tris, Hepes, MOPS), reducing agents (e.g., dithiothreitol, 2-mercaptoethanol), and denaturing compounds (e.g., 8 M urea, SDS, CHAPS) and for the elimination of contaminants such as nucleic acids (e.g., DNases), are added to the extraction buffer. The use of protease inhibitors, such as PMSF or EDTA, is strongly recommended. Most of the buffers used for protein extraction contain compounds that are not fully compatible with MS and thus a subsequent cleaning is required. However, a large number of the conventional protein extraction buffers are compatible with electrophoretic detection approaches. The best extraction method has to be experimentally determined for each type of sample and different mechanisms are available that can be used alone or in combination: organic solvents or detergents; enzymatic extraction; liquid nitrogen; mechanical disruption (e.g., use of ULTRA-TURRAX®, blenders, manual grinding with mortar or pestle, Ballotini beads); sonication; and compression/expansion. The method must be compatible with the measurement method that is going to be used afterwards, and also with the amount of material that is going to be processed. If measurements are not being performed immediately, protein samples should be stored in a way that proteolysis and protein aggregation and modification are minimized, freezing in liquid nitrogen (−196°C) or at −80°C is recommended.

Sample purification can commonly include one or several of the following processes [27]: (1) procedures for clarifying and concentrating proteins, such as protein precipitation (e.g., ammonium sulfate, TCA/acetone, chloroform), centrifugation, filtration (e.g., dialysis, ultrafiltration), lyophilization; (2) procedures for enrichment, depletion, or fractionation: for reduce sample complexity and increase the proportion of the target proteins. Chromatography (e.g., RP; SCX; SEC) and PAGE (1-DE or 2-DE) are the most commonly applied fractionation techniques, and also used is off-gel isoelectric focusing, ultracentrifugation, and phase partitioning. For the depletion of high-abundance proteins, or enrichment in the interesting ones, commercial immunoaffinity depletion kits are available, although generally for the removal of albumin and some other human plasma proteins, this reflecting the difficulty faced in food proteomics in comparison to clinical proteomics in terms of method development. In this sense, the use of bead-based combinatorial peptide ligand libraries (CPLL) such as ProteoMiner has worked well for the compression of the dynamic range of the protein concentration in food samples [30–33]. Since the use of CPLL enhances the signal of trace components up to four orders of magnitude, it represents a useful tool for studying matrices with very low concentration of proteins, such as commercial beverages [34,35]. Depletion and fractionation procedures offer significant potential for improving the overall sensitivity of MS analysis; however, these strategies suffer from the limitations of low sample throughput since the fractions obtained have to be measured individually and potentially poor sample recovery, due to non-specific binding of the target proteins to the antibody columns used for depletion or to high-abundance proteins (e.g., albumin), can lead to a partial or total loss of the analytes and to an underestimation of protein concentration [36].

A critical step in bottom-up proteomics is protein digestion, which is performed after electrophoresis isolation of the protein ("in-gel" digestion) for gel-based approaches, or directly on the protein mixture without previous fractionation ("in-solution" digestion). Many proteolytic enzymes can be used for protein digestion, although trypsin offers some advantages, as it mainly yields peptides of optimum amino acid composition, charge state, and homogeneous fragmentation by collision induced dissociation (CID) for MS analysis due to its cleavage specificity, and for these reasons is the most commonly used protease. Since typical digestion times range from 6 to 24 h, protein digestion is the most time-consuming step in the proteomic workflow. To accelerate protein digestion, and also to try to increase the digestion efficiency, alternatives to the standard protocols have been proposed, such as the use of microwaves, infrared radiation, high-intensity focused ultrasounds, modified trypsin, and immobilized trypsin (e.g., in columns or nanoparticles) [37,38].

Gel-based or MS-based approaches

Although nowadays there is a general tendency within the proteomics community to move to gel-free workflows, the fact remains that electrophoresis has been extensively used and continues to be used in food authentication studies. Since data provided by protein electrophoresis is often complementary to data obtained from LC-MS-based approaches, we consider that protein electrophoresis, mainly 2-DE, still has a role in food-related studies.

SDS–PAGE, IEF, and 2DE protein electrophoretic patterns have been extensively used for the development of methods aimed at the detection and authentication of the species present in food products [11], and IEF is the only official validated method for species identification according to the Association of Official Analytical Chemists (AOAC, 1995).

However, electrophoretic methodologies for food authentication show some limitations, such as the analysis of hydrophobic and poorly soluble proteins, limited dynamic range of the available detection methods, high complexity, and dynamic range of the proteins in the sample, degradation of proteins due to food processing, closely related species not well resolved, misidentification when analyzing samples with a mixture of several species, or the need for reference samples. For these reasons, food authentication methods using MS have gained much attention recently.

MS is the most common method of choice for the high-throughput identification, characterization, and quantification of proteins, allowing the analysis of thousands of proteins in one experiment. From the analysis of the mass spectra obtained, information about the peptide sequence and abundance can be obtained by applying a great variety of bioinformatic tools.

One important factor when deciding how to perform the analysis in the mass spectrometer is the method for the acquisition of the MS/MS data, depending on the nature and the stage of the research. The approach known as data-dependent analysis (DDA) is the most popular acquisition mode when the objective is to analyze the highest possible number of proteins present in the sample. However, detected peptides are biased towards those coming from the highest abundance proteins. To avoid this, the data-independent analysis (DIA) acquisition mode can be used. In this mode, instead of selecting one specific ion for fragmentation, all peptide ions present in the mass analyzer are fragmented. This approach is currently growing in popularity due to recent advances introduced in MS instruments, such as elevated-energy MS (MS^e) [39] and sequential window acquisition of all theoretic mass spectra (SWATH) [40] acquisition modes.

When the protein or peptide is known, a targeted proteomics analysis can be done using an SRM analysis [22]. This LC-MS/MS acquisition mode in combination with complete protein digestion is considered the gold standard method for the quantification of proteins due to its high sensitivity and specificity.

Foodomics

Foodomics, the application of systems biology approaches in food science by using and integrating a variety of omics tools (e.g., proteomics, genomics, metabolomics, transcriptomics, lipidomics, microbiomics) [41], can greatly improve comprehensive knowledge on food quality by identifying markers for several features, such as food contaminants, food origin, and food authentication [42]. Interactomics and network analysis provide reference proteins and genes maps that can act as the base for further studies [43]. This can help to evaluate the supramolecular complexities, such as the formation of protein complexes or the interaction of proteins with other molecules [44]. Since these changes may introduce alterations in food functional properties, these integrated approaches could be used to assess the technological processes undergone by a food product [45]. Functional proteomics and systems biology can also help with the selection of which protein fragment should be chosen as a marker for a specific condition [46]. For example, protein fragments overlapping with linear epitopes are preferable if the same fragment is to be used in MS and immunochemical methods. Another example is the investigation of conserved protein domains among different species, which can extend the analytical opportunities (e.g., a peptide marker of crab tropomyosin may be used to detect shrimp meat).

Many tools are available in order to build and visually explore protein interaction networks according to regularly updated databases. Most of these algorithms, such as ingenuity pathway analysis (Qiagen, Redwood City, CA, USA), Cytoscape [47] or Pathway Studio [48], work on the basis of a web page where the list of interesting proteins can be uploaded and searched for their annotations in databases (these can be their own built-in databases or publicly available databases such as Gene Ontology (GO), [49], KEGG [50] or BIND [51]) and mapping them to known biological pathways and gene ontology terms. A comprehensive list of pathway and interaction database can be downloaded from http://www.pathguide.org/ (accessed Nov. 2015). Proteins are classified according to their functional roles, cell localization, biological process, and interaction. From this information, pathway and network information can be obtained. Although customized pathways and databases can be created in some of these tools, all of them use human and model organisms› databases. Therefore, work is needed to extend the applicability of these approaches to food technology related studies. For instance, maize and rice protein networks are included in Pathway Studio, and pathways for chicken, cow, wild pig, and *Saccharomyces cerevisiae* are included in the Reactome database [52].

APPLICATIONS IN FOOD AUTHENTICATION

In the following sections, we describe studies that have used proteomic technologies for the assessment of food authentication in different kinds of foodstuffs. Although species authentication is the area of greatest activity, proteomics is also being applied to some other food authentication issues such as breed identification, geographic origin, undeclared addition of plant/animal protein material and type of material added (e.g., milk/milk proteins, blood/its constituents), production method (wild/farmed), technological processing, proportion of ingredients used, and detection of GMOs (Fig. 2). In Table 1, the most recent proteomics-based studies are compiled.

Figure. 2. Current challenges associated with food authenticity (inner circle) suitable to be overcome by proteomic approaches (outer circle).

Table 1. Proteomic approaches to assess authenticity of food products published in the last six years.

Food	Main technique	Purpose of analysis	Target	Reference
Milk and dairy products	IEF	Milk adulteration	Caseins	[55]
	MALDI-TOF MS protein/peptide profiling	Milk adulteration	Low Mr. proteins (< 25 kDa)	[58]
		Milk adulteration and thermal treatment	Several milk proteins	[59]
		Milk adulteration	Caseins	[63]
		Milk adulteration	Caseins and β-lactoglobulin	[64]
		Thermal treatment in milk	Casein phosphopeptides	[60]
	2-DE + PMF and MALDI-TOF MS Peptide profiling	Adulteration of fresh milk with powder milk	Modified whey proteins and/or caseins	[68]
	MS/MS DDA + PFF	Adulteration of milk powder with pea and soy proteins	Glycinin, β-conglycinin, legumin, vicilin	[66]
	SRM	Cheese adulteration	Caseins	[65]

Meat	2-DE	Differentiation of meat species (cattle, pork, chicken, turkey, duck and goose)	Myosin light chains	[71,72]
		Differentiation of two pig breeds	Muscle proteins	[84]
		Differentiation of two cattle breeds	Liver proteins	[85]
	2-DE DIGE	Differentiation of two Norwegian breeds	Muscle water-soluble proteins	[83]
	MALDI-TOF MS Peptide profiling	Species identification (32 mammal species)	Collagen	[76]
	MS/MS DDA	Detection of chicken meat in meat mixes	Myosin light chain 3	[77]
		Detection of horse meat in meat mixes	Myosin and myoglobin	[78]
	pSRM	Detection of beef, horse, pork, and lamb meat in meat mixes	Myoglobin	[82]
	SRM	Detection of horse and pork in beef	Troponin, myosin, myoglobin, hemoglobin, pyruvate kinase	[79]
		Detection of cooked pork meat	Albumin and lactate dehydrogenase	[80]
		Detection of beef, pork, horse, and lamb meat in meat mixes	Myoglobin	[81]
Shellfish	Native IEF	Differentiation of shrimp species	Sarcoplasmic calcium-binding proteins	[90]
	2-DE	Discrimination of two scallop populations	Mantle proteins	[94]
	2-DE and PMF	Differentiation of shrimp species	Arginine kinase	[95]
	MS/MS DDA + PFF	Differentiation of shrimp species	Arginine kinase	[96]
		Pandalus borealis markers	Arginine kinase	[97]
	pSRM	Differentiation of shrimp species	Arginine kinase	[98]
	MALDI-TOF MS protein profiling library	Differentiation of shrimp species, geographic origin, and fresh/frozen state	Muscle proteins	[99]

Fish	2-DE	Differentiation of tuna species	Muscular proteins	[106,107]
	2-DE, PMF, and MS/MS	Discrimination of two river fish species	Triose phosphate isomerase	[111]
	MS/MS spectral library	Differentiation of fish species	Muscle proteins	[116]
	pSRM	Differentiation of hake species	Parvalbumins	[113]
Wine	MALDI-TOF MS fingerprinting	Discrimination of white wine varieties	Wine proteins	[119]
		Classification of Croatian white wines	Wine protein and peptides	[120]
	MS/MS DDA + PFF	Addition of fining agents in white wines	Caseins	[30]
		Addition of fining agents in red wines	Caseins	[122]
		Ingredient verification for commercial aperitif	Soluble proteins	[34]
	MS/MS DDA + PFF and XIC	Addition of fining agents in white wines	Caseins	[124,125]
	MS/MS DIA	Addition of fining agents in white wines	Casein, lysozyme and ovalbumin	[126]
	SRM	Addition of fining agents in white wines	Lysozyme	[123]
Honey	MALDI-TOF MS protein profiling	Geographical origin	Water-soluble honey proteins	[148]
	2-DE	Floral origin	Water-soluble honey proteins	[149]
Gelatin	MS/MS DDA + PFF	Species used	Collagen	[150]

GMOs	2-DE	Comparison of GM and non-GM maize	Kernel proteome	[138]
			Leaf proteome	[139]
			Grain proteome	[140]
			Leaf proteome	[141]
		Comparison of GM and non-GM soybean	Seed proteome	[133,134]
		Comparison of GM and non-GM common bean	Grain proteome	[142]
		Comparison of GM and non-GM potato	Leaf proteome	[135]
			Tuber proteome	[135]
	2-DE DIGE	Comparison of GM and non-GM rice	Seed proteome	[143]
	LC-MS/MS with isobaric labeling	Comparison of GM and non-GM rice	Seed proteome	[144]
	LC-MS XIC	GM and non-GM tomato varieties	Fruit proteome	[145]

Milk and dairy products

Adulteration of cheese with less expensive milk is one of the main problems in the dairy industry. The reference method for determining the presence of cow milk in ewe and/or goats cheese is based on the detection of the bands for cow γ2- and γ3-caseins on an IEF gel [53] and has proven successful in verifying the presence of cow casein [54,55]. In general, this reference method is performed as a qualitative technique using certified reference materials with 0% and 1% cow›s milk. At these levels, the marker bands for bovine milk ($\gamma_{2\,-}$ and γ_3-caseins) are not visible when staining the gel with Coomassie blue; therefore, a milk sample having at least 5% cow's milk has to be run in parallel to ascertain the position of these bands on the gel [55]. Moreover, for cheese samples below 5% cow's milk, the marker bands could be not detected using the most common staining method. Other problems reported in relation to this method are interfering bands causing false-positive responses and the need for an excellent electrophoretic separation of γ-caseins [54]. The method can yield ambiguous results in some cases, and more resolute and sensitive techniques should be used to complement it.

MS [1]-based (as opposed to MS/MS-based) protein profiling, using either MALDI-TOF or ESI-MS, has been extensively investigated for the possible detection of fraudulent practices in the dairy industry. The use of the MS protein profiling method provides a shorter analysis time due to a faster sample preparation step that skips protein digestion.

MALDI-TOF protein profiling was used for the investigation of the addition of bovine milk to ewe and buffalo milk [56] and in the adulteration of mozzarella cheese [57], using whey proteins as biomarkers. The same technique has been used to reveal adulteration of donkey and goat milk with cow, ewe, and buffalo milk at levels down to 0.5% [58]. Recently, Sassi et al., using a similar method, identified markers of bovine, water buffalo, ovine, and goat milk, but also markers of thermal treatments [59]. MALDI-TOF MS has also been used to detect the addition of UHT milk to raw or pasteurized milk with an LOD of 10% [60]. ESI-MS has also been applied to milk adulteration. The protein β-lactoglobulin was used for the detection and quantification of bovine milk in goat [61] caprine or ovine milk [62] by means of HPLC-ESI-MS or capillary electrophoresis-MS. The presence of cow milk was detected at levels not lower than 5%.

MS^1-based peptide profiling incorporates a digestion step in the sample preparation workflow. Comparisons between sample groups are then based on specific peptides instead of proteins. Casein tryptic peptides were used for detecting and quantifying milk adulteration by means of MALDI-TOF and LC-ESI-MS analysis, which were able to detect as low as 0.5% of bovine, ovine, buffalo, and caprine milks in milk mixtures [63]. In this work, synthetic peptide analogues of the natural specific peptides were used, which increased accuracy and specificity. Calvano et al. [64] described a method combining in-solution digestion of whole milk samples and MALDI-TOF analysis of the tryptic digests which was able to detect a 5% level of cow and goat milk adulteration.

Although these MS^1-based methods perform well in simple matrices such as milk, specificity could be compromised when a more complex background matrix is analyzed. In these cases, the extra specificity that MS/MS-based methods can achieve by providing amino acid sequence information could help in preventing false positives and negatives and in achieving better accuracy in quantitative assays. Guarino et al. (2010) [65] reported an LC-MS/MS method for the detection and quantification of sheep's milk in goat's and cow's cheese. After plasmin digestion of the caseins extracted from the cheese samples, one peptide specific for sheep was selected. An SRM method targeting this marker peptide was then optimized and the analytical performance of the method was reported. For quantification, calibration curves were built using cheeses containing different percentages of the three types of milk. Using this approach, the authors reported the detection of as low as 2% of sheep›s milk in goat›s and cow›s cheeses. MS/MS analysis was also applied to the detection of soy and pea proteins added to skimmed milk powder (SMP) [66]. The masses from the peptides showing a differential behavior between adulterated SMP and control SMP samples were obtained by MS, and then the product ion spectra corresponding to these peptides were generated and the peptides and proteins identified by database searching. The source of the adulteration (pea or soy) was therefore identified. However, no different varieties/cultivars of soy and pea were used to check for intraspecies variability. Moreover, a method based on ultra-high-performance liquid chromatography (UHPLC) and UV detection has been published that outperformed this MS/MS-based method since it was able to detect adulteration of skimmed milk powder with soy, pea, rice, and hydrolyzed wheat protein isolates in the same analysis [67].

The addition of reconstituted milk (powder milk) to fresh milk is forbidden in some circumstances, such as the production of Protected Designation of origin cheeses [5], making it necessary to develop methodologies for detecting this kind of adulteration. Although the adulterant material, namely, the powder derivative, would have the same composition and even the same proteins as liquid milk, protein modification can occur during milk powder production due to the high temperatures used, and therefore proteomics methodologies could help to detect this illegal practice. Calvano et al. [68]detected differences among liquid and powder milk using 2-DE, and lactosylated diagnostic peptides for powdered milk were identified by MALDI MS. By detecting these marker peptides in milk samples adulterated with powder milk subjected to in-solution digestion of the proteins, as little as 1% adulteration could be detected.

Meat

As far back as 1993, ESI-MS of intact proteins was highlighted as a potential method for meat speciation [69]. Although the method was able to differentiate between pork, beef, sheep, and horse origin, and the target proteins, namely, hemoglobin and myoglobin, showed a high degree of heat stability, only the pure commercial proteins were used and no actual meat material was tested. In a subsequent study, this time using ESI-MS/MS on intact myoglobin extracted from beef and pork, and commercial proteins from horse and sheep, Ponce-Alquicira and Taylor [70] were able to differentiate sheep and beef from each other and from horse and pork, but instrument resolution was not enough to differentiate horse and pork. This issue is likely to be overcome with current array of high-resolution mass spectrometers available.

Montowska and Pospiec [71,72] found interspecies differences in 2-DE protein patterns between cattle, pig, chicken, turkey, duck, and goose, in both raw meat and processed products. Some of the proteins were stable during meat aging and resistant to thermal processing, and some of them could even be identified in highly processed products such as fermented sausages and were therefore proposed as suitable markers. However, a validation study focusing on amino acid sequence information from these stable proteins would be necessary in order to establish a high-throughput targeted MS/MS-based method, such as SRM, for the differentiation of the species. This requirement is strengthened by the fact that for duck and goose there is still little protein sequence information in current databases.

Adulteration of meat has also been investigated by means of MS/MS. Leitner et al. [73] reported a method for detecting soybean proteins added to processed meat products using 2D-LC-MS/MS. The presence of unique peptides from a glycinin subunit was consistently found in the four different commercial heat-processed meat products (from pork, chicken, turkey, and beef). Grundy et al. [74,75] were able to detect as low as 5% bovine or porcine blood-based gelling agents spiked in different food matrices (minced pork, beef, chicken, lamb, tuna, and cod) by the SRM detection of bovine-specific fibrinopeptides. The method was validated by analyzing 18 commercial ready-meal products with various coatings and sauces. Fibrinopeptides could not be detected in spiked cod, and the authors suggested a possible cleavage by a cod peptidase to explain it.

Buckley et al. [76] present a robust method of analyzing genus-specific collagen peptides extracted from bone fragments in 32 different mammal species, identifying a total of 92 peptide markers that could be used for species identification, for example, in processed food and animal feed. Cattle, pig, sheep, and chicken meat and processed bone meal samples were also analyzed. To reduce intraspecies variation, different individuals were sampled, for sheep (n = 10) and goat (n = 7), where different breeds were considered (two and three breeds, respectively). To simplify the analysis, on-tip C18 solid phase extraction instead of LC, before MALDI-TOF MS/MS, was used. Sentandreu et al. [77] developed an LC-MS/MS methodology, with in-solution isoelectric focusing fractionation, for detecting chicken in meat mixes. The extracted ion chromatograms for two chicken-specific peptides from myosin light chain 3 were used to calculate the amount of the peptides after normalization with spiked synthetic peptides. Even though the MS acquisition method and instrument used, namely, DDA and LCQ IT, are not the most sensitive choices, the method reported allowed the detection of as low as 0.5% chicken meat in pork meat and can be applied to both raw and cooked meats. The use of stable isotope labeled peptides as an internal standard for normalization assures the highest accuracy in these kinds of assays. Processed and raw horse meats were detected in meat mixtures at 0.5% levels using heat-stable species-specific peptides, also using a DDA method [78]. Von Bargen et al. [79] developed an SRM method for the detection of horse and pork in beef. After identification of the biomarker peptides by a shotgun MS/MS-based approach, peptides specific to horse and pig were included in an SRM assay capable of detecting as low as 0.55% horse or pork contamination in a beef matrix, or 0.13% pork contamination in beef when an MS^3 method was used. Different raw meat samples were analyzed (n = 21), although there was no validation in processed meats, where sensitivity is expected to be reduced, was addressed. Four thermostable peptides were suggested as markers for pork detection in cooked meat products using SRM [80], although only pork meat samples were analyzed, and therefore the method was not tested on mixed commercial meat products. SRM and pSRM have also been used for detection of beef, pork, horse, and lamb myoglobin in meat mixes [81,82], although quantitative data presented in the studies was preliminary. Moreover, some of the monitored peptides were shared by other species such as donkey, goat, buffalo, dog, and rabbit.

It is also possible to identify animals between individual breeds on the basis of proteomic analyses using the 2-DE technique. Using DIGE, differences in protein abundance between two different Norwegian pig breeds, Landrace and Duroc, were investigated [83]. A total of 94 proteins were significantly changed in their relative abundance according to ANOVA and multivariate statistical analysis, and 50 of them were identified by PMF and PFF analysis. Variability due to age was addressed, since three age groups were compared. In a similar study [84], 25 protein spots were found to be differentially expressed in two pig breads, Meishan and Large White. This study used silver-stained gels instead of DIGE technology but had less statistical power due to the reduced number of biological replicates (3 per group). Several proteins (n = 14) were identified by MALDI-TOF PMF, and seven of them were validated using quantitative real time-PCR (qRT-PCR). In another study, comparing cattle breeds of Holstein and Chianina differences in the expression levels of 39 proteins were observed using 2-DE with Coomassie staining [85]. However, since protein extracts were obtained from liver, results cannot be extrapolated to muscle tissue or "meat." In all these studies assessing inter-breed variability, differences due to environment were not addressed, since all the samples within a breed come from the same farm and were fed the same diet.

Shellfish

Seafood products, including shellfish and fish, are among the most internationally traded food commodities. Because of the great phenotypic similarity between related species, different kinds of intentional or inadvertent adulteration can occur, such as the substitution of a highly valuable species for a similar but cheaper one, or the labeling of an aquaculture-obtained product as fished wild. Therefore, legislation has appeared all over the world in order to guarantee authenticity and traceability in the trading of these products. For example, European legislation [7] advises that seafood products should be labeled, indicating the commercial designation of the species, the production method (wild or farmed), and the geographic area in which the shellfish or fish was caught or farmed. Therefore, the proper analytical methodologies should exist in order to assess these requirements.

In the pioneer studies by An et al. [86,87], it was demonstrated that electrophoretic protein profiles could be used for differentiating very closely related species. Although limited to only three shrimp species, and no information about the sampling was provided, they were able to differentiate them using SDS–PAGE or IEF of the proteins from muscle samples. SDS–PAGE and IEF gels have also been used for the differentiation of shrimp from crab meat and from fish, lobster, and cephalopod meat, respectively [88,89]. However, in all these studies, it was shown that there was no single extraction method and electrophoretic system valid for all applications: SDS–PAGE and IEF of the sarcoplasmic, water-soluble proteins, worked well for differentiating raw samples, while SDS extracts performed better for the cooked muscle. Moreover, no extensive sampling was done, and gels were analyzed visually, without any kind of statistical analysis.

The water-soluble muscular proteins were used in a study for differentiating shrimp and prawn species of commercial interest using native IEF with subsequent identification of the species-specific proteins by LC-MS/MS [90]. This study represents the most complete electrophoretic study described to date for the identification of closely related species since the 14 shrimp and prawn species studied were unambiguously identified, and different populations were included for some of the species, therefore assessing intraspecies variability. The p*I* of the marker protein bands was reported using image analysis software and thus the results can be used as a reference database for the identification of unknown samples. No cooked samples were analyzed, but the fact that the species-specific proteins were identified as heat-stable sarcoplasmic calcium-binding proteins (SCPs) extends the potential application of the method to heat-processed and ready-to-eat products, although it has yet to be proven. Moreover, SCPs have been established as crustacean allergens [91], so a method targeting this group of proteins would have an impact on food safety in addition to food authenticity.

2-DE has also been used for studying quantitative differences between two marine mussel species [92]. Several protein spots (n = 37) were found to be differentially expressed by the computer-assisted analysis of silver-stained 2-DE gels, although only 15 of them were identified by means of PMF or PFF analysis. However, the authors noted that the differences observed could be caused by environmental variations in addition to genetic variability since each species originated from a different geographical location. In a subsequent study, the same authors reported species-specific peptides for the three European marine mussel species by combining 2-DE protein isolation, PMF analysis of prominent protein spots and product ion peptide *de novo* sequencing [93]. The specific peptides could also be detected using SIM, which is known to improve sensitivity but can lack specificity in some instances. Although only four biological replicates from each of the three species were considered, the study represented the first application of a complete bottom-up proteomics approach to species authentication.

In a recent study, 2-DE was used for the discovery of potential biomarkers for discriminating between two populations of the great scallop [94]. Several protein spots ($n = 38$) were found to be differentially expressed, although only 11 of them could be identified by MS/MS analysis. The study considered the implications of physiological and temporal variability since animals of similar sizes were collected at two different time points and comprised a similar development state for all the scallops. Since no technical replicates were reported (which are generally needed when 2-DE gels are stained with Coomassie blue to address biases due to gel-to-gel variability), the results should be treated with caution until a verification experiment with a different technique and including a higher number of samples is performed. However, this preliminary study is a good example of the potential of proteomics to discriminate populations living in contrasting environments.

The sarcoplasmic protein arginine kinase (AK) was proposed as a biomarker for *Decapoda* species identification due to its variability found in 2-DE profiles [95]. PMF spectra of AK were used as a marker for the authentication of six shrimp species of commercial relevance. Although the number of biological replicates was relatively low, this study represents the first study dealing with species differentiation by MS-based proteomics, where data were statistically analyzed instead of the common visually reporting of differences between species. Specifically, a PMF spectra clustering method was developed that can help in the classification of unknown samples by comparing with reference samples. In a subsequent study, several species-specific peptides from AK that can serve as diagnostic peptides for the identification of the seven most commercial shrimp and prawn species were identified and characterized by means of *de novo* sequencing of the corresponding product ion spectra [96,97]. Although samples from different populations/geographical regions were used for some of the analyzed species, possible variability due to seasonal changes was not addressed. Since the sequences of the species-specific peptides were reported, they can be used as a reference for the identification of unknown samples. Moreover, identified species-specific peptides opened the way to develop fast and easy-to-use methods for the sensitive detection of the species, such as immunoassays via antibody development, or targeted MS assays. Actually, a pSRM assay for the identification of the seven most commercial shrimp species, targeting those previously characterized peptides, was reported and validated using commercial samples [98]. This method includes very fast sample preparation using high-intensity focused ultrasound and the complete analysis including sample processing and pSRM takes no > 90 min. This can be further reduced to < 45 min if the LC is skipped and samples are analyzed offline since this approach resulted in the same outcome as when using the online LC. One extra advantage of pSRM methods over other targeted MS methods, such as SIM or SRM, is that, in pSRM, the product ion spectra of the target peptides are registered, yielding a higher specificity. To increase specificity of SRM assays, most QqQ now include the option of acquiring a triggered product ion scan, although it has not yet been commonly used.

Finally, Salla and Murray [99] have reported a method for the MALDI-TOF MS protein profiling of shrimp muscle proteins. Unknown samples (n = 74) were compared to a mass spectra library previously generated using reference specimens from six shrimp species. The commercial software MALDI BioTyper (Bruker Daltonics, MA, USA) was used for library generation and mass spectra comparison and 97% accuracy for species level matching was reported. The method was able to correctly classify shrimps from several geographic origins for two of the species, and with different size and fresh/frozen state for one of the species. The great advantage of this method is the speed of analysis due to the fast sample preparation step that avoids protein digestion, but the main drawback of this type of analysis based on only one stage of mass spectrometry analysis (MS^1 or MS) is that only the mass of the analyte is obtained. Thus, since the peptide or protein is not being fragmented and no product ion spectra are being obtained, no structural information is derived. The comparison in this approach is purely made on pattern recognition, and since no unique biomarker is reported, an extensive validation of such approach should be essential.

Fish

As for shellfish, classical electrophoretic techniques such as SDS–PAGE and IEF have been extensively used for fish species authentication. In an interlaboratory study carried out by nine laboratories, the performance of optimized SDS–PAGE and urea-IEF standard operation procedures was evaluated for the identification of ten commercial fish species after cooking [100]. It was concluded that both methods performed differently depending on the species, and therefore the use of both methods was advised for species differentiation. The need for proper reference samples to run in parallel to the unknown sample being analyzed is one of the main issues associated with electrophoretic methodologies. As an alternative, it has been proposed to build a library of reference gel images stored at public repositories, such as the Gel Library from the US Food and Drug Administration (http://www.fda.gov/Food/FoodScienceResearch/RFE/ucm219129.htm, accessed November 2015), based on the Official Method 980.16 (AOAC, 1995). Due to this, Bossier and Cooreman [101] built a computerized databank of IEF protein profiles from 17 flatfish species and checked it for the authentication of 17 commercial fish fillets. Interspecimen variability was assessed for some of the species by analyzing up to three specimens, although no information about different populations or geographic regions was reported. The authors concluded that intraspecies similarity was larger than interspecies similarity, although it has to be noted that this could due to samples from one species being collected from the same geographic region. Renon et al. [102]used the IEF sarcoplasmic protein pattern to distinguish between swordfish, blue marlin and Mediterranean spearfish, in fresh, frozen, and cold-smoked fillets, although only one biological replicate for swordfish and marlin, and two for spearfish, were used.

Since the 1990s, differences in the qualitative 2-DE profiles of water-soluble muscular proteins have been used for the discrimination between fish species such as gadoid fishes [103], flat fishes [104] and hake species [105]. In these studies, several specimens from distant fishing banks were investigated for some of the species in order to assess for intraspecies variability, and three to four biological replicates for each species or geographical location were considered. Technical replicates were included and gel images were computer processed, thus increasing the power of the analysis. Since the differential proteins found were tentatively identified as parvalbumins, a group of heat-resistant sarcoplasmic proteins, it was claimed that the methodologies described could be used for the identification of those fish species in heat-treated products. Nucleoside diphosphate kinase A (NDK A) was also identified as a marker protein for the identification of hake species. Some other proteins that have demonstrated potential for discriminating fish species by 2-DE are triose phosphate isomerase, pyruvate kinase, troponin T, and beta-enolase, which has been proposed as potential markers for the tuna species *Thunnus thynnus*, *Thunnus alalunga*, *Thunnus albacares*, and *Thunnus. obesus*, respectively [106,107]. However, further research, including a higher number of biological replicates from the different population/origins and technical replicates for assessing the inter-gel variability, would be needed to validate the results obtained. The myosin light chain (MLC) isoforms were proposed for differentiation of cod, saithe, haddock, mackerel, and capelin [108], although no sampling information was reported. Interestingly, MLCs also enabled the differentiation of breeding stock, showing observable differences in two stocks of Arctic char (*Salvelinus alpinus*). 2-DE also achieved the differentiation of five puffer fish species using urea-soluble protein extracts from the muscle of three specimens of each species [109], although the potential species-specific proteins were not further characterized and thus not identified. Martínez et al. [110] were able to differentiate wild from farmed cod using 2-DE. Several spots were found to have different electrophoretic mobility, and some other spots were absent in wild but present in farmed cod samples. However, the spots were not sequenced so the proteins were not identified. The main drawback of this study was the lack of statistical analysis, since electrophoretic profiles were only checked visually. Moreover, no source of intra-group variability, such as population or development-related differences, was assessed.

PMF and PFF provide some advantages over 2-DE since they allow the identification of the protein, which is useful for the direct comparison and classification of samples. Combining 2-DE with PMF and PFF, differences in the muscular proteins from two commercially important river fish species from the genus *Sperata* were investigated [111]. Several species-specific proteins (n = 11) were identified, and triosephosphate isomerase isoforms were proposed as specific markers. Using a similar approach, the suitability of the PMF spectra of the proteins in the parvalbumin fraction for the classification of ten hake species was tested [112]. Six species could be identified, while the other four species were grouped in two clusters. Two subspecies from one of the species could also be differentiated. The work was completed with the development of a pSRM assay targeting the differential parvalbumin peptides, using five reference specimens for each of the 11 commercial hake species studied [113]. The method differentiated the species, except four of them that are grouped in two clusters, and was able to distinguish between the two subspecies for one of the species. Ten commercial hake foodstuffs were used for validation, comprising only five of the species. Since parvalbumins are heat-resistant proteins, the methodology can be applied to processed and cooked products. The same authors, after 2-DE analysis, PMF, and PFF *de novo* sequencing of the protein NDK B, reported a set of specific peptides that can help in the classification of the 11 hake species [114]. However, only two of the species could be completely differentiated from the others, while the other nine species grouped in two clusters comprising several species each.

Mazzeo et al. [115] developed a MALDI-TOF protein profiling method able to obtain specific profiles for 25 different fish species. Signals observed around 11 kDa, and subsequently identified as parvalbumins, were selected as specific biomarkers. Commercial fish products (fillets and fish sticks) were also successfully analyzed. Although five biological replicates were used, variability due to population or geographical changes was not addressed. However, the work represents the highest number of species included in a fish authentication study, and possibly the fastest method to date, since the sample preparation step and the MALDI analysis take only few minutes.

Wulff et al. [116] demonstrated the suitability of a completely different proteome-wide approach for fish species differentiation. They first generated a reference product ion spectral library using muscle samples from 22 different species. "Unknown" samples, both fresh and heavily processed, could then be unambiguously identified by comparison to the spectral library. Although the data sets from the fresh "unknown" samples were replicates of the same samples that had been used for the library generation, 21 out of 25 heavily processed samples were correctly identified, therefore showing the potential of this direct spectral matching method. The method does not require genome or protein databases; it uses all acquired product ion data (therefore there is no need for selection or filtering of spectra) and it uses an automated and standardized workflow.

Wine

Traceability and quality control in the wine industry have attracted much attention in recent years. Proteomics-based techniques can be used to investigate methods for distinguishing and classifying different wines, such as different vintages or different grape varieties, but also to detect and quantify non-declared additions of non-wine proteins such as fining agents. Nevertheless, protein and peptide analysis techniques have not yet been fully exploited to study authenticity topics in wine and other alcoholic and non-alcoholic beverages. The main reason is that, in these kinds of samples, proteins are present only at trace levels due to filtering steps during industrial processing and the usual removal of the residual proteins from the finished product. Moreover, only a little amount of the proteins from the raw material survives the fermentation process and they tend to aggregate leading to sediments, so only small-size species or small fragments coming from the proteolysis of high-molecular-mass components remain after industrial manipulations [117]. The presence of exogenous proteins (e.g., yeast, bacteria, and fungi) further complicates the use of protein or peptide fingerprinting to identify frauds or authenticate origin, composition, varieties, or vintage. Therefore, sample preparation, mainly protein extraction and enrichment, is critical, affecting not only the number of proteins identified, but also reproducibility of results. The use of CPLLs for protein enrichment has been reported to highly increase not only the proteome coverage since > 100 grape proteins could be identified in wine, but also sensitivity (down to a few $ng{\cdot}mL^{-1}$ could be detected for some proteins) and reproducibility [117,118].

Chambery et al. [119] applied a peptide profile fingerprint approach using MALDI-TOF MS for the differentiation of three white wine varieties. The MALDI spectra revealed differences between wines after analysis with PCA and cluster analysis, but since only a few representative samples were analyzed, these differences cannot be considered as markers for the grape variety. Furthermore, one of the main drawbacks of protein/peptide profile methodologies of wine is that, since peptides are not identified, error in the classification of a wine sample can be induced because of the presence of exogenous proteins, such as those coming from yeast or fungi. Therefore, some exogenous peaks can appear in the MS spectrum when analyzing one bottle of wine but may not appear in another from the same vintage or grape variety. An intra-sample study in which a series of bottles of the same wine group are check for fingerprint pattern changes should be done to overcome this issue. In a recent study [120], MALDI-TOF fingerprinting followed by unsupervised statistical methods showed potential for wine type classification, although the method was not able to distinguish between all the 33 Croatian white wine types that were included in the analysis. LC-MS/MS following CPLL protein capture was used to prove the genuineness of a commercial liqueur, stated to be of plant origin [34]. Proteins from the species declared in the product label were identified, although proteins from other vegetable species were also found but not identified due to lack of database entries for most of plants.

From July 2012, wine fining agents must be declared on the wine label if above 0.25 mg·L^{-1}[6,121]. Casein and lysozyme are the two most popular fining agents added to wine in order to remove the residual proteins remaining in the finished product, and they are also major food allergens from milk and egg, respectively. Therefore, reliable methods enabling the detection of these added proteins at trace levels are needed. Casein has been detected by DDA, with previous enrichment using combinatorial peptide ligand libraries and a further isolation of the target protein casein by SDS–PAGE, with LOD values as little as 1 μg·L^{-1} and 3.8 μg·L^{-1}, for white and red wines, respectively [30,122]. However, the complexity of the sample treatment was not ideally suited to high-throughput analysis. Moreover, as stated in Section 2.5, since information about the target protein/peptides is known, the use of targeted MS analysis is recommended for better specificity and sensitivity. Cryar et al. [123] reported an SRM method targeting three tryptic peptides from lysozyme. Using stable isotopic dilution (SID) experiments for the quantification of 1 ppm lysozyme in white wine, an uncertainty as low as 3% was reached. Monaci et al. [124,125] were able to detect casein in white wines at 100 μg·mL^{-1} using LC-MS/MS DDA on a Q-TOF mass spectrometer, or at 1.6 μg·mL^{-1} when the production ion spectra was generated on a high-resolution Orbitrap mass spectrometer. When using high-energy collision dissociation (HCD) DIA in a high-resolution instrument, the extracted ion chromatograms of eight specific peptides allowed the multiplex determination of casein, lysozyme, and ovalbumin in white wines with LODs ranging from 0.4 to 1.1 μg·mL^{-1}[126]. Multiplexing the three proteins in the same assay allows the simultaneous detection of traces of caseinate and egg-white powder potentially remaining in white wines upon fining.

GMOs

More than 40 countries have recently adopted labeling regulations regarding GM food [127]. For instance, EU regulation mandates labeling of food and feed containing a 0.9% threshold for GM ingredients [4]. In the USA, although the FDA has refused to require labeling of GM foods since 1992, the development of a GM food labeling legislation is being reconsidered due to recent decisions by some food companies to require the labeling of all GMO products sold in its stores and to the development of GM animals for consumption [128]. Therefore, analytical methods able to detect and quantify the use of GMOs in food and feed are required in order to enforce legislation. Profiling techniques such as metabolomics and proteomics are being tested in order to identify changes in the composition of genetically engineered crops [129,130]. Non-targeted proteomics allows the simultaneous comparison of hundreds of plant components without the need for previous knowledge of their identity, allowing the identification of newly expressed or over-expressed proteins after the genetic modification of the organism. 2-DE has extensively been used for investigating changes between GMOs and their non-GMO isogenic varieties or non-transgenic comparators in tomato [131], wheat [132], soybean [133,134], potato [135,136], maize [137–141], and common bean [142]. Remarkably, although the DIGE technique would help improve the low inter-experimental reproducibility achieved using 2-DE approaches, it has scarcely been used for comparative studies regarding GMOs. A study using DIGE did not find alterations in protein expression caused by GM events in rice as compared with natural variation and conventional breeding [143], although a gel-free approach, using iTRAQ labeling revealed differential expression between GM and wild-type rice in 103 proteins [144]. In a recent study, Mora et al. [145] reported a label-free LC-MS workflow for the relative quantification of proteins between GM and non-GM tomato varieties. Quantification was based on the intensities of the extracted ion chromatograms for the tryptic peptides matching a protein, and normalization with a spiked exogenous protein was performed to account for run-to-run variation. However, all of these are discovery studies, meaning that they focused on the identification of the differences between the GMO and the corresponding non-GMO. Further research is required on the development of methods targeting these differences for the detection and quantification of GMOs in actual food products at the low levels as dictated by legislation. Although protein expression could potentially be used to differentiate GMOs from wild-type crops, permitted or licensed GMO's have well documented insertions/deletions in their genome making detection via genomic approaches more specific. The current targeted genetic approaches are both faster and more cost effective than the proteomic methods investigated to date.

Other foodstuffs

Classical protein electrophoretic techniques have been used for the authentication of other food commodities, such as legumes and ginseng herbs. IEF followed by chemometric data analysis was able to differentiate between four dry bean cultivars [146]. Yeh et al. [147], used a simple SDS–PAGE protocol, to demonstrate differences in some proteins between Oriental and American ginseng that can be useful for the differentiation of the two varieties.

Protein profiling using MALDI-TOF was used for the determination of the geographical origin of honey from Hawaiian bees [148]. Several samples from Hawaii origin (n = 16) were used as the references for building a database library of protein mass spectra barcodes, and 38 commercial samples, including 15 labeled Hawaii origin, were used to test the method. Four non-Hawaiian commercial samples yielded a correlation coefficient in the same range as the Hawaii-origin ones. However, principal component analysis (PCA) of protein fingerprints segregated honeys depending on the geographical origin, and those from Hawaii were clustered together. An attempt to classify honey varieties (chestnut, acacia, sunflower, eucalyptus, and orange) combining CPLL enrichment and 2-DE analysis was unsuccessful since no plant proteins were identified [149].

Zhang et al. [150] identified differential peptides for bovine and porcine gelatins using LC-MS/MS after tryptic digestion, although future experiments should be done using gelatins from different geographic areas. Since protein modifications such as proline hydroxylation can interfere with peptide identification, the product ion data obtained should be verified for correct identification, making the method less amenable to automation.

FUTURE TRENDS

Although the use of protein/peptide biomarkers by means of proteomic technologies is still limited in assessing food authenticity as compared to other well-established methodologies such as immnunoassays or DNA-based analysis, it represents a promising alternative because of its robustness, sensitivity, multiplexing capacity, and high-throughput and discriminating power. As in DNA-based analyses, when using PFF-based methods, discrimination is made at the sequence level, making it possible to differentiate between closely related species. Moreover, peptide-based methods can overcome one of the major drawbacks of DNA methods, that is the degradation of DNA in highly processed samples [76], since (i) marker peptides can be quite stable against processing, (ii) modifications in amino acidic sequence due to food processing (non-enzymatic PTMs) can be monitored, and (iii) heat-stable proteins can be selected as targets. This fact should encourage researchers to develop standard proteomics-based protocols to assess food authenticity. However, in future studies, efforts to address some important considerations that are lacking in most of the studies published to date should be considered. These include the careful reporting of technical details as required by the existing standards for proteomics approach, the inclusion of a proper number of representative samples (biological replicates) and technical replicates, the use of powerful and adequate statistical analysis, and the development and implementation of more powerful MS equipment and bioinformatic tools. As inferred by most of the studies compiled in this review, the major problem regarding the use of proteomics for food authentication is protein extraction from complex matrices and low concentration of target proteins; therefore, further efforts have to be made at the extraction and enrichment steps in order to increase sensitivity and reproducibility of the assays. Another limiting factor that food analysis laboratories may face is the restricted availability of reference materials, considering the wide range of possible analytes, concentration, and matrix combinations of the real samples to be analyzed [151].

Since pre-analytical factors have a large impact on the output of proteomic studies, a more detailed description of sample treatment factors such as processing and sample storage are needed. In order to increase the reproducibility of results, standardization of sample treatment and a careful reporting of technical details is also necessary. The use of MIAPEs should be a standard for all kinds of studies using proteomic technologies, not only for those clinically related. Correct and complete sampling should be done, in order to assess for changes caused by (i) food processing and (ii) intra-group variability of the specimens used in the foodstuff, such as geographical origin or population changes (intraspecies variability), season changes over the year, development status (age), physiological rhythms, or pathological conditions. Different statistical methods have to be assessed to find the best discriminators that meet the expected requirements in terms of classification and differentiation. Power analysis should be included in all analyses in order to check the validity of the study and to design future experiments.

Furthermore, future improvements in high-resolution MS instruments will increase sensitivity and specificity of the assays. Targeted MS/MS acquisition methods can help to fill the gap between preliminary discovery studies and final validation experiments where complete sampling is necessary to address all the possible sources of variability. The development of validated SRM LC-MS methods, combined with the use of certified reference materials as standards, may help to provide reliable analytical results that can be comparable among different laboratories. In addition to SRM, some other recently developed MS acquisition methods promise much for the future. For instance, data-independent analysis acquisition (DIA) modes, such as SWATH and MS^e, allow the quantitative analysis of low abundance peptides in the same run that is used for protein identification, improving the analytical coverage and merging the initial discovery and the subsequent validation steps in one platform. Finally, the development of new bioinformatic tools and the application of existing ones to food proteomes and matrices will play a key role in the implementation of integrative biology and systems biology in food technology and nutrition areas. A collaborative effort should be made by the scientific community for the sharing of publicly available databases and data sets and for the development of certified reference materials and standard operating procedures.

REFERENCES

1. J.C. Moore, J. Spink, M. Lipp, Development and application of a database of food ingredient fraud and economically motivated adulteration from 1980 to 2010, J. Food Sci. 77 (2012) R118–R126.
2. J. Spink, D.C. Moyer, Defining the public health threat of food fraud, J. Food Sci. 76 (2011) R157–R163.
3. E. Parliament, European Council, Regulation (EC) No 178/2002 of the European Parliament and of the Council of 28 January 2002 laying down the general principles and requirements of food law, establishing the European Food Safety Authority and laying down procedures in matters of food safety, Off. J. Eur. Communities L31 (2002) 1–24.
4. E. Parliament, European Council, Regulation (EC) No 1830/2003 of 22 September 2003 concerning the traceability and labelling of genetically modified organisms and the traceability of food and feed products produced from genetically modified organisms and amending Directive 2001/18/EC, Off. J. Eur. Union L268 (2003) 24–28.
5. E. Parliament, European Council, Regulation (EU) No 1151/2012 of the European Parliament and of the Council of 21 November 2012 on quality schemes for agricultural products and foodstuffs, Off. J. Eur. Union L343 (2012) 1–29.
6. European Commission, Commission Implementing Regulation (EU) No 579/2012 of 29 June 2012 amending Regulation (EC) No 607/2009 laying down certain detailed rules for the implementation of Council Regulation (EC) No 479/2008 as regards protected designations of origin and geographical indications, traditional terms, labelling and presentation of certain wine sector products, Off. J. Eur. Union L171 (2012) 4–6.
7. E. Parliament, European Council, Regulation (EU) No 1379/2013

of the European Parliament and of the Council of 11 December 2013 on the common organisation of the markets in fishery and aquaculture products, amending Council Regulations (EC) No 1184/2006 and (EC) No 1224/2009 and repealin, Off. J. Eur. Union L354 (2013) 1–21.

8. U.S. Food and Drug Administration, US Federal Food, Drug, and Cosmetic Act. Chapter IV: Food, 2014 (http://www.fda.gov/RegulatoryInformation/Legislati. http:// www.fda.gov/RegulatoryInformation/Legislation/ FederalFoodDrugandCosmeticActFDCAct/FDCActChapterIVFood/default.htm (accessed November 1, 2015)).

9. European Parliament, European Parliament Resolution of 14 January 2014 on the Food Crisis, Fraud in the Food Chain and the Control thereof (2013/2091(INI)). P7_TA(2014)0011, 2014.

10. S.A. Drivelos, C.A. Georgiou, Multi-element and multi-isotope-ratio analysis to determine the geographical origin of foods in the European Union, Trends Anal. Chem. 40 (2012) 38–51.

11. I. Ortea, A. Pascoal, B. Cañas, J.M. Gallardo, J. Barros-Velázquez, P. Calo-Mata, Food authentication of commercially-relevant shrimp and prawn species: from classical methods to foodomics, Electrophoresis 33 (2012) 2201–2211.

12. E. Cubero-Leon, R. Peñalver, A. Maquet, Review on metabolomics for food authentication, Food Res. Int. 60 (2014) 95–107.

13. A. Pandey, M. Mann, Proteomics to study genes and genomes, Nature 405 (2000) 837–846.

14. D.J.C. Pappin, P. Hojrup, A. Bleasby, Rapid identification of proteins by peptide mass fingerprinting, Curr. Biol. 3 (1993) 327–332.

15. J.K. Eng, A.L. McCormack, R.J. Yates III, An approach to correlate tandem mass spectral data of peptides with amino acid sequences in a protein database, J. Am. Soc. Mass Spectrom. 5 (1994) 976–989.

16. F.W. McLafferty, K. Breuker, M. Jin, X. Han, G. Infusini, H. Jiang, et al., Top-down MS, a powerful complement to the high capabilities of proteolysis proteomics, FEBS J. 274 (2007) 6256–6268.

17. Y. Zhao, O.N. Jensen, Modification-specific proteomics: strategies for characterization of post-translational modifications using enrichment techniques, Proteomics 9 (2009) 4632–4641.

18. S. Clerens, J.E. Plowman, J.M. Dyer, Food proteomics: mapping modifications, in: J.L. Heazlewood, C.J. Petzold (Eds.),Proteomic Appl. Biol., InTech, Rijeka, 2012.
19. M. Pischetsrieder, R. Baeuerlein, Proteome research in food science, Chem. Soc. Rev. 38 (2009) 2600–2608.
20. A. Panchaud, M. Affolter, P. Moreillon, M. Kussmann, Experimental and computational approaches to quantitative proteomics: status quo and outlook, J. Proteome 71 (2008) 19–33.
21. J.S. Minden, S.R. Dowd, H.E. Meyer, K. Stühler, Difference gel electrophoresis, Electrophoresis 30 (2009) S156–S161.
22. S. Gallien, E. Duriez, B. Domon, Selected reaction monitoring applied to proteomics, J. Mass Spectrom. 46 (2011) 298–312.
23. L. Kiemer, G. Cesareni, Comparative interactomics: comparing apples and pears? Trends Biotechnol. 25 (2007) 448–454.
24. S. Serim, U. Haedke, S.H.L. Verhelst, Activity-based probes for the study of proteases: recent advances and developments, ChemMedChem 7 (2012) 1146–1159.
25. P.M. Angel, R.M. Caprioli, Matrix-assisted laser desorption ionization imaging mass spectrometry: in situ molecular mapping, Biochemistry 52 (2013) 3818–3828.
26. C. Wu, A.L. Dill, L.S. Eberlin, R.G. Cooks, D.R. Ifa, Mass spectrometry imaging under ambient conditions, Mass Spectrom. Rev. 32 (2012) 218–243.
27. D. Martínez-Maqueda, B. Hernández-Ledesma, L. Amigo, B. Miralles, J.A. GómezRuiz, Extraction/fractionation techniques for proteins and peptides and protein digestion, in: F. Toldrá, L.M.L. Nollet (Eds.), Proteomics Foods Princ. Appl, Springer, New York 2013, pp. 21–50.
28. C. Taylor, N. Paton, K. Lilley, P.-A. Binz, R. Julian, A. Jones, et al., The minimum information about a proteomics experiment (MIAPE), Nat. Biotechnol. 25 (2007) 887–893.
29. S. Martínez-Bartolomé, P.A. Binz, J.P. Albar, The minimal information about a proteomics experiment (MIAPE) from the proteomics standards initiative, Methods Mol. Biol. 1072 (2014) 765–780.

30. A. Cereda, A.V. Kravchuk, A. D'Amato, A. Bachi, P.G. Righetti, Proteomics of wine additives: mining for the invisible via combinatorial peptide ligand libraries, J. Proteome 73 (2010) 1732–1739.

31. E. Fasoli, A. D'Amato, A.V. Kravchuk, A. Citterio, P.G. Righetti, In-depth proteomic analysis of non-alcoholic beverages with peptide ligand libraries, J. Proteome 74 (2011) 1080–1090.

32. C. Esteve, A. D'Amato, M.L. Marina, M.C. García, A. Citterio, P.G. Righetti, Identification of olive (Olea europaea) seed and pulp proteins by nLC-MS/MS via combinatorial peptide ligand libraries, J. Proteome 75 (2012) 2396–2403.

33. C. Esteve, A. D'Amato, M.L. Marina, M.C. García, P.G. Righetti, Identification of avocado (Persea americana) pulp proteins by nano-LC-MS/MS via combinatorial peptide ligand libraries, Electrophoresis 33 (2012) 2799–2805.

34. M.J. Lerma-García, A. D'Amato, E. Fasoli, E.F. Simó-Alfonso, P.G. Righetti, According to the CPLL proteome sheriffs, not all aperitifs are created equal! Biochim. Biophys. Acta 1844 (2014) 1493–1499.

35. C. Cilindre, E. Fasoli, A. D'Amato, G. Liger-Belair, P.G. Righetti, It's time to pop a cork on champagne's proteome! J. Proteome 105 (2014) 351–362.

36. T. Shi, D. Su, T. Liu, K. Tang, D.G. Camp II, W.-J. Quian, et al., Advancing the sensitivity of selected reaction monitoring-based targeted quantitative proteomics, Proteomics 12 (2012) 1074–1092.

37. Q. Chen, T. Liu, G. Chen, Highly efficient proteolysis accelerated by electromagnetic waves for peptide mapping, Curr. Genomics 12 (2011) 380–390.

38. J.L. Capelo, R. Carreira, M. Diniz, L. Fernandes, M. Galesio, C. Lodeiro, et al., Overview on modern approaches to speed up protein identification workflows relying on enzymatic cleavage and mass spectrometry-based techniques, Anal. Chim. Acta 650 (2009) 151–159.

39. S.J. Geromanos, J.P.C. Vissers, J.C. Silva, C.A. Dorschel, G.-Z. Li, M.V. Gorenstein, et al., The detection, correlation, and comparison of peptide precursor and product ions from data independent LC-MS with data dependant LC-MS/MS, Proteomics 9 (2009) 1683–1695.

40. L.C. Gillet, P. Navarro, S. Tate, H. Röst, N. Selevsek, L. Reiter, et al., Targeted data extraction of the MS/MS spectra generated by data-independent acquisition: a new concept for consistent and accurate proteome analysis, Mol. Cell. Proteomics 11 (2012) (O111.016717).

41. A. Cifuentes, Food analysis and foodomics, J. Chromatogr. A 1216 (2009) 7109.

42. C. Zheng, A. Chen, System biological research on food quality for personalized nutrition and health using foodomics techniques: a review, J. Food Nutr. Res. 2 (2014) 608–616.

43. A. D'Alessandro, L. Zolla, A. Scaloni, The bovine milk proteome: cherishing, nourishing and fostering molecular complexity. An interactomics and functional overview, Mol. BioSyst. 7 (2011) 579–597.

44. G. Picariello, G. Mamone, F. Addeo, P. Ferranti, Novel mass spectrometry-based applications of the "Omic" sciences in food technology and biotechnology, Food Technol. Biotechnol. 50 (2012) 286–305.

45. G. Mamone, G. Picariello, S. Caira, F. Addeo, P. Ferranti, Analysis of food proteins and peptides by mass spectrometry-based techniques, J. Chromatogr. A 1216 (2009) 7130–7142.

46. P. Minkiewicz, M. Darewicz, A. Iwaniak, J. Sokolowska, P. Starowicz, J. Bucholska, et al., Common amino acid subsequences in a universal proteome-relevance for food science, Int. J. Mol. Sci. 16 (2015) 20748–20773.

47. M.S. Cline, M. Smoot, E. Cerami, A. Kuchinsky, N. Landys, C. Workman, et al., Integration of biological networks and gene expression data using Cytoscape, Nat. Protoc. 2 (2007) 2366–2382.

48. A. Nikitin, S. Egorov, N. Daraselia, I. Mazo, Pathway Studio—the analysis and navigation of molecular networks, Bioinformatics 19 (2003) 2155–2157.

49. M. Ashburner, C.A. Ball, J.A. Blake, D. Botstein, H. Butler, J.M. Cherry, et al., Gene ontology: tool for the unification of biology. The Gene Ontology Consortium, Nat. Genet. 25 (2000) 25–29.

50. M. Kanehisa, The KEGG database, Novartis Found. Symp. 42–46 (2002).

51. C. Alfarano, C.E. Andrade, K. Anthony, N. Bahroos, M. Bajec, K. Bantoft, et al., The biomolecular interaction network database and related tools 2005 update, Nucleic Acids Res. 33 (2005) 418–424.

52. D. Croft, G. O'Kelly, G. Wu, R. Haw, M. Gillespie, L. Matthews, et al., Reactome: a database of reactions, pathways and biological processes, Nucleic Acids Res. 39 (2011) D691–D697.

53. European Commission, Regulation (EC) No 273/2008 of 5 March 2008 laying down detailed rules for the application of Council Regulation (EC) No 1255/1999 as regards methods for the analysis and quality evaluation of milk and milk products, Off. J. Eur. Union L88 (2008) 53–61.

54. H.K. Mayer, Milk species identification in cheese varieties using electrophoretic, chromatographic and PCR techniques, Int. Dairy J. 15 (2005) 595–604.

55. J. Špoljarić, N. Mikulec, D. Plavljanić, B. Radeljević, J. Havranek, N. Antunac, Proving the adulteration of ewe and goat cheeses with cow milk using the reference method of isoelectric focusing of γ-casein, Mljekarstvo 63 (2013) 115–121.

56. R. Cozzolino, S. Passalacqua, S. Salemi, P. Malvagna, E. Spina, D. Garozzo, Identification of adulteration in milk by matrix-assisted laser desorption/ionization time-of- flight mass spectrometry, J. Mass Spectrom. 36 (2001) 1031–1037.

57. R. Cozzolino, S. Passalacqua, S. Salemi, D. Garozzo, Identification of adulteration in water buffalo mozzarella and in ewe cheese by using whey proteins as biomarkers and matrix-assisted laser desorption/ionization mass spectrometry, J. Mass Spectrom. 37 (2002) 985–991.

58. F. Di Girolamo, A. Masotti, G. Salvatori, M. Scapaticci, M. Muraca, L. Putignani, A sensitive and effective proteomic apprach to identify she-donkey›s and goat›s milk adulteration by MALDI-TOF MS fingerprinting, Int. J. Mol. Sci. 15 (2014) 13697–13719.

59. M. Sassi, S. Arena, A. Scaloni, MALDI-TOF-MS platform for integrated proteomic and peptidomic profiling of milk samples allows rapid detection of food adulterations, J. Agric. Food Chem. 63 (2015) 6157–6171, http://dx.doi.org/10.1021/acs. jafc.5b02384.

60. G. Pinto, S. Caira, M. Cuollo, O. Fierro, M.A. Nicolai, L. Chianese,

et al., Lactosylated casein phosphopeptides as specific indicators of heated milks, Anal. Bioanal. Chem. 402 (2012) 1961–1972.

61. R.-K. Chen, L.-W. Chang, Y.-Y. Chung, M.-H. Lee, Y.-C. Ling, Quantification of cow milk adulteration in goat milk using high-performance liquid chromatography with electrospray ionization mass spectrometry, Rapid Commun. Mass Spectrom. 18 (2004) 1167–1171.

62. L. Műller, P. Barták, P. Bednář, I. Fryšová, J. Ševčík, K. Lemr, Capillary electrophoresis-mass spectrometry—a fast and reliable tool for the monitoring of milk adulteration, Electrophoresis 29 (2008) 2088–2093.

63. M. Cuollo, S. Caira, O. Fierro, G. Pinto, G. Picariello, F. Addeo, Toward milk speciation through the monitoring of casein proteotypic peptides, Rapid Commun. Mass Spectrom. 24 (2010) 1687–1696.

64. C.D. Calvano, C. De Ceglie, A. Monopoli, C.G. Zambonin, Detection of sheep and goat milk adulterations by direct MALDI–TOF MS analysis of milk tryptic digests, J. Mass Spectrom. 47 (2012) 1141–1149.

65. C. Guarino, F. Fuselli, A. La Mantia, L. Longo, A. Faberi, R.M. Marianella, Peptidomic approach, based on liquid chromatography/electrospray ionization tandem mass spectrometry, for detecting sheep›s milk in goat›s and cow›s cheeses, Rapid Commun. Mass Spectrom. 24 (2010) 705–713.

66. J.H.G. Cordewener, D.M.A.M. Luykx, R. Frankhuizen, E. B.M.G., Untargeted LC-QTOF mass spectrometry method for the detection of adulterations in skimmed milk powder, J. Sep. Sci. 32 (2009) 1216–1223.

67. J.E. Jablonski, J.C. Moore, J.M. Harnly, Nontargeted detection of adulteration of skim milk powder with foreign proteins using UHPLC-UV, J. Agric. Food Chem. 62 (2014) 5198–5206.

68. C.D. Calvano, A. Monopoli, P. Loizzo, M. Faccia, C.G. Zambonin, Proteomic approach based on MALDI-TOF MS to detect powdered milk in fresh cow›s milk, J. Agric. Food Chem. 61 (2013) 1609–1617.

69. A.J. Taylor, R. Linforth, O. Weir, T. Hutton, B. Green, Potential of electrospray massspectrometry for meat pigment identification,

Meat Sci. 33 (1993) 75–83.

70. E. Ponce-Alquicira, A.J. Taylor, Extraction and ESI-CID-MS/MS analysis of myoglobins from different species, Food Chem. 69 (2000) 81–86.
71. M. Montowska, E. Pospiech, Myosin light chain isoforms retain their species-specific electrophoretic mobility after processing, which enables differentiation between six species: 2DE analysis of minced meat and meat products made from beef, pork and poultry, Proteomics 12 (2012) 2879–2889.
72. M. Montowska, E. Pospiech, Species-specific expression of various proteins in meat tissue: proteomic analysis of raw and cooked meat and meat products made from beef, pork and selected poultry species, Food Chem. 136 (2013) 1461–1469.
73. A. Leitner, F. Castro-Rubio, M.L. Marina, W. Lindner, Identification of marker proteins for the adulteration of meat products with soybean proteins by multidimensional liquid chromatography-tandem mass spectrometry, J. Proteome Res. 5 (2006) 2424–2430.
74. H.H. Grundy, P. Reece, M.D. Sykes, J.A. Clough, N. Audsley, R. Stones, Screening method for the addition of bovine blood-based binding agents to food using liquid chromatography triple quadrupole mass spectrometry, Rapid Commun. Mass Spectrom. 21 (2007) 2919–2925.
75. H.H. Grundy, P. Reece, M.D. Sykes, J.A. Clough, N. Audsley, R. Stones, Method to screen for the addition of porcine blood-based binding products to foods using liquid chromatography/triple quadrupole mass spectrometry, Rapid Commun. Mass Spectrom. 22 (2008) 2006–2008.
76. M. Buckley, M. Collins, J. Thomas-Oates, J.C. Wilson, Species identification by analysis of bone collagen using matrix-assisted laser desorption/ionisation time-of-flight mass spectrometry, Rapid Commun. Mass Spectrom. 23 (2009) 3843–3854.
77. M.A. Sentandreu, P.D. Fraser, J. Halket, R. Patel, P.M. Bramley, A proteomic based approach for detection of chicken in meat mixes, J. Proteome Res. 9 (2010) 3374–3383.
78. A.J. Claydon, H.H. Grundy, A.J. Charlton, M.R. Romero, Identification of novel peptides for horse meat speciation in highly

processed foodstuffs, Food Addit. Contam. Part A 32 (2015) 1718–1729.

79. C. Von Bargen, J. Dojahn, D. Waidelich, H.-U. Humpf, J. Brockmeyer, New sensitive high-performance liquid chromatography–tandem mass spectrometry method for the detection of horse and pork in Halal beef, J. Agric. Food Chem. 61 (2013) 11986–11994.
80. S.A. Sarah, W.N. Faradalila, M.S. Salwani, I. Amin, S.A. Karsani, A.Q. Sazili, LC-QTOFMS identification of porcine-specific peptide in heat treated pork identifies candidate markers for meat species determination, Food Chem. 199 (2016) 157–164.
81. A. Watson, Y. Gunning, N.M. Rigby, M. Philo, E.K. Kemsley, Meat authentication via multiple reaction monitoring mass spectrometry of myoglobin peptides, Anal. Chem. 87 (2015) 10315–10322.
82. A. Ruiz Orduna, E. Husby, C.T. Yang, D. Ghosh, F. Beaudry, Assessment of meat authenticity using bioinformatics, targeted peptide biomarkers and high-resolution mass spectrometry, Food Addit. Contam. Part A Chem. Anal. Control Expo. 32 (2015) 1709–1717.
83. K. Hollung, H. Grove, E.M. Færgestad, M.S. Sidhu, P. Berg, Comparison of muscle proteome profiles in pure breeds of Norwegian Landrace and Duroc at three different ages, Meat Sci. 81 (2009) 487–492.
84. Y.J. Xu, M.L. Jin, L.J. Wang, A.D. Zhang, B. Zuo, D.Q. Xu, et al., Differential proteome analysis of porcine skeletal muscles between Meishan and Large White, J. Anim. Sci. 87 (2009) 2519–2527.
85. A.M. Timperio, A. D›Alessandro, L. Pariset, G.M. D›Amici, A. Valentini, L. Zolla, Comparative proteomics and transcriptomics analyses of livers from two different Bos taurus breeds: "Chianina and Holstein Friesian,", J. Proteome 73 (2009) 309–322.
86. H. An, M.R. Marshall, W.S. Otwell, C.I. Wei, Electrophoretic identification of raw and cooked shrimp using various protein extraction systems, J. Food Sci. 53 (1988) 313–318.
87. H. An, M.R. Marshall, W.S. Otwell, C.I. Wei, Species identification of raw and boiled shrimp by a urea gel isoelectric focusing technique, J. Food Sci. 54 (1989) 233–236.
88. T. Civera, E. Parisi, The use of sodium dodecyl sulphate

polyacrylamide gel electrophoresis for the identification of raw and cooked shrimps and crabs, Ital. J. Food Sci. 3 (1991) 149–157.

89. H. Rehbein, Food control by applied biochemistry of marine organisms: comparison of proteins and metabolites from fish and invertebrate muscle, Helgoländer Meeresun. 49 (1995) 747–757.

90. I. Ortea, B. Cañas, P. Calo-Mata, J. Barros-Velázquez, J.M. Gallardo, Identification of commercial prawn and shrimp species of food interest by native isoelectric focusing, Food Chem. 121 (2010) 569–574.

91. K. Shiomi, Y. Sato, S. Hamamoto, H. Mita, K. Shimakura, Sarcoplasmic calciumbinding protein: identification as a new allergen of the black tiger shrimp Penaeus monodon, Int. Arch. Allergy Immunol. 146 (2008) 91–98.

92. J.L. López, A. Marina, J. Vázquez, G. Álvarez, A proteomic approach to the study of the marine mussels Mytilus edulis and M. galloprovincialis, Mar. Biol. 141 (2002) 217–223.

93. J.L. López, A. Marina, G. Álvarez, J. Vázquez, Application of proteomics for fast identification of species-specific peptides from marine species, Proteomics 2 (2002) 1658–1665.

94. S. Artigaud, R. Lavaud, J. Thébault, F. Jean, Ø. Strand, T. Strohmeier, et al., Proteomic based comparison between populations of the great scallop, Pecten maximus, J. Proteome 105 (2014) 164–173.

95. I. Ortea, B. Cañas, P. Calo-Mata, J. Barros-Velázquez, J.M. Gallardo, Arginine kinase peptide mass fingerprinting as a proteomic approach for species identification and taxonomic analysis of commercially relevant shrimp species, J. Agric. Food Chem. 57 (2009) 5665–5672.

96. I. Ortea, B. Cañas, J.M. Gallardo, Mass spectrometry characterization of speciesspecific peptides from arginine kinase for the identification of commercially relevant shrimp species, J. Proteome Res. 8 (2009) 5356–5362.

97. A. Pascoal, I. Ortea, J.M. Gallardo, B. Cañas, J. Barros-Velázquez, P. Calo-Mata, Species identification of the northern shrimp (Pandalus borealis) by polymerase chain reaction-restriction fragment length polymorphism and proteomic analysis, Anal. Biochem. 421 (2012) 56–67.

98. I. Ortea, B. Cañas, J.M. Gallardo, Selected tandem mass spectrometry

ion monitoring for the fast identification of seafood species, J. Chromatogr. A. 1218 (2011) 4445–4451.

99. V. Salla, K.K. Murray, Matrix-assisted laser desorption ionization mass spectrometry for identification of shrimp, Anal. Chim. Acta 794 (2013) 55–59.

100. M. Etienne, M. Jérome, J. Fleurence, H. Rehbein, R. Kündiger, R. Mendes, et al., Identification of fish species after cooking by SDS–PAGE and urea IEF: a collaborative study, J. Agric. Food Chem. 48 (2000) 2653–2658.

101. P. Bossier, K. Cooreman, A databank able to be used for identifying and authenticating commercial flatfish (Pleuronectiformes) products at the species level using isoelectric focusing of native muscle proteins, Int. J. Food Sci. Technol. 35 (2000) 563–568.

102. P. Renon, C. Bernardi, R. Malandra, P.A. Biondi, Isoelectric focusing of sarcoplasmic proteins to distinguish swordfish, blue marlin and Mediterranean spearfish, Food Control 16 (2005) 473–477.

103. C. Piñeiro, J. Barros-Velázquez, C.G. Sotelo, R.I. Pérez-Martín, J.M. Gallardo, Two-dimensional electrophoretic study of the water-soluble protein fraction in white muscle of gadoid fish species, J. Agric. Food Chem. 46 (1998) 3991–3997.

104. C. Piñeiro, J. Barros-Velázquez, C.G. Sotelo, J.M. Gallardo, The use of two-dimensional electrophoresis in the characterization of the water-soluble protein fraction of commercial flat fish species, Zeitschrift Für Leb. Und-Forsch. A 208 (1999) 342–348.

105. C. Piñeiro, J. Vázquez, A. Marina, J. Barros-Velázquez, J.M. Gallardo, Characterization and partial sequencing of species-specific sarcoplasmic polypeptides from commercial hake species by mass spectrometry following two-dimensional electrophoresis, Electrophoresis 22 (2001) 1545–1552.

106. T. Pepe, M. Ceruso, A. Carpentieri, I. Ventrone, A. Amoresano, A. Anastasio, Proteomics analysis for the identification of three species of Thunnus, Vet. Res. Commun. 34 (2010) S153–S155.

107. T. Pepe, M. Ceruso, A. Carpentieri, I. Ventrone, A. Amoresano, A. Anastasio, et al., Differentiation of four tuna species by two-dimensional electrophoresis and mass spectrometric analysis, in: J.L. Heazlewood, C.J. Petzold (Eds.),Proteomic Appl. Biol., InTech,

2012.

108. I. Martinez, T.J. Friis, Application of proteome analysis to seafood authentication, Proteomics 4 (2004) 347–354.

109. T.-Y. Chen, C.-Y. Shiau, C.-I. Wei, D.-F. Hwang, Preliminary study on puffer fish proteome-species identification of puffer fish by two-dimensional electrophoresis, J. Agric. Food Chem. 52 (2004) 2236–2241.

110. I. Martínez, R. Slizyte, E. Dauksas, High resolution two-dimensional electrophoresis as a tool to differentiate wild from farmed cod (Gadus morhua) and to assess the protein composition of klipfish, Food Chem. 102 (2007) 504–510.

111. S.K. Barik, S. Banerjee, S. Bhattacharjee, D. Gupta S.K., S. Mohanty, et al., Proteomic analysis of sarcoplasmic peptides of two related fish species for food authentication, Appl. Biochem. Biotechnol. 171 (2013) 1011–1021.

112. M. Carrera, B. Cañas, P. Piñeiro, J. Vázquez, J.M. Gallardo, Identification of commercial hake and grenadier species by proteomic analysis of the parvalbumins fraction, Proteomics 6 (2006) 5278–5287.

113. M. Carrera, B. Cañas, D. López-Ferrer, C. Piñeiro, J. Vázquez, J.M. Gallardo, Fast monitoring of species-specific peptide biomarkers using high-intensity-focused-ultrasound-assisted tryptic digestion and selected MS/MS ion monitoring, Anal. Chem. 83 (2011) 5688–5695.

114. M. Carrera, B. Cañas, C. Piñeiro, J. Vázquez, J.M. Gallardo, De novo mass spectrometry sequencing and characterization of species-specific peptides from nucleoside diphosphate kinase B for the classification of commercial fish species belonging to the family Merlucciidae, J. Proteome Res. 6 (2007) 3070–3080.

115. M.F. Mazzeo, B. de Giulio, G. Guerriero, G. Ciarcia, A. Malorni, G.L. Russo, et al., Fish authentication by MALDI-TOF mass spectrometry, J. Agric. Food Chem. 56 (2008) 11071–11076.

116. T. Wulff, M.E. Nielsen, A.M. Deelder, F. Jessen, M. Palmblad, Authentication of fish products by large-scale comparison of tandem mass spectra, J. Proteome Res. 12 (2013) 5253–5259.

117. P.G. Righetti, A. D›Amato, E. Fasoli, E. Boschetti, In taberna

quando sumus: a drunkard›s cakewalk through wine proteomics, Food Technol. Biotechnol. 50 (2012) 253–260.

118. F. Di Girolamo, G.P. Righetti, M. Soste, Y. Feng, P. Picotti, Reproducibility of combinatorial peptide ligand libraries for proteome capture evaluated by selected reaction monitoring, J. Proteome 89 (2013) 215–226.

119. A. Chambery, G. del Monaco, A. Di Maro, A. Parente, Peptide fingerprint of high quality Campania white wines by MALDI-TOF mass spectrometry, Food Chem. 113 (2009) 1283–1289.

120. D. Resetar, M. Marchetti-Deschmann, G. Allmaier, J.P. Katalinic, S. Kraljevic, Matrix assisted laser desorption ionization mass spectrometry linear time-of-flight method for white wine fingerprinting and classification, Food Control 64 (2016) 157–164.

121. European Council, Council Regulation (EC) No 1234/2007 of 22 October 2007 establishing a common organisation of agricultural markets and on specific provisions for certain agricultural products (Single CMO Regulation), Off. J. Eur. Union L299 (2007) 1–149.

122. A. D›Amato, A.V. Kravchuk, A. Bachi, P.G. Righetti, Noah›s nectar: the proteome content of a glass of red wine, J. Proteome 73 (2010) 2370–2377.

123. A. Cryar, C. Pritchard, W. Burkitt, M. Walker, G. O›Connor, M. Quaglia, A mass spectrometry-based reference method for the analysis of lysozyme in wine and the production of certified reference materials, J. Assoc. Public Anal. 40 (2012) 77–80.

124. L. Monaci, I. Losito, F. Palmisano, A. Visconti, Identification of allergenic milk proteins markers in fined white wines by capillary liquid chromatography– electrospray ionization-tandem mass spectrometry, J. Chromatogr. A 1217 (2010) 4300–4305.

125. L. Monaci, I. Losito, F. Palmisano, A. Visconti, Reliable detection of milk allergens in food using a high-resolution, stand-alone mass spectrometer, J. AOAC Int. 94 (2011) 1034–1042.

126. L. Monaci, I. Losito, E. De Angelis, R. Pilolli, A. Visconti, Multi-allergen quantification of fining-related egg and milk proteins in white wines by high-resolution mass spectrometry, Rapid Commun. Mass Spectrom. 27 (2013) 2009–2018.

127. G.P. Gruere, S.R. Rao, A review of international labeling policies

of genetically modified food to evaluate India›s proposed rule, AgBioforum 10 (2007) 51–64.

128. A.E. Wohlers, Labeling of genetically modified food: closer to reality in the United States? Politics Life Sci. 32 (2013) 73–84.
129. V. García-Cañas, C. Simó, C. León, E. Ibáñez, A. Cifuentes, MS-based analytical methodologies to characterize genetically modified crops, Mass Spectrom. Rev. 30 (2011) 396–416.
130. A. Valdés, C. Simó, C. Ibáñez, V. García-cañas, Foodomics strategies for the analysis of transgenic foods, Trends Anal. Chem. 52 (2013) 2–15.
131. D. Corpillo, G. Gardini, A.M. Vaira, M. Basso, S. Aime, G.P. Accotto, et al., Proteomics as a tool to improve investigation of substantial equivalence in genetically modi- fied organisms: the case of a virus-resistant tomato, Proteomics 4 (2004) 193–200.
132. F. Scossa, D. Laudencia-Chingcuanco, O.D. Anderson, W.H. Vensel, D. Lafiandra, R. D›Ovidio, et al., Comparative proteomic and transcriptional profiling of a bread wheat cultivar and its derived transgenic line overexpressing a low molecular weight glutenin subunit gene in the endosperm, Proteomics 8 (2008) 2948–2966.
133. A.R. Brandão, H.S. Barbosa, M.A. Arruda, Image analysis of two-dimensional gel electrophoresis for comparative proteomics of transgenic and non-transgenic soybean seeds, J. Proteome 73 (2010) 1433–1440.
134. H.S. Barbosa, S.C. Arruda, R.A. Azevedo, M.A. Arruda, New insights on proteomics of transgenic soybean seeds: evaluation of differential expression of enzymes and proteins, Anal. Bioanal. Chem. 402 (2012) 299–314.
135. C. Goulet, M. Benchabane, R. Anguenot, F. Brunelle, M. Khalf, D. Michaud, A companion protease inhibitor for the protection of cytosol-targeted recombinant proteins in plants, Plant Biotechnol. J. 8 (2010) 142–154.
136. M. Khalf, C. Goulet, J. Vorster, F. Brunelle, R. Anguenot, I. Fliss, et al., Tubers from potato lines expressing a tomato Kunitz protease inhibitor are substantially equivalent to parental and transgenic controls, Plant Biotechnol. J. 8 (2010) 155–169.
137. L. Zolla, S. Rinalducci, P. Antonioli, P.G. Righetti, Proteomics as a

complementary tool for identifying unintended side effects occurring in transgenic maize seeds as a result of genetic modifications, J. Proteome Res. 7 (2008) 1850–1861.

138. E. Barros, S. Lezar, M.J. Anttonen, J.P. Van Dijk, R.M. Röhlig, E.J. Kok, et al., Comparison of two GM maize varieties with a near-isogenic non-GM variety using transcriptomics, proteomics and metabolomics, Plant Biotechnol. J. 8 (2010) 436–451.

139. G.M. Balsamo, G.C. Cangahuala-Inocente, J.B. Bertoldo, H. Terenzi, A.C. Arisi, Proteomic analysis of four Brazilian MON810 maize varieties and their four non-genetically-modified isogenic varieties, J. Agric. Food Chem. 59 (2011) 11553–11559.

140. A. Coll, A. Nadal, M. Rossignol, P. Puigdomènech, M. Pla, Proteomic analysis of MON810 and comparable non-GM maize varieties grown in agricultural fields, Transgenic Res. 4 (2011) 939–949.

141. S.Z. Agapito-Tenfen, M.P. Guerra, O.G. Wikmark, R.O. Nodari, Comparative proteomic analysis of genetically modified maize grown under different agroecosystems conditions in Brazil, Proteome Sci. 11 (2013) 46.

142. G.M. Balsamo, P.A. Valentim-Neto, C.S. Mello, A.C.M. Arisi, Comparative proteomic analysis of two varieties of genetically modified (GM) Embrapa 5.1 common bean (Phaseolus vulgaris L.) and their non-GM counterparts, J. Agric. Food Chem. 63 (2015) 10569–10577.

143. C.Y. Gong, Q. Li, H.T. Yu, Z. Wang, T. Wang, Proteomics insight into the biological safety of transgenic modification of rice as compared with conventional genetic breeding and spontaneous genotypic variation, J. Proteome Res. 11 (2012) 3019–3029.

144. J. Luo, T. Ning, Y. Sun, J. Zhu, Y. Zhu, Q. Lin, et al., Proteomic analysis of rice endosperm cells in response to expression of hGM-CSF, J. Proteome Res. 8 (2009) 829–837.

145. L. Mora, P.M. Bramley, P.D. Fraser, Development and optimisation of a label-free quantitative proteomic procedure and its application in the assessment of genetically modified tomato fruit, Proteomics 13 (2013) 2016–2030.

146. R.M. Ruiz, R. Ortiz, P. Álvarez, Dry bean cultivar characterisation by isoelectric focusing electrophoresis in polyacrylamide gel, J. Sci. Food Agric. 81 (2001) 1126–1131.

147. J.M. Yeh, J.W.-Y. Ong, K.Y.-L. Yap, Proteomic analysis of ginseng, ICBPE 2006—Proc. Int. Conf. Biomed. Pharm. Eng. 2006, pp. 459–460.

148. J. Wang, M.M. Kliks, W. Qu, S. Jun, G. Shi, Q.X. Li, Rapid determination of the geographical origin of honey based on protein fingerprinting and barcoding using MALDI TOF MS, J. Agric. Food Chem. 57 (2009) 10081–10088.

149. F. Di Girolamo, A. D›Amato, P.G. Righetti, Assessment of the floral origin of honey via proteomic tools, J. Proteome 12 (2012) 3688–3693.

150. G.F. Zhang, T. Liu, Q. Wang, L. Chen, J.D. Lei, J. Luo, et al., Mass spectrometric detection of marker peptides in tryptic digests of gelatin: a new method to differentiate between bovine and porcine gelatin, Food Hydrocoll. 23 (2009) 2001–2007.

151. H. Emons, A. Held, F. Ulberth, Reference materials as crucial tools for quality assurance and control in food analysis, Pure Appl. Chem. 78 (2006) 135–143.

CHAPTER

6

COMMON AMINO ACID SUBSEQUENCES IN A UNIVERSAL PROTEOME—RELEVANCE FOR FOOD SCIENCE

Piotr Minkiewicz, Małgorzata Darewicz, Anna Iwaniak, Jolanta Sokołowska, Piotr Starowicz, Justyna Bucholska and Monika Hrynkiewicz

Department of Food Biochemistry, University of Warmia and Mazury in Olsztyn, Plac Cieszyński 1, Olsztyn-Kortowo 10-726, Poland

ABSTRACT

A common subsequence is a fragment of the amino acid chain that occurs in more than one protein. Common subsequences may be an object of interest for food scientists as biologically active peptides, epitopes, and/or protein markers that are used in comparative proteomics. An individual bioactive fragment, in particular the shortest fragment containing two or three amino acid residues, may occur in many protein sequences. An individual linear epitope may also be present in multiple sequences of precursor proteins. Although recent recommendations for prediction of allergenicity and cross-reactivity include not only sequence identity, but also similarities in secondary and tertiary structures surrounding the common fragment, local sequence identity may be used to screen protein sequence databases for potential allergens *in silico*. The main weakness of the screening process is that it overlooks allergens and cross-reactivity cases without identical fragments corresponding to linear epitopes. A single peptide may also serve as a marker of a group of allergens that belong to the same family and, possibly, reveal cross-reactivity. This review article discusses the benefits for food scientists that follow from the common subsequences concept.

Keywords: allergens; biologically active peptides; biomarkers; epitopes; databases

INTRODUCTION

Food science is being rapidly integrated with other areas, such as chemistry, biology, medicine or pharmacology. The ideas, methods and concepts originating in the above fields are applied to solve food-related problems. The concept of common subsequences creates new opportunities for analyzing problems in selected areas of interest related to food science.

A universal proteome [1] is defined as the set of all existing proteins. It is also referred to as a proteomosphere [2] or the protein universe [3].

A common subsequence is a fragment of the amino acid chain that occurs in more than one protein. Such a fragment may be regarded as a motif, *i.e.*, a reproducible pattern in a protein sequence that is attributed to a specific biological function [4]. Common subsequences may constitute continuous motifs. The role of short, continuous motifs in the biological functions of proteins and immunology constitutes the domain of peptidology [5]. Such short subsequences may play important roles as fragments of entire proteins (e.g., as epitopes responsible for interactions between proteins and antibodies) or after release through proteolytic enzymes (e.g., as hormones). In the latter case, short fragments may constitute "cryptides", peptides that are encrypted in protein sequences, are inactive inside the protein chain and are activated after enzymatic release. This definition of bioactive peptides was introduced by Schlimme and Meisel [6] for products of food protein hydrolysis.

The shortest peptides containing two or three amino acid residues are especially interesting for food scientists because they pass from the gastrointestinal tract to the blood [7,8]. Affinity for small intestine isoforms of the oligopeptide transporter (target ID: CHEMBL4605) is emphasized in the ChEMBL database [9,10] as a standard property of dipeptides and tripeptides.

Compounds that are used as drugs or potential drugs are characterized by low molecular weight [11,12]. The criteria for the selection of potential drugs, referred to as the "Rule of five", include molecular weight, number of H-bond donors, number of H-bond acceptors and hydrophobicity measured as the logarithm of the octanol/water partition coefficient [11]. The above parameters should not exceed the following values: molecular weight—500 Da, number of H-bond donors—5, number of H-bond acceptors—10, logarithm of the octanol/water partition coefficient (CLogP)—5 [11]. The shortest peptides that fulfill the above criteria are annotated in chemical databases, such as PubChem [13,14], ChemSpider [15,16] or ChEMBL, as potential objects of interest in pharmacology. Those criteria may also be applied to select potentially bioactive food components.

Every protein may be a source of biologically active dipeptides and tripeptides. The shortest sequences are most consistent with the hypothesis formulated by Karelin *et al.* [17], which states that all existing proteins are precursors of peptides whose biological activity is revealed after release.

Common subsequences are also used in comparative proteomics on the assumption that homologous proteins (which possess a common ancestor) can release similar sets of peptides during proteolysis. Comparative proteomics supports the search for non-sequenced proteins based on the presence of fragments representative of the protein family. Peptides are identified by mass spectrometry to detect a protein family that contains the same or similar fragments [2,18]. Peptides identified in such experiments should be characterized by the greatest possible length.

This review article discusses various aspects of common short fragments in proteins using the example of bioactive peptides, epitopes and protein biomarkers. The presented examples include proteins and peptides originating from organisms that are major food resources (e.g., wheat, cattle, chickens, fish) or microorganisms utilized in the food industry (yeasts).

BIOLOGICALLY ACTIVE PEPTIDES

Biologically active peptides are involved in the regulation of many processes in living organisms. They may be produced in the body by synthesis or hydrolysis of precursors (endogenous peptides) or supplied with food (exogenous peptides). Peptides of the latter category constitute valuable components of functional foods, *i.e.*, foods with defined biological activity. Functional foods may support conventional treatments of selected diseases. Hypertension is the best known example of diseases that can be effectively mitigated by diet.

The most recent review article describing the state of the art in proteomics and peptidomics research relating to both categories of bioactive peptides was published by Dallas *et al.* [19]. Peptides and proteins can be identified with the use of standard proteomics or peptidomics techniques involving mass spectrometry [19,20]. Two questions have to be answered when a peptide or a protein is identified in an organism, cell, tissue or food product: which biologically active fragments are present in the analyzed protein or peptide and what are the possible precursors of the analyzed peptide? To answer the first question, data has to be interpreted in a manner similar to the top-down approach in proteomics [20]. This protocol begins with a search for the short fragments of protein sequence. To answer the second question, the peptide sequence should be used as a query, and the database of protein or peptide sequences should be searched for longer sequences containing the analyzed fragment.

The exemplary results of "top-down mimicking" search are presented in Figure 1 and Table 1. The sequence of protein from yeast (*Saccharomyces cerevisiae*), a microorganism that is broadly applied in food technology, was used as a query.

```
  MTKNFIVTLKKNTPDVEAKKFLDSVHHAGGSIVHEFDIIKGYTIKVPDVLHLNKLKEKHNDVIENVEEDKEVHTN
1    NF            EAKKF       AGGS        KGY   VP    HLNKLKEK      IE  VE   KEV
2      IV                              IV      II        VL
3         LK       EAK     VHH                            LHL   LK
4                            HA                  VP
5                     KF           EF
6                     KF           EF
7         KK        KK                                                  EE
```

Figure 1. Location of biologically active fragments in the sequence of yeast (*Saccharomyces cerevisiae*, strain ATCC 204508/S288c) protease B inhibitor 2 (Accession No P0CT04 in the UniProt Knowledgebase [21,22]). (1) angiotensin I-converting enzyme (ACE) inhibitors; (2) glucose uptake stimulators; (3) antioxidant fragments; (4) dipeptidyl peptidase IV inhibitors; (5) calmodulin-dependent phosphodiesterase 1 inhibitors; (6) renin inhibitors; (7) fragments with other activities (see Table 1). Bioactive fragments were found with the use of the BIOPEP search engine [23,24] where the protein sequence was the query. The search was performed in May 2014.

Table 1. Reference data for biologically active fragments of yeast (*Saccharomyces cerevisiae*, strain ATCC 204508/S288c) protease B inhibitor 2, indicated in Figure 1.

ID [a]	Sequence [b]	Activity	Primary Resource [c]	Reference
3379	AKK	ACE inhibitor	Muscle of fish of the genus Sardina [d]	[25]
3532	GY	ACE inhibitor	Synthetic	[26]
7587	VP	ACE inhibitor	Synthetic	[26]
7600	AG	ACE inhibitor	Synthetic	[26]
7602	HL	ACE inhibitor	Synthetic	[27]
7604	KG	ACE inhibitor	Synthetic	[26]
7607	GS	ACE inhibitor	Synthetic	[26]
7616	GG	ACE inhibitor	Synthetic	[27]
7623	EA	ACE inhibitor	Synthetic	[26]
7654	NKL	ACE inhibitor	Wakame (*Undaria pinnatifida*) [d]	[28]

7683	NF	ACE inhibitor	Garlic (*Allium sativum*) [d]	[28]
7692	KF	ACE inhibitor	Garlic (*Allium sativum*) [d]	[28]
7693	KL	ACE inhibitor	Wakame (*Undaria pinnatifida*) [d]	[28]
7698	NK	ACE inhibitor	Wakame (*Undaria pinnatifida*) [d]	[28]
7827	IE	ACE inhibitor	Bovine (*Bos taurus*) milk [d]	[29]
7828	EV	ACE inhibitor	Bovine (*Bos taurus*) milk [d]	[29]
7829	VE	ACE inhibitor	Bovine (*Bos taurus*) milk [d]	[29]
7832	LN	ACE inhibitor	Bovine (*Bos taurus*) milk [d]	[29]
7840	EK	ACE inhibitor	Bovine (*Bos taurus*) milk [d]	[29]
7841	KE	ACE inhibitor	Bovine (*Bos taurus*) milk [d]	[29]
8320	VL	Glucose uptake stimulating	Bovine (*Bos taurus*) whey [d]	[30]
8322	IV	Glucose uptake stimulating	Bovine (*Bos taurus*) whey [d]	[30]
8325	II	Glucose uptake stimulating	Bovine (*Bos taurus*) whey [d]	[30]
8329	EE	Vasoactive substance release stimulating	Soybean (*Glycine max*) [d]	[31]
3305	LH	Antioxidant	Soybean (*Glycine max*) [d]	[32]
3317	HL	Antioxidant	Soybean (*Glycine max*) [d]	[32]
3319	HH	Antioxidant	Soybean (*Glycine max*) [d]	[32]
7794	VHH	Antioxidant	Chicken (*Gallus gallus*) egg [d]	[33]
7995	LHL	Antioxidant	Synthetic	[34]
8130	EAK	Antioxidant	Bonito (*Katsuwonus pelamis*) [d]	[35]
8217	LK	Antioxidant	Chicken (*Gallus gallus*) egg [d]	[36]
3751	KK	Bacterial permease ligand	Synthetic	[37]

3181	VP	Dipeptidyl peptidase IV inhibitor	Rat (*Rattus norvegicus*)	[38]
3184	HA	Dipeptidyl peptidase IV inhibitor	Rat (*Rattus norvegicus*)	[38]
8249	KF	CaMPDE inhibitor	Pea (*Pisum sativum*) [d]	[39]
8250	EF	CaMPDE inhibitor	Pea (*Pisum sativum*) [d]	[39]
8248	KF	Renin inhibitor	Pea (*Pisum sativum*) [d]	[39]
8251	EF	Renin inhibitor	Pea (*Pisum sativum*) [d]	[39]

[a] ID number in the BIOPEP database; [b] Sequence given in a single-letter code; [c] Source from which the peptide was isolated for the first time; [d] Organism used as a food resource. Abbreviations used in Table 1: ACE—angiotensin I-converting enzyme; CaMPDE—calmodulin-dependent phosphodiesterase 1.

Among the bioactive peptides shown in Table 1, angiotensin I-converting enzyme (EC 3.4.15.1) inhibitors are most abundant. In the BRENDA database [40,41] the recommended name of the enzyme is peptidyl-dipeptidase A. The enzyme participates in the release of angiotensin II, a peptide that causes vasoconstriction. ACE inhibitors may thus lower blood pressure *in vivo* [42]. They are the most extensively studied class of bioactive peptides from food [42,43,44]. Renin (EC 3.4.23.15) inhibitors are also involved in blood pressure regulation. Renin releases the peptide angiotensin I from its precursor, angiotensinogen. Angiotensin I is inactive, but it is a substrate for conversion to the vasoconstrictor angiotensin II. Renin inhibitors pose an alternative to ACE inhibitors. They attract the interest of researchers as drugs [45] as well as bioactive components of functional foods that prevent hypertension [46,47]. Some peptides, including KF (Table 1), are capable of inhibiting the angiotensin-converting enzyme as well as renin. Peptides with sequences VP and HA are inhibitors of dipeptidyl peptidase IV (EC 3.4.14.5). This enzyme participates in the hydrolysis of the insulinotropic hormone, glucagon-like peptide 1. Due to this function, enzyme inhibitors can be used in the treatment of type II diabetes. Inhibitors of dipeptidyl peptidase IV may be used as anti-diabetic drugs [48] and components of functional foods designed for the treatment of diabetes [49]. Peptides with sequences KF and EF inhibit calmodulin-dependent phosphodiesterase 1 (EC 3.1.4.17). In the BRENDA database, the recommended name of the enzyme is 3′,5′-cyclic-nucleotide phosphodiesterase. The enzyme is involved in the metabolism of cyclic adenosine 3,5-monophosphate (cAMP) and regulation of cellular processes mediated by this compound. Inhibitors of 3′,5′-cyclic-nucleotide phosphodiesterase constitute potential treatment for cancer [50], inflammatory [51,52], autoimmune [51], and neurological diseases [52]. Antioxidant peptides from food, in particular short-chain peptides, are considered helpful in the prevention of oxidative damage [53]. Food components that stimulate glucose uptake (including peptides) are recommended for athletes [54].

Several examples of "top-down mimicking" database searches are shown in Table 2. Simulated proteolysis *in silico* of proteins from the human digestive tract [55] is included. All other examples presented in Table 2 are related to food. The BIOPEP database [23,24] was used in all cases, and query peptide or protein sequences were longer than the target sequences. The target sequences were short peptides (usually dipeptides and tripeptides) summarized in the database. Peptide sequences used as queries were obtained by mass spectrometry.

The examples presented in Table 2 account for only *in silico* and *in silico* together with experimental research. The second option involves mass spectrometry, followed by database screening. In addition to the examples shown in Table 2, short sequences (dipeptides and tripeptides) were also matched exactly in the database. In a food experiment conducted by Barba de la Rosa *et al.* [56], bioactive dipeptides and tripeptides were identified in hydrolysates of amaranthus proteins. In food science, *in silico* research also involved proteolysis simulations that seemed to be the weak point of experiments. The results of *in silico* and *in vitro* studies were compared to demonstrate differences between predicted and experimentally obtained patterns of proteolysis. The observed differences included both the absence of the predicted peptides [57] and the presence of peptides that were not expected to be released by enzymes with known specificity [58]. A successful prediction of an antimicrobial peptide released from casein by proteolysis has been recently described by Guinane *et al.* [59]. The noted differences could be explained by the fact that the specificity of proteolytic enzymes may be affected by reaction conditions, changes in protein structure and possible interactions with other compounds in the reaction environment.

An example of "bottom-up mimicking" (query sequence shorter than the target) search results is presented in Table 3.

Table 2. Examples of protocols involving the search for shorter fragments in sequences of proteins or peptides relevant for food and/or nutrition sciences.

Database Search Application	Reference
Location of short, bioactive fragments in sequences of peptides released during hydrolysis of bovine and trout meat proteins in the porcine digestive tract. Peptides used as query sequences were identified by mass spectrometry.	[60]
Location of bioactive fragments in sequences of rapeseed proteins. Protein sequences from UniProt were used as queries.	[61]
Location of bioactive fragments in sequences of bovine meat proteins. Protein sequences from UniProt were used as queries.	[62]
Location of short, bioactive fragments in sequences of peptides released during hydrolysis of fish sarcoplasmic proteins. Peptides used as query sequences were identified by mass spectrometry.	[63]
Location of bioactive fragments in sequences of cereal proteins. Protein sequences from UniProt were used as queries.	[64]
The BIOPEP database was used to determine the profiles of potential biological activity of salmon proteins. Some of the predicted peptides were identified in protein hydrolysates by liquid chromatography and mass spectrometry.	[58]
Location of bioactive fragments in sequences of proteins from the human digestive tract, followed by proteolysis simulation by digestive proteolytic enzymes. Protein sequences from UniProt were used as queries.	[55]
Location of bioactive fragments in sequences of amaranthus proteins. Protein sequences from UniProt were used as queries.	[65]

Table 3. Proteins containing fragment PANLPWGSSNV with an ACE inhibitory activity [66] (ID 49468 in the PepBank database [67,68]). The UniProt Knowledgebase [21,22] was screened with the BLAST program [69,70] with the use of the above sequence as a query and screening parameters described by Minkiewicz *et al.* [71]. The search was performed in May 2014.

No	Protein Name	Entry Name in UniProtKB	Organism [a]
1.	Uncharacterized protein	TR:W4ZV89_WHEAT	*Triticum aestivum* (4565)
2.	Glyceraldehyde-3-phosphate dehydrogenase	SP:G3P3_YEAST	*Saccharomyces cerevisiae* (strain ATCC 204508/S288c) (559292)
3.	Glyceraldehyde-3-phosphate dehydrogenase	TR:A6ZUK2_YEAS7	*Saccharomyces cerevisiae* (strain YJM789) (307796)
4.	Glyceraldehyde-3-phosphate dehydrogenase	TR:B3LI45_YEAS1	*Saccharomyces cerevisiae* (strain RM11-1a) (285006)
5.	Glyceraldehyde-3-phosphate dehydrogenase	TR:B5VJD4_YEAS6	*Saccharomyces cerevisiae* (strain AWRI1631) (545124)
6.	Glyceraldehyde-3-phosphate dehydrogenase	TR:C8Z985_YEAS8	*Saccharomyces cerevisiae* (strain Lalvin EC1118 / Prise de mousse) (643680)
7.	Glyceraldehyde-3-phosphate dehydrogenase	TR:E7KD02_YEASA	*Saccharomyces cerevisiae* (strain AWRI796) (764097)
8.	Glyceraldehyde-3-phosphate dehydrogenase	TR:E7KP33_YEASL	*Saccharomyces cerevisiae* (strain Lalvin QA23) (764098)
9.	Glyceraldehyde-3-phosphate dehydrogenase	TR:E7LUX3_YEASV	*Saccharomyces cerevisiae* (strain VIN 13) (764099)
10.	Glyceraldehyde-3-phosphate dehydrogenase	TR:E7NI37_YEASO	*Saccharomyces cerevisiae* (strain FostersO) (764101)
11.	Glyceraldehyde-3-phosphate dehydrogenase	TR:E7Q4A2_YEASB	*Saccharomyces cerevisiae* (strain FostersB) (764102)
12.	Glyceraldehyde-3-phosphate dehydrogenase	TR:E7QF80_YEASZ	*Saccharomyces cerevisiae* (strain Zymaflore VL3) (764100)
13.	Glyceraldehyde-3-phosphate dehydrogenase	TR:G2WES0_YEASK	*Saccharomyces cerevisiae* (strain Kyokai no. 7/NBRC 101557) (721032)
14.	Tdh3p	TR:H0GGT7_9SACH	*Saccharomyces cerevisiae* x *Saccharomyces kudriavzevii* VIN7 (1095631)
15.	Uncharacterized protein	TR:J7S7S3_KAZNA	*Kazachstania naganishii* (strain ATCC MYA-139/BCRC 22969/CBS 8797/CCRC 22969/KCTC 17520/NBRC 10181/NCYC 3082) (1071383)

16.	Tdh3p	TR:N1P2H7_YEASC	*Saccharomyces cerevisiae* (strain CEN.PK113-7D) (889517)
17.	Tdh3p	TR:W7PUI3_YEASX	*Saccharomyces cerevisiae* R008 (1182966)
18.	Tdh3p	TR:W7RBG4_ YEASX	*Saccharomyces cerevisiae* P283 (1177187)

[a] Defined by the Latin name and NCBI taxonomic identifier [72,73] (in parentheses).

The query peptide originates from yeasts (*Saccharomyces cerevisiae*) [66]. All proteins presented in Table 3 belong to the Glyceraldehyde/ Erythrose phosphate dehydrogenase family (Signature IPR020831 in the InterPro classification system [74,75]). The data shown in Table 3 illustrate the possibility of the same fragment occurring in homologous proteins from various microbial species and strains. This phenomenon is noted when the bioactive fragment occurs in a strongly conserved part of the protein chain. The observation that a biologically active peptide may possess more than one precursor is emphasized in the AHTPDB database of antihypertensive peptides [43,44].

Peptide LAPSLPGKPKPD (BIOPEP ID: 8388; 8547; 8548; 8550) may serve as an example of a peptide with a single known precursor. It was found (26 February 2015) only in the sequence of visual system homeobox 2 (Entry name in UniProt: VSX2_CHICK) from chicken (*Gallus gallus*) eggs. Similar fragments containing 8–10 of the 12 amino acid residues in the above peptide were found in sequences of five microbial enzymes annotated in the UniProt database. Peptide LAPSLPGKPKPD is multifunctional. It acts as an inhibitor of angiotensin I-converting enzyme (EC 3.4.15.1), dipeptidyl peptidase IV (EC 3.4.14.5), α-glucosidase (EC 3.2.1.20), and it has antioxidant properties [76]. A BLAST [69,70] search (19 March 2015) revealed that complete sequences of visual system homeobox 2 proteins are available for mammals, reptiles and fish. Partial sequences of rock pigeon (*Columba livia*; protein: TR:I1TED5_ COLLI), turkey (*Meleagris gallopavo*; protein: TR:G1NJ43_MELGA) and mallard (*Anas platyrhynchos* protein: TR:U3J596_ANAPL) proteins were characterized by 98%–100% identity with chicken visual system homeobox 2 proteins. Therefore, a more comprehensive list of bird proteins could reveal a higher number of potential precursors of peptide LAPSLPGKPKPD. To date, the chicken genome and proteome have been studied most extensively in birds due to the significance of chicken as a food source. Chicken egg proteins are also studied as a source of peptides with various biological activities [77].

LINEAR EPITOPES

Epitopes attract the interest of researchers dealing with three fields of science: allergology, immunochemical analysis methods (ELISA) and vaccinology. The first two areas also capture the interest of food scientists due to the prevalence of food allergies and the broad application of immunochemical methods in food analysis. Epitopes are defined as protein fragments responsible for interactions with the immune system (antibodies, B cells, T cells). Epitopes are divided into two classes: sequential (linear) epitopes which are continuous fragments of the primary protein structure, and conformational epitopes which are formed by neighboring amino acid residues on the surface of the antigen, but do not form a continuous fragment in the primary structure. Role of spatial structure of epitopes is recently emphasized even in the case of linear ones [78,79].

The standard protocol for the search of linear epitopes covers protein hydrolysis or synthesis of protein chain fragments, followed by experimental detection of interactions between specific peptides and antibodies of allergy sufferers. Albrecht and co-workers [78] pointed out that interactions between antibodies and fragments of synthetic proteins do not always lead to interactions with the same fragment encrypted in the entire protein sequence. The spatial structure of a short peptide may differ from the structure of the same peptide that is a part of a larger molecule. On the other hand, an example of a protein modified by insertion of a linear epitope and interaction with immunoglobulin E has been described [78].

Some allergenic proteins, such as caseins, which are major milk proteins, do not form compact or well-established spatial structures. In this case, allergenic properties may be retained under denaturing conditions (e.g., after heating, a process that is commonly applied in food processing) [80,81]. The presence of proteins and protein fragments that do not form a well-defined structure ("naturally denatured" proteins) is a relatively common phenomenon [82,83]. In the case of naturally denatured proteins or protein fragments, interactions between short peptide fragments and antibodies of allergenic patients imply interactions between entire proteins and, consequently, cross-reactivity of proteins containing the same fragment which is recognized as an epitope.

The experimental criteria for allergenic proteins recommended by the International Union of Immunological Societies (IUIS) were summarized by Breiteneder and Chapman [84]. An allergenic protein should meet a number of biochemical criteria, such as a known sequence and posttranslational modification pattern (if applicable), purification to homogeneity or near homogeneity, determination of basic physicochemical properties (molecular weight, isoelectric point) and production of monoclonal or monospecific antibodies that interact with the allergen. Immunological criteria of allergenicity include comparisons of the prevalence of serum IgE antibodies in as many patients as possible (at least 50 are recommended), determination of allergenic activity (e.g., skin tests), reducing IgE binding capacity of the allergenic extract after allergen removal (e.g., by immunoabsorption) and detection of IgE binding ability of a recombinant allergen. Proteins whose sequences have not been confirmed experimentally, but predicted based on sequence or structure analysis, are sometimes regarded as allergens *in silico*. The simplest method of determining *in silico* allergenicity and predicting cross-reactivity involves a comparison of the sequence of the analyzed protein with experimentally confirmed allergens. A protein is a potential allergen if it contains a fragment with at least six to eight amino acid residues which are identical to the fragment of a known allergen, or a fragment of at least 80 amino acid residues with a minimum 35% identity with a fragment of the known allergen [85,86,87]. The presence of common sequential epitopes (as long as possible) seems to increase the likelihood of cross-reactivity [88,89]. Common sequential epitopes are usually present in homologous proteins, *i.e.*, proteins that have a common ancestor and belong to the same family. Homologous proteins possess similar amino acid sequences and similar structure. Protein families are classified based on the presence of characteristic domains that are attributed to protein functions [90]. Protein families are described in domain databases such as InterPro [74,75] and Pfam [91,92]. The AllFam database of allergen families [93,94] was developed based on the protein classification system found in the Pfam database. Fragments containing five amino acid residues are the smallest units that interact with the immune system [95]. The distribution of pentapeptides in the universal proteome was analyzed by *in silico* studies [95,96,97]. Tools for comparing protein sequences and identifying common pentapeptides and other common motifs have been recently developed [98,99,100,101].

In this review article, the possible distribution of epitopes across protein families is discussed based on 4 epitopes from wheat (*Triticum aestivum*) ω5-gliadin (UniProt entry name: Q402I5_WHEAT) [102]. Modifications of those epitopes have been proposed to significantly decrease gliadin allergenicity [103]. The discussed epitopes have the following amino acid sequences: QQFPQQQ (IEDB ID 52028), QQIPQQQ (IEDB ID 52043), QQLPQQQ (IEDB ID 52066) and QQYPQQQ (IEDB ID 52180). Numbers in parentheses indicate ID numbers in the Immune Epitope Database (IEDB) [104,105]. The search protocol was identical to that whose results are presented in Table 3. Wheat and related species are among the most commonly used resources in the food industry. Gliadins and their homologs from other cereals belong to the best known group of food proteins.

Due to space constraints, the table containing a list of proteins and fragments identical to gliadin epitopes is presented in the form of a supplement. The supplement includes proteins identified "at protein level" (e.g., by mass spectrometry) as well as amino acid sequences translated from the known nucleotide sequences (putative or identified at transcript level). It also contains a table of protein families defined according to the InterPro database (with links to records of particular domains), species annotated by their Latin names and taxonomic identifiers (with link to annotations of species on the UniProt website) and proteins annotated based on their entry names in UniProt (with links to particular records). Four peptides have been listed based on amino acid sequences in a single-letter code and chemical identifiers: SMILES [106], InChI and InChIKeys [107]. As previously noted [108] peptide structures annotated with the SMILES code may be used as input in many cheminformatics programs. They are used in specialized peptide databases such as AHTPDB [44]. Links to peptide databases deploying chemical codes (SMILES) are available on the MetaComBio website [109,110]. InChI and InChIKey offer more advantages in comparison with SMILES. Few versions of annotation of the same structure are possible using the last code. SMILES requires special search engines, whereas InChI and especially InChIKey are more versatile and may be used as queries in popular search engines such as Google™ [111]. Peptide structures described with InChI and InChIKey are thus more effective in identifying datasets (such as the supplement to this publication) than sequences written in a single-letter code. Peptides are commonly annotated with chemical codes in chemical databases such as ChEMBL, ChemSpider and PubChem. InChiKeys are also used in the BRENDA database [40,41]. The amino acid sequences presented in the supplement were translated from amino acid sequences in FASTA format into chemical codes using the Open Babel program [112,113].

The number of proteins containing the above-mentioned sequences is high due to the fact that repeated glutamine residues belong to the most common motifs in known protein sequences. For instance, according to the database associated with the Tachyon program, the fragment containing five glutamine residues is present in more than half a million sequences [98,99].

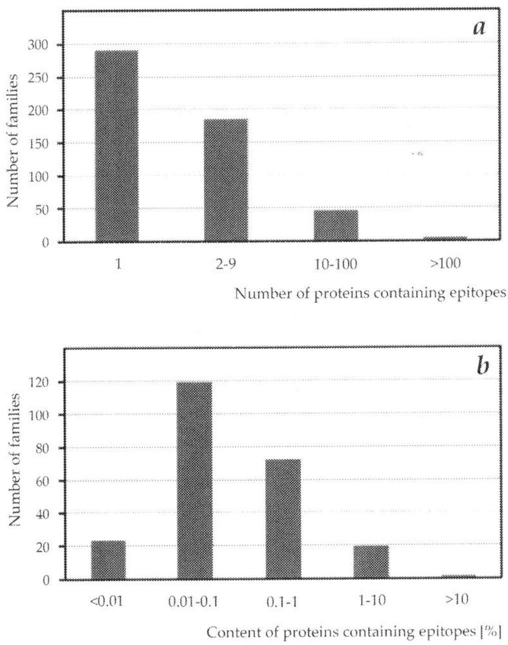

Figure 2. Distribution of proteins containing at least one of the four IgE-binding epitopes of ω5-gliadin [102] across protein families. (**a**) distribution based on the number of proteins containing epitopes in the family; (**b**) percentage

content of proteins with epitopes in the family. The data for families containing at least two proteins with epitopes are shown in **b**.

Heptapeptides are distributed across hundreds of protein families (Figure 2), although in most cases, only several proteins in a family contain at least one epitope. Epitopes are most abundant in four protein families: gliadin, alpha/beta (IPR001376) (208 proteins), gliadin /low molecular weight glutenin (IPR001954) (199 proteins), bifunctional trypsin/alpha-amylase inhibitor helical domain (IPR013771) (199 proteins) and bifunctional inhibitor/plant lipid transfer protein/seed storage helical domain (IPR016140) (197 proteins). These families are characterized by a high ratio of epitope-containing sequences to proteins at 12.72%, 9.57%, 5.63% and 2.79%, respectively. They contain proteins that are closely related to ω5-gliadin, the first discovered precursor of the discussed epitopes.

In our previous publications, we discussed the distribution of specific epitopes at the level of protein families based on the presence of the corresponding domains rather than individual proteins [71,114]. Such choice is explained by the fact that the number of known protein sequences grows rapidly, but the number of known domains remains much more stable [3]. The observed increase in the number of protein families can be attributed to the discovery of new multidomain families. The probability that a protein contains an epitope together with a domain defining its family and function seems to be a more reliable determinant of epitope distribution, and it can be used as a prognostic tool. That probability may be expressed as the percentage of proteins within a family (with a domain defining the family), which contain an epitope (or another fragment of interest). A family may be defined in accordance with the rules of InterPro, Pfam or AllFam databases. Proteins belonging to the same family possess similar sequences, spatial structure and, consequently, similar physico-chemical properties (e.g., solubility under various conditions, susceptibility to thermal denaturation) that affect behavior during food processing. Proteins belonging to the same family can be expected to have a similar pattern of bioactive fragments.

There are two possible patterns of distribution of fragments that are identical to known epitopes. The hexapeptides from Baltic cod (*Gadus morhua* subsp. *callarias*) parvalbumin are distributed randomly across many protein families. None of them is present in other parvalbumins [71]. The epitopes from shrimp (*Farfantepenaeus aztecus*) tropomyosin, which contain 10–15 amino acid residues, occur in homologs of their precursor (other tropomyosins). Only one fragment with five residues is broadly distributed across various protein families. Two fragments containing eight amino acid residues were present in several proteins that did not belong to the tropomyosin family [114].

The traditional system for the classification of allergens includes the route of exposure (ingestion, inhalation, injection or contact), although several different routes can exist for the same allergen [115]. Routes of exposure for particular allergens are annotated e.g. in Allergome database [116,117]. Allergens are also divided into the following groups: food, indoor, outdoor and injected [84]. Organisms that synthesize proteins containing fragments identical to known epitopes may be classified based on the possibility of human contact [71,114]. Species that synthesize proteins containing fragments identical to linear epitopes may be divided into the following categories: animal and plant species relevant for the food industry and/or agriculture (mainly edible), microorganisms that are useful and potentially useful for the food industry and/or agriculture (e.g., used in biotechnological processes), human symbionts and commensals (e.g., gut microorganisms) as well as human pathogens and parasites [71]. The first two groups may be interesting from the point of view of food safety. A more detailed classification has been proposed in an article describing the distribution of fragments identical to shrimp (*Farfantepenaeus aztecus*) tropomyosin epitopes [114]. Invertebrates that synthesize tropomyosins containing fragments identical to the above epitopes belong to the following categories: edible invertebrates (crustaceans and molluscs), human parasites (e.g., worms), parasites of edible animals and plants (potential food contaminants), as well as species that come into contact with humans by other ways (indoor organisms such as dust mites and invertebrates cultured in laboratories, such as *Caenorhabditis elegans* or *Drosophila melanogaster*).

The enclosed supplement contains information about organisms belonging to all of the above categories (not only edible organisms or organisms used in food technology). The most abundant protein families are found in wheat and other cereals. Fragments identical to wheat gliadin epitopes were also found in the proteins of edible birds (e.g., *Gallus gallus*) or fish (*Takifugu rubripes*, *Oreochromis niloticus*). Yeast (*Saccharomyces cerevisiae*) is a model microorganism that is used in the food industry and contains proteins with fragments identical to the epitopes from wheat gliadin. Darewicz *et al.* [118] reported local sequence similarity between proteomes of wheat (*Triticum aestivum*) and yeast (*Saccharomyces cerevisiae*), where the yeast proteome contained short sequence fragments similar to celiac-toxic peptides. Vojdani and Tarash [119] observed interactions between yeast proteins and antibodies of patients suffering from celiac disease. Proteins from those species revealed cross-reactions with the human immune system.

Database screening may produce results that go beyond the area of interest in food science. The resulting data could also be interesting from the point of view of biological and medical sciences. *Candida albicans* is an example of a commensal microorganism that is ubiquitous in the human gut, but may cause opportunistic infections. This microorganism may be thus classified in two categories as a commensal and a pathogen [120]. *Candida albicans* proteins contain fragments identical to fragments of wheat (Table S1) as well as cod parvalbumins [71]. Protein sequences of the human parasite *Trichinella spiralis* contain short subsequences that are identical to the fragments of three allergenic proteins: wheat gliadin (Table S1), cod parvalbumin [71] and shrimp tropomyosin [114]. Fragments identical to wheat gliadin epitopes are also present in human protein sequences. Kanduc described [96] the degree of identity between allergenic epitopes (e.g., from food allergens) and human proteins at the pentapeptide level. Amino acid subsequences identical to parvalbumin and tropomyosin fragments were are also found in the human proteome [71,114]. Human tropomyosin, which contains a fragment identical to the shrimp allergen, is regarded as an autoantigen [121].

PEPTIDES RELEVANT AS ALLERGEN MARKERS

Mass spectrometry is the recommended method for identifying and determining allergenic proteins in foods. Some of them are major food components. The presence of such proteins in food products may also result from contamination or adulteration. Qualitative and quantitative analyses of food allergens are based on the identification of peptides representative of allergens and considered as allergen markers (signatures) [122,123,124]. Peptides are usually released by trypsin (EC 3.4.21.4), an enzyme that is widely used in proteomics [122,123,124,125]. Recent experiments conducted with the use of mass spectrometry were described by Pilolli *et al.* [126], Gomaa and Boye [127] and by Posada-Ayala *et al.* [128]. Koeberl *et al.* [124] discussed the advantages of mass spectrometry for allergen analysis, including short time of analysis and the option of identifying multiple allergens in a single analysis. Some authors [129,130] recommended protein fragments overlapping with linear epitopes as markers for mass spectrometry. In this approach, the same fragments can be used in mass spectrometry and immunochemical methods, which is an unquestioned advantage.

The unique character of peptides (presence of a fragment in a single precursor) is emphasized as a major advantage in analyses that rely on the identification of protein fragments [131]. The rapid increase in the number of protein sequences annotated in UniProt [21,22], NCBI [132,133] or Allergome [116,117] makes this recommendation increasingly difficult to fulfill. Peptides used as markers may be released from multiple precursors, as demonstrated in the examples in Table 4.

Both peptides listed in Table 4 are fragments of multiple proteins from several species. The criteria for the choice of protein markers [131] indicate that an identified peptide may originate from proteins that are unlikely to be present in foods. Such proteins may originate from species that are not edible, wild, not used as sources of industrially processed foods or inhabit limited areas. In Table 4, such species are represented by wild birds from south-east Asia: *Gallus lafayetii*and *Gallus sonneratii*. The milk of the yak (*Bos mutus*) and water buffalo (*Bubalus bubalis*) as well as quail (*Coturnix coturnix japonica*) eggs are used as food, but they are less popular than cow milk and chicken eggs. The likelihood that yak or buffalo milk proteins will be identified in food products depends on their geographical origin. Peptide FFVAPFPEVFGK may indicate the presence of bovine α_{s1}-casein in products originating from Europe as well as the presence of yak or buffalo proteins in products from central or Southeast Asia.

α_{s1}-Casein from *Bos mutus* and lysozyme C from *Coturnix coturnix japonica* are not annotated in Allergome (as checked 28 April 2015), but they share linear epitopes with bovine (*Bos taurus*) α_{s1}-casein and chicken (*Gallus gallus*) lysozyme C, respectively. Both proteins can be classified as allergens *in silico* according to criteria that are based on local sequence identity, including common fragments containing at least six to eight amino acid residues [85,86] or common, experimentally found epitopes [89].

Proteins which are the precursors of peptide FFVAPFPEVFGK belong to the following families: casein (IPR001588) and α_{s1}-casein (IPR026999) in the InterPro database [74,75]; casein (PF00363) in the Pfam database [91,92] and alpha/beta casein (AF065) in the AllFam database [93,94]. Proteins containing the FESNFNTQATNR fragment belong to families: Glyco_hydro_22 (IPR001916), Glyco_hydro_22_CS (IPR019799), Glyco_hydro_22_lys (IPR000974), Lysozyme-like_dom (IPR023346) and Lysozyme_C (IPR030056) in the InterPro database, Lys (PF00062) in Pfam database and C-type lysozyme/alpha-lactalbumin family (AF016) in the AllFam database. In both cases, the group of precursors of a given peptide marker includes only a part of the protein family. The same applies to group markers predicted *in silico*[134] as well as common epitopes [71,114]. The group of proteins identified or determined in a single marker (signature) peptide should be precisely defined and updated to track the increase in the number of known protein sequences.

The presence of conserved fragments in a family creates new analytical opportunities. The same fragment may be present in proteins with and without a known sequence. The strategy that relies on local identity or similarity between sequenced and non-sequenced proteins is referred to as comparative proteomics [2,18]. Numerous edible organisms have not been subject to extensive studies aimed at protein sequencing to date. Edible insects, emerging as novel food resources [135,136], could also constitute a source of such proteins. Arthropod tropomyosins are allergens. Tropomyosins from various arthropods contain many identical fragments [114]. It is likely that selected peptides—markers of crustacean tropomyosins—may be used to detect allergens in insects. This could also apply to allergens from other sources.

Table 4. Proteins containing fragments FFVAPFPEVFGK and FESNFNTQATNR, used as markers of α_{s1}-casein from milk and lysozyme from eggs, respectively [126]. The UniProt Knowledgebase [21,22] was screened with the BLAST program [69,70] with the use of the above sequence as a query and screening parameters described by Minkiewicz *et al.* [71]. The search was performed in April 2015.

No	Entry Name in UniProtKB	Allergome Annotation	Organism [a]
Peptide (R)FFVAPFPEVFGK [b]—marker of α_{s1}-casein			
1.	CASA1_BOVIN	Bos d 9.0101; Code 10197	*Bos taurus* (9913)
2.	CASA1_BUBBU	Bub b 8; Code 1259	*Bubalus bubalis* (89462)
3.	G3C8Y4_BUBBU		*Bubalus bubalis* (89462)
4.	B5B3R8_BOVIN	Bos d 9; Code 2734	*Bos taurus* (9913)
5.	L8I5S0_9CETA		*Bos mutus (Bos grunniens)* (72004)
6.	G3C8Y5_BUBBU		*Bubalus bubalis* (89462)
7.	Q4F6X6_BUBBU		*Bubalus bubalis* (89462)
Peptide (K)FESNFNTQATNR [c]—marker of lysozyme C			
1.	LYSC_CHICK	Gal d 4.0101; Code 3294	*Gallus gallus* (9031)
2.	LYSC_COTJA		*Coturnix coturnix japonica* (93934)
3.	B8YK77_GALLA	Gal la 4; Code 9143	*Gallus lafayetii* (9032)
4.	B8YK75_GALSO	Gal so 4; Code 9144	*Gallus sonneratii* (9033)
5.	B8YK79_CHICK	Gal d 4; Code 362	*Gallus gallus* (9031)
6.	B8YJP1_CHICK	Gal d 4; Code 362	*Gallus gallus* (9031)
7.	B8YJN9_CHICK	Gal d 4; Code 362	*Gallus gallus* (9031)
8.	B8YJT7_CHICK	Gal d 4; Code 362	*Gallus gallus* (9031)

[a] Defined by the Latin name and NCBI taxonomic identifier [72,73] (in parentheses); [b] Fragment preceded by arginine residue in the sequences of all proteins annotated in the Table. The preceding residue (in parentheses) was included in the query sequence; [c] Fragment preceded by lysine residue in the sequences of all proteins annotated in the Table. The preceding residue (in parentheses) was included in the query sequence.

MASS SPECTROMETRY AS A TOOL FOR EXPERIMENTAL IDENTIFICATION OF COMMON SUBSEQUENCES

Experimental proteomics or peptidomics studies (relating to food and nutrition) require the identification of peptide sequences. Mass spectrometry is a popular identification tool. The significance of mass spectrometry in research into proteins and their fragments was emphasized and extensively discussed in several reviews [7,19,122,123,124,125,137,138,139,140]. Almost all peptide sequences listed in databases and discussed in bioinformatics studies, including in this review article, were identified by mass spectrometry. There are no special mass spectrometric techniques that support the search for common subsequences. The question "Is this subsequence common?" requires bioinformatics tools.

Several practical applications of mass spectrometry in food peptide analyses are presented in Table 5.

Table 5. Selected applications of mass spectrometry for the identification of food peptides.

Aim of the Experiment	Mass Spectrometry Technique	Separation Method	Reference
Identification of Angiotensin I-converting enzyme (ACE) inhibitory peptides released during simulated gastrointestinal digestion of salmon (*Salmo salar*) muscles	ESI-IT-MS/MS, SRM	RP-HPLC, low TFA concentration in mobile phase	[58]
Detection and quantitative determination of peptides that are markers of bovine (*Bos taurus*) casein and chicken (*Gallus gallus*) egg proteins	ESI-MS/MS, SRM	RP-HPLC	[126]
Detection and quantitative determination of peptides that are markers of mustard allergen Sin a 1 in foods	ESI-MS/MS, SRM	RP-HPLC	[128]
Identification of peptides from peanut (*Arachis hypogaea*) allergens	nano-ESI Q-TOF MS/MS	capillary RP-HPLC	[129]

Identification of peptides from soybean (*Glycine max*) allergens	MALDI-TOF and MALDI-TOF-TOF	RP-HPLC	[130]

Abbreviations used in Table 5: ESI: electrospray ionization; IT: ion trap; MALDI: Matrix-Assisted Laser Desorption Ionization; MS: mass spectrometry; MS/MS: tandem mass spectrometry; Nano-ESI: nanoelectrospray; Q-TOF: quadrupole-time-of-flight; RP-HPLC: reversed-phase high-performance liquid chromatography; SRM: selected reaction monitoring; TFA: trifluoroacetic acid; TOF: time of flight.

Mass spectrometry protocols used for peptide identification include peptide fragmentation to determine the complete or partial sequence. Various tandem mass (MS/MS) techniques are used for this purpose, including triple quadrupole or ion trap. Electrospray (ESI) is the most popular peptide ionization method, and matrix-assisted laser desorption (MALDI) is also commonly used. Reversed-phase high-performance liquid chromatography with a water/acetonitrile mobile phase is usually applied as a method for on-line separation in combination with mass spectrometry. Trifluoroacetic acid (TFA), used as the third mobile phase component causes quenching electrospray ionization. Protocols with low TFA concentration in the mobile phase are thus developed. Formic acid may be also used as a mobile phase component. Formic acid produces mass spectra of excellent quality, but the quality of the resulting chromatograms is low. MALDI-MS is applied off line with RP-HPLC without any restrictions concerning trifluoroacetic acid concentrations. The MALDI ionization technique is more resistant to the presence of inorganic salts than ESI. On-line capillary electrophoresis with mass spectrometry is also used in proteomics and peptidomics [139]. The selected reaction monitoring (SRM) method supports quantitative analyses of peptides. It involves measurements of peak intensity corresponding to selected fragment ion or ions from the peptide of interest [126,128]. The SRM method involving more fragment ions from a single peptide may be used for peptide identification [58].

FINAL REMARKS

Common amino acid subsequences occur in numerous proteins. This phenomenon should be taken into consideration in food science. The point for discussion is: Are common subsequences "friends" or "foes" of food scientists?

In biologically active peptides, the presence of common subsequences creates new opportunities for experimental design. Experiments involving bioactive peptides may be designed to search for new active compounds or known compounds in peptide mixtures [7]. The first strategy involves the separation of peptide mixtures into fractions, measurements of peptide activity, determination of amino acid sequences in active fraction compounds and confirmation of biological activity with the use of synthetic peptides. In the second strategy, peptides are identified be screening databases based on identified sequences as queries or by predicting peptide release. Examples of such experiments are summarized in Table 2. The search for novel active peptides and the existing databases may be significantly enhanced by high-throughput screening of peptide libraries. An experiment of the type has been recently described by Lan *et al.* [141]. They constructed a library of 367 dipeptides and screened it for compounds inhibiting dipeptidyl peptidase IV. The active peptides discovered via similar experiments may be found in food protein sequences and identified among products of their hydrolysis. Chanput *et al.* [142] predicted the biological activity of tripeptides, determined their location in protein sequences and simulated proteolysis. In regard to longer peptides which contain at least six amino acid residues, protein databases may be screened with the use of peptide sequences as queries to identify novel precursors and sources of bioactive peptides. Such protocols may support research into novel resources that can be potentially used in the production of functional foods.

The presence of common linear epitopes was recommended as a criterion of allergenicity and cross-reactivity prediction [89]. Recent recommendations to consider proteins as allergens *in silico* are more rigorous, and they account for the spatial structure of epitopes, even if they are sequential [79]. We can achieve consensus that protein database screening using epitope sequences as query may serve for construction of preliminary lists of potential allergens. They can be thus subjected to structure modeling *in silico* and finally to experiments aimed on fulfilling criteria summarized in a review published by Breiteneder and Chapman [84]. The structure and properties of the closest neighbor may be taken into account for the shortest sequences that are regarded as epitopes (pentapeptides). Common epitopes are particularly often found in conserved protein families such as tropomyosins [114]. This protein family is also characterized by conserved spatial structure. The presence of common linear epitopes is emphasized in databases such as the BIOPEP database of allergenic proteins and their epitopes [134] and Immune Epitope Database [105]. The latter contains a program that searches for common epitopes in a user-defined set of protein or peptide sequences. As previously noted [89,114] cross-reactivity between the allergens is also possible without identical epitopes or any identical fragments. This is a weak point of allergenicity and cross-reactivity predictions based on common subsequences.

In the context of the search for allergen markers, the presence of common subsequences may be considered as a weakness that obstructs the identification of a peptide signature of a unique protein. Despite the above, common subsequences create new opportunities for finding peptides that are markers of more than one protein. A single peptide may be a marker of a group of cross-reacting peptides. It would be interesting to use a single peptide as a marker of proteins with both known and unknown sequence according to the paradigm of comparative proteomics [18]. Chassaigne *et al.* [129] and Cucu *et al.* [130] recommend the use of peptide markers that overlap with epitopes, and such analyses would also create new opportunities. The same fragment can be used as a marker of a group of proteins identified by mass spectrometry and a marker of the same group of proteins detected by immunochemical methods.

The preparation and presentation of data relating to peptide sequences analyzed in single experiments or multiple precursors of single peptides may be fraught with problems. The process of updating major databases by insertion of hundreds of bioactive peptide sequences from a single protein chain or hundreds of precursors of single peptides may be difficult in real time. Publication of data in datasets such as the enclosed supplement creates useful opportunities for researchers. It is recommended that such datasets contain references or links to major databases (UniProt, Allergome, BIOPEP, IEDB, *etc.*) to provide as much information as possible in compact form. Data may be published in the form of supplements to articles or as separate datasets that are uploaded for instance on the websites of the authors' institutions. The use of chemical codes (InChI, InChiKey) for encoding peptides, in particular short peptides containing two or three amino acid residues, as recommended by Southan [111] may make finding of such datasets easier.

In this review we discussed the benefits for food scientist that follow from the use of common subsequences in the universal proteome and their relevance for food science. This phenomenon seems to be well known in biologically active peptides, where it has been discussed in the example of common epitopes, but it has not yet been analyzed in fragments that are protein markers.

Supplementary Information

Supplementary materials can be found at http://www.mdpi.com/1422-0067/16/09/20748/s1.

Acknowledgments

This publication was financed by the University of Warmia and Mazury in Olsztyn.

Conflict of Interest

The authors declare no conflict of interest.

REFERENCES

1. Kusalik, A.; Trost, B.; Bickis, M.; Fasano, C.; Capone, G.; Kanduc, D. Codon number shapes peptide redundancy in the universal proteome composition. *Peptides* 2009, *30*, 1940–1944. .
2. Shevchenko, A.; Valcu, C.-M.; Jungueira, M. Tools for exploiting proteomosphere. *J. Proteom.* 2009, *72*, 137–144. .
3. Levitt, M. Nature of the protein universe. *Proc. Natl. Acad. Sci. USA* 2009, *106*, 11079–11084. .
4. Liu, Z.P.; Wu, L.Y.; Wang, Y.; Zhang, X.S.; Chen, L. Bridging protein local structures and protein functions. *Amino Acids* 2008, *35*, 627–650. .
5. Lucchese, G.; Stufano, A.; Trost, B.; Kusalik, A.; Kanduc, D. Peptidology: Short amino acid modules in cell biology and immunology. *Amino Acids* 2007, *33*, 703–707. .

6. Schlimme, E.; Meisel, H. Bioactive peptides derived from milk proteins. Structural, physiological and analytical aspects. *Nahrung* 1995, *39*, 1–20. .
7. Minkiewicz, P.; Dziuba, J.; Darewicz, M.; Iwaniak, A.; Dziuba, M.; Nałęcz, D. Food peptidomics. *Food Technol. Biotechnol.* 2008, *46*, 1–10. .
8. Wang, L.; Wang, Q.; Qian, J.; Liang, Q.; Wang, Z.; Xu, J.; He, S.; Ma, H. Bioavailability and bioavailable forms of collagen after oral administration to rats. *J. Agric. Food Chem.* 2015, *63*, 3752–3756. .
9. ChEMBL Database. Available online: https://www.ebi.ac.uk/chembldb/ (accessed on 1 May 2015).
10. Bento, A.P.; Gaulton, A.; Hersey, A.; Bellis, L.J.; Chambers, J.; Davies, M.; Krüger, F.A.; Light, Y.; Mak, L.; McGlinchey, S.; *et al.* The ChEMBL bioactivity database: An update. *Nucleic Acids Res.* 2014, *42*, D1083–D1090. .
11. Lipinski, C.A.; Lombardo, F.; Dominy, B.W.; Feeney, P.J. Experimental and computational approaches to estimate solubility and permeability in drug discovery and development settings. *Adv. Drug Delivery Rev.* 1997, *23*, 3–25. .
12. Reymond, J.-L.; Awale, M. Exploring chemical space for drug discovery using the chemical universe database. *ACS Chem. Neurosci.* 2012, *3*, 649–657. .
13. PubChem Database. Available online: https://pubchem.ncbi.nlm.nih.gov/ (accessed on 1 May 2015).
14. Wang, Y.; Suzek, T.; Zhang, J.; Wang, J.; He, S.; Cheng, T.; Shoemaker, B.A.; Gindulyte, A.; Bryant, S.H. PubChem BioAssay: 2014 update. *Nucleic Acids Res.* 2014, *42*, D1075–D1082. .
15. ChemSpider Database. Available online: http://www.chemspider.com/Default.aspx (accessed on 1 May 2015).
16. Pence, H.E.; Williams, A. Chemspider: An online chemical information resource. *J. Chem. Educ.* 2010, *87*, 1123–1124. .
17. Karelin, A.A.; Blischenko, E.Y.; Ivanov, V.T. A novel system of peptidergic regulation. *FEBS Lett.* 1998, *428*, 7–12. .
18. Shevchenko, A.; Sunyaev, S.; Loboda, A.; Shevchenko, A.; Bork, P.; Ens, W.; Standing, K.G. Charting the proteomes of organism

with unsequenced genomes by MALDI-quadrupole time-of-flight mass spectrometry and BLAST homology searching. *Anal. Chem.* 2001, *73*, 1917–1926. .

19. Dallas, D.C.; Guerrero, A.; Parker, E.A.; Robinson, R.C.; Gan, J.; German, J.B.; Barile, D.; Lebrilla, C.B. Current peptidomics: Applications, purification, identification, quantification, and functional analysis. *Proteomics* 2015, *15*, 1026–1038. .
20. Catherman, A.D.; Skinner, O.S.; Kelleher, N.L. Top down proteomics: Facts and perspectives. *Biochem. Biophys. Res. Commun.* 2014, *445*, 683–693. .
21. UniProtKB Website. Available online: http://www.uniprot.org/help/uniprotkb (accessed on 1 May 2015).
22. UniProt Consortium. UniProt: A hub for protein information. *Nucleic Acids Res.* 2015, *43*, D204–D212. .
23. BIOPEP Database. Available online: http://www.uwm.edu.pl/biochemia/index.php/pl/biopep (accessed on 1 May 2015).
24. Minkiewicz, P.; Dziuba, J.; Iwaniak, A.; Dziuba, M.; Darewicz, M. BIOPEP database and other programs for processing bioactive peptide sequences. *J. AOAC Int.* 2008, *91*, 965–980. . [PubMed]
25. Matsufuji, H.; Matsui, T.; Seki, E.; Osajima, K.; Nakashima, M.; Osajima, Y. Angiotensin I-converting enzyme inhibitory peptides in an alkaline proteinase hydrolysate derived from sardine muscle. *Biosci. Biotechnol. Biochem.*1994, *58*, 2244–2245. .
26. Cheung, H.-S.; Wang, F.-L.; Ondetti, M.A.; Sabo, E.F.; Cushman, D.W. Binding of peptide substrates and inhibitors of angiotensin-converting enzyme. *J. Biol. Chem.* 1980, *255*, 401–407. . [PubMed]
27. Cushman, D.W. Angiotensin converting enzyme inhibitors: Evolution of a new class of antihypertensive drugs. In *Mechanisms of Action and Clinical Implications*; Horovitz, Z.P., Ed.; Urban & Schwarzenberg: Munich, Germany, 1981; p. 19. .
28. Meisel, H.; Walsh, D.J.; Murray, B.; FitzGerald, R.J. ACE inhibitory peptides. In *Nutraceutical Proteins and Peptides in Health and Disease*; Mine, Y., Shahidi, F., Eds.; In CRC Taylor & Francis Group: Boca Raton, FL, USA; London, UK; New York, NY, USA, 2006; pp. 269–315. .

29. Van Platerink, C.J.; Janssen, H.-G.M.; Haverkamp, J. Application of at-line two-dimensional liquid chromatography-mass spectrometry for identification of small hydrophilic angiotensin I-inhibiting peptides in milk hydrolysates. *Anal. Bioanal. Chem.* 2008, *391*, 299–307. .

30. Morifuji, M.; Koga, J.; Kawanaka, K.; Higuchi, M. Branched-chain amino acid-containing dipeptides, identified from whey protein hydrolysates, stimulate glucose uptake in L6 myotubes and isolated skeletal muscles. *J. Nutr. Sci. Vitaminol.* 2009, *55*, 81–86. .

31. Ringseis, R.; Motthes, B.; Lehmann, V.; Becker, K.; Schöps, R.; Ulbrich-Hofmann, R.; Eder, K. Peptides and hydrolysates from casein and soy protein modulate the release of vasoactive substances from human aortic endothelial cells. *Biochim. Biophys. Acta Gen. Subj.* 2005, *1721*, 89–97. .

32. Chen, H.-M.; Muramoto, K.; Yamauchi, F.; Nokihara, K. Antioxidant activity of designed peptides based on the antioxidant peptide isolated from digests of a soybean protein. *J. Agric. Food Chem.* 1996, *44*, 2619–2623. .

33. Mine, Y. Egg proteins and peptides in human health-chemistry, bioactivity and production. *Curr. Pharm. Des.*2007, *13*, 875–884. .

34. Saito, K.; Jin, D.-H.; Ogawa, T.; Muramoto, K.; Hatakeyama, E.; Yasuhara, T.; Nokihara, K. Antioxidative properties of tripeptide libraries prepared by the combinatorial chemistry. *J. Agric. Food Chem.* 2003, *51*, 3668–3674. .

35. Suetsuna, K. Separation and identification of antioxidant peptides from proteolytic digest of dried bonito. *Nippon Suisan Gakkaishi* 1999, *65*, 92–96, (In Japanese, Abstract in English). .

36. Huang, W.-Y.; Majumder, K.; Wu, J. Oxygen radical absorbance capacity of peptides from egg white protein ovotransferrin and their interactions with phytochemicals. *Food Chem.* 2010, *123*, 635–641. .

37. Sleigh, S.H.; Tame, J.R.H.; Dodson, E.J.; Wilkinson, A.J. Peptide binding in OppA, the crystal structure of the periplasmic oligopeptide-binding protein in the unliganded form and in complex with lysillysine. *Biochemistry*1997, *36*, 9747–9758. .

38. Bella, A.M.; Erickson, R.H., Jr.; Kim, Y.S. Rat intestinal brush

border membrane dipeptidyl-aminopeptidase IV: Kinetic properties and substrate specificities of the purified enzyme. *Arch. Biochem. Biophys.* 1982, *218*, 156–162. .

39. Li, H.; Aluko, R.E. Identification and inhibitory properties of multifunctional peptides from pea protein hydrolysate. *J. Agric. Food Chem.* 2010, *58*, 11471–11476. .
40. BRENDA Database. Available online: http://www.brenda-enzymes.org/ (accessed on 1 May 2015).
41. Chang, A.; Schomburg, I.; Placzek, P.; Jeske, L.; Ulbrich, M.; Xiao, M.; Sensen, C.W.; Schomburg, D. BRENDA in 2015: Exciting developments in its 25th year of existence. *Nucleic Acids Res.* 2015, *43*, D439–D446. .
42. Iwaniak, A.; Minkiewicz, P.; Darewicz, M. Food-originating ACE inhibitors, including antihypertensive peptides, as preventive food components in blood pressure reduction. *Compr. Rev. Food Sci. Food Saf.* 2014, *13*, 114–134. .
43. AHTPDB Database. Available online: http://crdd.osdd.net/raghava/ahtpdb/ (accessed on 1 May 2015).
44. Kumar, R.; Chaudhary, K.; Sharma, M.; Nagpal, G.; Chauhan, J.S.; Singh, S.; Gautam, A.; Raghava, G.P.S. AHTPDB: A comprehensive platform for analysis and presentation of antihypertensive peptides. *Nucleic Acids Res.* 2015, *43*, D956–D962. .
45. Azizi, M.; Ménard, J. Renin inhibitors and cardiovascular and renal protection: An endless quest? *Cardiovasc. Drugs Ther.* 2013, *27*, 145–153. .
46. Takahashi, S.; Gotoh, T.; Hata, K.; Tokiwano, T.; Yoshizawa, Y.; Hiwatashi, K.; Ogasawara, H.; Hori, K. Renin inhibitors in foodstuffs: Structure-function relationship. *J. Biol. Macromol.* 2014, *14*, 71–84. .
47. Udenigwe, C.C.; Mohan, A. Mechanisms of food protein-derived antihypertensive peptides other than ACE inhibition. *J. Funct. Foods* 2014, *8C*, 45–52. .
48. Juillerat-Jeanneret, L. Dipeptidyl peptidase IV and its inhibitors: Therapeutics for type 2 diabetes and what else. *J. Med. Chem.* 2014, *57*, 2197–2212. .

49. Ojeda, M.J.; Cereto-Massagué, A.; Valls, C.; Pujadas, G. DPP-IV, an important target for antidiabetic functional food design. In *Foodinformatics. Applications of Chemical Information to Food Chemistry*; Martinez-Mayorga, K., Medina-Franco, J.L., Eds.; Springer International Publishing AG: Cham, Switzerland, 2014; pp. 177–212. .
50. Fajardo, A.M.; Piazza, G.A.; Tinsley, H.N. The role of cyclic nucleotide signaling pathways in cancer: Targets for prevention and treatment. *Cancers* 2014, *6*, 436–458. .
51. Martinez, A.; Gil, C. cAMP-specific phosphodiesterase inhibitors: Promising drugs for inflammatory and neurological diseases. *Expert Opin. Ther. Pat.* 2014, *24*, 1311–1321. .
52. Miller, M.S. Phosphodiesterase inhibition in the treatment of autoimmune and inflammatory diseases: Current status and potential. *J. Recept. Ligand Channel Res.* 2015, *8*, 19–30. .
53. Freitas, A.C.; Andrade, J.C.; Silva, F.M.; Rocha-Santos, T.A.P.; Duarte, A.C.; Gomes, A.M. Antioxidative peptides: Trends and perspectives for future research. *Curr. Med. Chem.* 2013, *20*, 4575–4594. .
54. Ormsbee, M.J.; Bach, C.W.; Baur, D.A. Pre-exercise nutrition: The role of macronutrients, modified starches and supplements on metabolism and endurance performance. *Nutrients* 2014, *6*, 1782–1808. .
55. Dave, L.A.; Montoya, C.A.; Rutherfurd, S.M.; Moughan, P.J. Gastrointestinal endogenous proteins as a source of bioactive peptides—An *in silico* study. *PLoS ONE* 2014, *9*, e98922. .
56. Barba de la Rosa, A.P.; Barba Montoya, A.; Martínez-Cuevas, P.; Hernández-Ledesma, B.; León-Galván, M.F.; de León-Rodríguez, A.; González, C. Tryptic amaranth glutelin digests induce endothelial nitric oxide production through inhibition of ACE: Antihypertensive role of amaranth peptides. *Nitric Oxide* 2010, *23*, 106–111. .
57. Chatterjee, A.; Kanawjia, S.K.; Khetra, Y.; Saini, P. Discordance between *in silico* & *in vitro* analyses of ACE inhibitory & antioxidative peptides from mixed milk tryptic whey protein hydrolysate. *J. Food Sci. Technol.* 2015. .
58. Darewicz, M.; Borawska, J.; Vegarud, G.E.; Minkiewicz, P.;

Iwaniak, A. Angiotensin I-converting enzyme (ACE) inhibitory activity and ACE inhibitory peptides of salmon (*Salmo salar*) protein hydrolysates obtained by human and porcine gastrointestinal enzymes. *Int. J. Mol. Sci.* 2014, *15*, 14077–14101. .

59. Guinane, C.M.; Kent, R.M.; Norberg, S.; O'Connor, P.M.; Cotter, P.D.; Hill, C.; Fitzgerald, G.F.; Stanton, C.; Ross, R.P. Generation of the antimicrobial peptide caseicin A from casein by hydrolysis with thermolysin enzymes. *Int. Dairy J.* 2015, *49*, 1–7. .

60. Bauchart, C.; Morzel, M.; Chambon, C.; Mirand, P.P.; Reynès, C.; Buffière, C.; Rémond, D. Peptides reproducibly released by *in vivo* digestion of beef meat and trout flesh in pigs. *Br. J. Nutr.* 2007, *98*, 1187–1195. .

61. Wanasundara, J.P.D. Proteins of *Brassicaceae* oilseeds and their potential as a plant protein source. *Crit. Rev. Food Sci. Nutr.* 2011, *51*, 635–677. .

62. Minkiewicz, P.; Dziuba, J.; Michalska, J. Bovine meat proteins as potential precursors of biologically active peptides—A computational study based on the BIOPEP database. *Food Sci. Technol. Int.* 2011, *17*, 39–45. .

63. Carrera, M.; Cañas, B.; Gallardo, J.M. The sarcoplasmic fish proteome: Pathways, metabolic networks and potential bioactive peptides for nutritional inferences. *J. Proteom.* 2013, *78*, 211–220. .

64. Cavazos, A.; González de Mejía, E. Identification of bioactive peptides from cereal storage proteins and their potential role in prevention of chronic diseases. *Compr. Rev. Food Sci. Food Saf.* 2013, *12*, 364–380. .

65. Montoya-Rodríguez, A.; Gómez-Favela, M.A.; Reyes-Moreno, C.; González de Mejía, E.; Milán-Carrillo, J. Identification of bioactive peptide sequences from amaranth (*Amaranthus hypochondriacus*) seed proteins and their potential role in the prevention of chronic diseases. *Compr. Rev. Food Sci. Food Saf.* 2015, *14*, 139–158. .

66. Kohama, Y.; Nagase, Y.; Oka, H.; Nakagawa, T.; Teramoto, T.; Murayama, N.; Tsujibo, H.; Inamori, Y.; Mimura, T. Production of angiotensin-converting enzyme inhibitors from baker's yeast glyceraldehyde-3-phosphate dehydrogenase. *J. Pharmacobiodyn.* 1990, *13*, 766–771. .

67. PepBank Database. Available online: http://pepbank.mgh.harvard.edu/ (accessed on 1 May 2015).
68. Duchrow, T.; Shtatland, T.; Guettler, D.; Pivovarov, M.; Kramer, S.; Weissleder, R. Enhancing navigation in biomedical databases by community voting and database-driven text classification. *BMC Bioinform.* 2009, *10*, 317. .
69. BLAST Program. Available online: http://www.ebi.ac.uk/Tools/sss/wublast/ (accessed on 1 May 2015).
70. Altschul, S.F.; Madden, T.L.; Schäffer, A.A.; Zhang, J.; Zhang, Z.; Miller, W.; Lipman, D.J. Gapped BLAST and PSI-BLAST: A new generation of protein database search programs. *Nucleic Acids Res.* 1997, *25*, 3389–3402. .
71. Minkiewicz, P.; Bucholska, J.; Darewicz, M.; Borawska, J. Epitopic hexapeptide sequences from Baltic cod parvalbumin beta (allergen Gad c 1) are common in the universal proteome. *Peptides* 2012, *38*, 105–109. .
72. NCBI Taxonomy Website. Available online: http://www.ncbi.nlm.nih.gov/taxonomy (accessed on 1 May 2015).
73. Federhen, S. Type material in the NCBI Taxonomy Database. *Nucleic Acids Res.* 2015, *43*, D1086–D1098. .
74. InterPro Website. Available online: http://www.ebi.ac.uk/interpro/ (accessed on 1 May 2015).
75. Mitchell, A.; Chang, H.-Y.; Daugherty, L.; Fraser, M.; Hunter, S.; Lopez, R.; McAnulla, C.; McMenamin, C.; Gift, N.; Pesseat, S.; *et al.* The InterPro protein families database: The classification resource after 15 years. *Nucleic Acids Res.* 2015, *43*, D213–D221. .
76. Zambrowicz, A.; Eckert, E.; Pokora, M.; Bobak, Ł.; Dąbrowska, A.; Szołtysik, M.; Trziszka, T.; Chrzanowska, J. Antioxidant and antidiabetic activities of peptides isolated from a hydrolysate of an egg-yolk protein by-product prepared with a proteinase from Asian pumpkin (*Cucurbita ficifolia*). *RSC Adv.* 2015, *5*, 10460–10467. .
77. Eckert, E.; Zambrowicz, A.; Pokora, M.; Polanowski, A.; Chrzanowska, J.; Szołtysik, M.; Dąbrowska, A.; Różański, H.; Trziszka, T. Biologically active peptides derived from egg proteins. *World's Poultry Sci. J.* 2013, *69*, 375–386. .

78. Albrecht, M.; Kühne, Y.; Ballmer-Weber, B.K.; Becker, W.-M.; Holzhauser, T.; Lauer, I.; Reuter, A.; Randow, S.; Falk, S.; Wangorsch, A.; *et al.* Relevance of IgE binding to short peptides for the allergenic activity of food allergens. *J. Allergy Clin. Immunol.* 2009, *124*, 328–336. .

79. Dall'Antonia, F.; Pavkov-Keller, T.; Zangger, K.; Keller, W. Structure of allergens and structure based epitope predictions. *Methods* 2014, *66*, 3–21. .

80. Ferreira, F.; Hawranek, T.; Gruber, P.; Wopfner, N.; Mari, A. Allergic cross-reactivity: From gene to the clinic. *Allergy* 2004, *59*, 243–267. .

81. Bartuzi, Z. The molecular traits of food allergens (Molekularne cechy alergenów pokarmowych). *Post. Dermatol. Alergol.* 2009, *26*, 310–312. (In Polish). .

82. Habchi, J.; Tompa, P.; Longhi, S.; Uversky, V.N. Introducing protein intrinsic disorder. *Chem. Rev.* 2014, *114*, 6561–6588. .

83. Van der Lee, R.; Buljan, M.; Lang, B.; Weatheritt, R.J.; Daughdrill, G.W.; Dunker, A.K.; Fuxreiter, M.; Gough, J.; Gsponer, J.; Jones, D.T.; *et al.* Classification of intrinsically disordered regions and proteins. *Chem. Rev.* 2014, *114*, 6589–6631. .

84. Breiteneder, H.; Chapman, M.D. Allergen nomenclature. In *Allergens and Allergen Immunotherapy*, 5th ed.; Lockey, R.F., Ledford, D.K., Eds.; CRC Press: Boca Raton, FL, USA, 2014; pp. 37–49. .

85. Bindslev-Jensen, C.; Sten, E.; Earl, L.K.; Crevel, R.W.R.; Bindslev-Jensen, U.; Hansen, T.K.; Skov, P.S.; Poulsen, L.K. Assessment of the potential allergenicity of ice structuring protein type III HPLC 12 using the FAO/WHO 2001 decision tree for novel foods. *Food Chem. Toxicol.* 2003, *41*, 81–87. .

86. Goodman, R.E. Practical and predictive bioinformatic methods for the identification of potentially cross-reactive protein matches. *Mol. Nutr. Food Res.* 2006, *50*, 655–660. .

87. Schein, C.H.; Ivanciuc, O.; Braun, W. Bioinformatic approaches to classifying allergens and predicting cross-reactivity. *Immunol. Allergy Clin. N. Am.* 2007, *27*, 1–27. .

88. Kleter, G.A.; Peijnenburg, A.A.C.M. Screening of transgenic

proteins expressed in transgenic food crops for the presence of short amino acid sequences identical to potential, IgE-binding linear epitopes of allergens. *BMC Struct. Biol.* 2002, *2*, 8. .

89. Minkiewicz, P.; Dziuba, J.; Gładkowska-Balewicz, I. Update of the list of allergenic proteins from milk based on local amino acid sequence identity with known epitopes from bovine milk proteins—A short report. *Pol. J. Food Nutr. Sci.* 2011, *61*, 153–158. .
90. Dessailly, B.H.; Redfern, O.C.; Cuff, A.; Orengo, C.A. Exploiting structural classifications for function prediction: Towards a domain grammar for protein function. *Curr. Opin. Struct. Biol.* 2009, *19*, 349–356. .
91. Pfam Database. Available online: http://pfam.xfam.org/ (accessed on 1 May 2015).
92. Finn, R.D.; Bateman, A.; Clements, J.; Coggill, P.; Eberhardt, R.Y.; Eddy, S.R.; Heger, A.; Hetherington, K.; Holm, L.; Mistry, J.; *et al.* The Pfam protein families database. *Nucleic Acids Res.* 2014, *42*, D222–D230. .
93. AllFam Database. Available online: http://www.meduniwien.ac.at/allergens/allfam/ (accessed on 1 June 2015).
94. Radauer, C.; Bublin, M.; Wagner, S.; Mari, A.; Breiteneder, H. Allergens are distributed into few protein families and possess a restricted number of biochemical functions. *J. Allergy Clin. Immunol.* 2008, *121*, 847–852. .
95. Kanduc, D. Pentapeptides as minimal functional units in cell biology and immunology. *Curr. Protein Pept. Sci.*2013, *14*, 111–120. .
96. Kanduc, D. Correlating low-similarity peptide sequences and allergenic epitopes. *Curr. Pharm. Des.* 2008, *14*, 289–295. .
97. Kanduc, D. Homology, similarity, and identity in peptide epitope immunodefinition. *J. Pept. Sci.* 2012, *18*, 487–494. .
98. Tachyon Program. Available online: http://tachyon.bii.a-star.edu.sg/index.action (accessed on 1 May 2015).
99. Tan, J.; Kuchibhatla, D.; Sirota, F.L.; Sherman, W.A.; Gattermayer, T.; Kwoh, C.Y.; Eisenhaber, F.; Schneider, G.; Maurer-Stroh, S. Tachyon search speeds up retrieval of similar sequences by several orders of magnitude. *Bioinformatics* 2012, *28*, 1645–1646. .

100. MimicMe Program. Available online: http://mimicme.uwaterloo.ca/ (accessed on 1 May 2015).
101. Petrenko, P.; Doxey, A.C. MimicMe: A web server for prediction and analysis of host-like proteins in microbial pathogens. *Bioinformatics* 2015, *31*, 590–592. .
102. Matsuo, H.; Morita, E.; Tatham, A.S.; Morimoto, K.; Horikawa, T.; Osuna, H.; Ikezawa, Z.; Kaneko, S.; Kohno, K.; Dekio, S. Identification of the IgE-binding epitope in omega-5 gliadin, a major wheat allergen in wheat-dependent exercise-induced anaphylaxis. *J. Biol. Chem.* 2004, *279*, 12135–12140. .
103. Abe, R.; Shimizu, S.; Yasuda, K.; Sugai, M.; Okada, Y.; Chiba, K.; Akao, M.; Kumagai, H.; Kumagai, H. Evaluation of reduced allergenicity of deamidated gliadin in a mouse model of wheat-gliadin allergy using an antibody prepared by a peptide containing three epitopes. *J. Agric. Food Chem.* 2014, *62*, 2845–2852. .
104. Immune Epitope Database. Available online: http://www.iedb.org/ (accessed on 1 May 2015).
105. Vita, R.; Overton, J.A.; Greenbaum, J.A.; Ponomarenko, J.; Clark, J.D.; Cantrell, J.R.; Wheeler, D.K.; Gabbard, J.L.; Hix, D.; Sette, A.; *et al.* The immune epitope database (IEDB) 3.0. *Nucleic Acids Res.* 2015, *43*, D405–D412. .
106. Weininger, D. SMILES, a chemical language and information system. 1. Introduction to methodology and encoding rules. *J. Chem. Inf. Comput. Sci.* 1988, *28*, 31–36. .
107. Heller, S.; McNaught, A.; Stein, S.; Tchekhovskoi, D.; Pletnev, I. InChI—The worldwide chemical structure identifier standard. *J. Cheminform.* 2013, *5*, 7. .
108. Iwaniak, A.; Minkiewicz, P.; Darewicz, M.; Protasiewicz, M.; Mogut, D. Chemometrics and cheminformatics in the analysis of biologically active peptides from food sources. *J. Funct. Foods* 2015, *16*, 334–351. .
109. MetaComBio Website. Available online: http://www.uwm.edu.pl/metachemibio/index.php/about-metacombio(accessed on 1 July 2015).
110. Minkiewicz, P.; Iwaniak, A.; Darewicz, M. Using internet databases for food science organic chemistry students to discover chemical

compound information. *J. Chem. Educ.* 2015, *92*, 874–876. .

111. Southan, C. InChI in the wild: An assessment of InChIKey searching in Google. *J. Cheminform.* 2013, *5*, 10. .

112. Open Babel Program. Available online: http://openbabel.org/wiki/Main_Page (accessed on 1 May 2015).

113. O'Boyle, N.M.; Banck, M.; James, C.A.; Morley, C.; Vandermeersch, T.; Hutchison, G.R. Open Babel: An open chemical toolbox. *J. Cheminform.* 2011, *3*, 33. .

114. Minkiewicz, P.; Sokołowska, J.; Darewicz, M. The occurrence of sequences identical with epitopes from the allergen Pen a 1.0102 among food and non-food proteins. *Pol. J. Food Nutr. Sci.* 2015, *65*, 21–29. .

115. Custovic, A. To what extent is allergen exposure a risk factor for the development of allergic disease? *Clin. Exp. Allergy* 2015, *45*, 54–62. .

116. Allergome Database. Available online: http://www.allergome.org/ (accessed on 1 May 2015).

117. Mari, A.; Rasi, C.; Palazzo, P.; Scala, E. Allergen databases: Current status and perspectives. *Curr. Allergy Asthma Rep.* 2009, *9*, 376–383. .

118. Darewicz, M.; Dziuba, J.; Minkiewicz, P. Computational characterisation and identification of peptides for *in silico*detection of potentially celiac-toxic proteins. *Food Sci. Technol. Int.* 2007, *13*, 125–133. .

119. Vojdani, A.; Tarash, I. Cross-reaction between gliadin and different food and tissue antigens. *Food Nutr. Sci.* 2013, *4*, 20–32. .

120. Yapar, N. Epidemiology and risk factors for invasive candidiasis. *Ther. Clin. Risk Manag.* 2014, *10*, 95–105. .

121. Mirza, Z.K.; Sastri, B.; Lin, J.J.C.; Amenta, P.S.; Das, K.M. Autoimmunity against human tropomyosin isoforms in ulcerative colitis—Localization of specific human tropomyosin isoforms in the intestine and extraintestinal organs. *Inflamm. Bowel Dis.* 2006, *12*, 1036–1043. .

122. Fæste, C.K.; Rønning, H.T.; Christians, U.; Granum, P.E. Liquid chromatography and mass spectrometry in food allergen detection. *J.*

Food Protect. 2011, *74*, 316–345. .

123. Cunsolo, V.; Muccilli, V.; Saletti, R.; Foti, S. Mass spectrometry in food proteomics: A tutorial. *J. Mass Spectrom.*2014, *49*, 768–784. .
124. Koeberl, M.; Clarke, D.; Lopata, A.L. Next generation of food allergen quantification using mass spectrometric systems. *J. Proteome Res.* 2014, *13*, 3499–3509. .
125. Tedesco, S.; Mullen, W.; Cristobal, S. High-throughput proteomics: A new tool for quality and safety in fishery products. *Curr. Protein Pept. Sci.* 2014, *15*, 118–133. .
126. Pilolli, R.; de Angelis, E.; Godula, M.; Visconti, A.; Monaci, L. Orbitrap™ monostage MS *versus* hybrid linear ion trap MS: Application to multi-allergen screening in wine. *J. Mass Spectrom.* 2014, *49*, 1254–1263. .
127. Gomaa, A.; Boye, J. Simultaneous detection of multi-allergens in an incurred food matrix using ELISA, multiplex flow cytometry and liquid chromatography mass spectrometry (LC–MS). *Food Chem.* 2015, *175*, 585–592. .
128. Posada-Ayala, M.; Alvarez-Llamas, G.; Maroto, A.S.; Maes, X.; Muñoz-Garcia, E.; Villalba, M.; Rodríguez, R.; Perez-Gordo, M.; Vivanco, F.; Pastor-Vargas, C.; *et al.* Novel liquid chromatography-mass spectrometry method for sensitive determination of the mustard allergen Sin a 1 in food. *Food Chem.* 2015, *183*, 58–63. .
129. Chassaigne, H.; Nørgaard, J.V.; van Hengel, A.J. Proteomics-based approach to detect and identify major allergens in processed peanuts by capillary LC-Q-TOF (MS/MS). *J. Agric. Food Chem.* 2007, *55*, 4461–4473. .
130. Cucu, T.; de Meulenaer, B.; Devreese, B. MALDI based identification of soybean protein markers—Possible analytical targets for allergen detection in processed foods. *Peptides* 2012, *33*, 187–196. .
131. Johnson, P.E.; Baumgartner, S.; Aldick, T.; Bessant, C.; Giosafatto, V.; Heick, J.; Mamone, G.; O'Connor, G.; Poms, R.; Popping, B.; *et al.* Current perspectives and recommendations for the development of mass spectrometry methods for the determination of allergens in foods. *J. AOAC Int.* 2011, *94*, 1026–1033. . [PubMed]
132. NCBI Proteins Database. Available online: http://www.ncbi.nlm.nih.gov/protein (accessed on 1 May 2015).

133. NCBI Resource Coordinators. Database resources of the National Center for Biotechnology Information. *Nucleic Acids Res.* 2015, *43*, D6–D17. .

134. Dziuba, M.; Minkiewicz, P.; Dąbek, M. Peptides, specific proteolysis products as molecular markers of allergenic proteins—*In silico* studies. *Acta Sci. Polon. Technol. Aliment.* 2013, *12*, 101–112. .

135. Belluco, S.; Losasso, C.; Maggioletti, M.; Alonzi, C.C.; Paoletti, M.G.; Ricci, A. Edible insects in a food safety and nutritional perspective: A critical review. *Compr. Rev. Food Sci. Food Saf.* 2013, *12*, 296–313. .

136. Mlcek, J.; Rop, O.; Borkovcova, M.; Bednarova, M. A comprehensive look at the possibilities of edible insects as food in Europe—A review. *Pol. J. Food Nutr. Sci.* 2014, *64*, 147–157. .

137. Carrasco-Castilla, J.; Hernández-Álvarez, A.J.; Jiménez-Martínez, C.; Gutiérrez-López, G.F.; Dávila-Ortiz, G. Use of proteomics and peptidomics methods in food bioactive peptide science and engineering. *Food Eng. Rev.* 2012, *4*, 224–243. .

138. Sauer, S.; Luge, T. Nutriproteomics: Facts, concepts, and perspectives. *Proteomics* 2015, *15*, 997–1013. .

139. Ibáñez, C.; Simó, C.; García-Cañas, V.; Cifuentes, A.; Castro-Puyana, M. Metabolomics, peptidomics and proteomics applications of capillary electrophoresis-mass spectrometry in foodomics: A review. *Anal. Chim. Acta*2013, *802*, 1–13. .

140. Sánchez-Rivera, L.; Martínez-Maqueda, D.; Cruz-Huerta, E.; Miralles, B.; Recio, I. Peptidomics for discovery, bioavailability and monitoring of dairy bioactive peptides. *Food Res. Int.* 2014, *63*, 170–181. .

141. Lan, V.T.T.; Ito, K.; Ohno, M.; Motoyama, T.; Ito, S.; Kawarasaki, Y. Analyzing a dipeptide library to identify human dipeptidyl peptidase IV inhibitor. *Food Chem.* 2015, *175*, 66–73. .

142. Chanput, W.; Nakai, S.; Theerakulkait, C. Introduction of new computer softwares for classification and prediction purposes of bioactive peptides: Case study in antioxidative tripeptides. *Int. J. Food Prop.* 2010, *13*, 947–959. .

CHAPTER

7

IDENTIFICATION OF MILK COMPONENT IN ANCIENT FOOD RESIDUE BY PROTEOMICS

Chuan Hong[1,2], Hongen Jiang[1,2], Enguo Lu¨ [3] , Yunfei Wu[4] , Lihai Guo[5] , Yongming Xie[5] , Changsui Wang[1,2], Yimin Yang[1,2]

[1] Laboratory of Human Evolution, Institute of Vertebrate Paleontology and Paleoanthropology, CAS, Beijing, China,

[2]Department of Scientific History and Archaeometry, Graduate University, Beijing, China,

[3] Xinjiang Institute of Archaeology, Xinjiang Uygur Autonomous Region, Urumchi, China,

[4] College of Life Science, Graduate University, CAS, Beijing, China,

[5] Asia Pacific Application Support Center, AB SCIEX, Shanghai, China

ABSTRACT

Background

Proteomic approaches based on mass spectrometry have been recently used in archaeological and art researches, generating promising results for protein identification. Little information is known about eastward spread and eastern limits of prehistoric milking in eastern Eurasia.

Methodology/Principal Finding

In this paper, an ancient visible food remain from Subeixi Cemeteries (cal. 500 to 300 years BC) of the Turpan Basin in Xinjiang, China, preliminarily determined containing 0.432 mg/kg cattle casein with ELISA, was analyzed by using an improved method based on liquid chromatography (LC) coupled with MALDI-TOF/TOF-MS to further identify protein origin. The specific sequence of bovine casein and the homology sequence of goat/sheep casein were identified.

Conclusions/Significance

The existence of milk component in ancient food implies goat/sheep and cattle milking in ancient Subeixi region, the furthest eastern location of prehistoric milking in the Old World up to date. It is envisioned that this work provides a new approach for ancient residue analysis and other archaeometry field.

INTRODUCTION

Along with the development of civilization, domesticated animals, such as cattle, goat, sheep and horse, became more and more important in farming practices. Series of reports have indicated that these animals were initially tamed for the production of meat, hide and bone, and later for a more diversified exploitation of their lifetime products and uses, including milk, wool, as well as riding, transportation or tillage [1]; and this process was entitled "Secondary Products Revolution (SPR)" by Andrew Sherratt [2], [3]. Among these secondary uses of livestock, dairying played a significant role in the agrarian civilization process of Europe, and also was crucial in the emergence and evolution of the nomadic culture of the Eurasian grasslands, due to the special nutritional value of dairy products [4], [5]. In consideration of this, much research focused on ancient dairying in order to provide more evidence about the progress and spread of animal domestication [6], [7].

The major approaches applied in research on dairying are as follows: (1) Lipid analysis, *e.g.*, the analysis of lipids preserved in pottery, which indicated that the exploitation of milk could be shifted to an earlier date, about in 7,000 BC [8]; (2) Immunological method, *e.g.*, Enzyme Linked Immunosorbent Assay (ELISA), applied in the identification of the presence of casein preserved in artisan materials [9]; (3) Zooarchaeological method, especially the harvest profile (age-at-death) of unearthed animal skeletons, *e.g.*, the quantity and the distribution of the age of slaughtered cattle unearthed in the Harappa relics in Pakistan showed that the mature cow had been raised for a longer time, which might have been used for the production of milk [10]; (4) The distribution of lactase persistence genes, *e.g.*, the ability of lactase persistence possibly emerged from people who had lived in the west of the Urals and the north of the Caucasus [11]. (5) Determining the weaning time of calves through the detection of nitrogen isotope [12].

However, in spite of the fact that cattle, sheep and goat can offer products with similar nutritional properties, the economic, social and cultural implications of using these animals for these purposes were very different. Determining the species of animal exploited for milk is also vital to understanding dairying [13]. Furthermore, zooarchaeological harvest profiles are the most useful method to observe general shifts of animal exploitation, but not to identify specific products [14]. Thus, the identification of the animal species related to milk residues has become a hotspot in international academic research. Proteomic methods have been recently applied in archeology and arts, thus opening a new window for residue analysis [15]–[23]. Proteomic approaches based on mass spectrometry have been particularly used to clearly discriminate sheep and goat from bovine alpha casein used in mural painting [18]. Milk additives (bovine milk, curd, and whey) in historical building materials have been identified via proteomic analysis and several sequences have been confirmed "*de novo*" by mass spectrometry [19]. These works imply that the proteomic method could be subsequently used to identify milk component in food residue and determine the precise origin of milk, which would help better understand prehistoric animal exploitation and subsistence shift according to SPR theory.

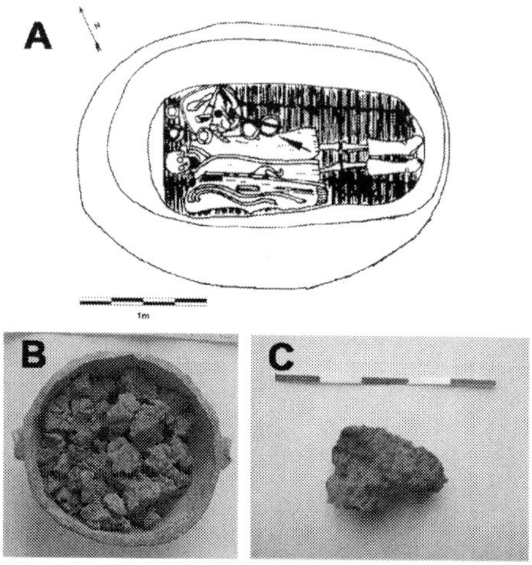

Figure 1. Samples chosen for archaeological analysis.

A) The line drawing of tomb where ancient residue existed. The bottom layer of III M27, showing the black food residues (arrow) in the pottery bowl (M27:9, adapted from XIA & MT, 2002); B) black residues in the pottery bowl; C) residues chosen for analysis. Scale bar=5 cm (photo b supplied by Prof. Enguo Lü).

The spread of domesticated cattle, sheep, and goats into ancient China came at least 5000, 4000, and 3700 years ago, respectively [24]; but there is little information about the spread of secondary products exploitation in eastern Asia, which has been largely ignored since SPR theory was suggested [25]. In the present study, a monoclonal ELISA technique has been used for the preliminary identification of cattle casein contained in an ancient black food residue. The positive result of ELISA analysis showed that the cattle casein concentration in the ancient food was 0.432 mg/kg, a little higher than the detection limit (0.12 mg/kg) and lower than the limit of quantification (0.5 mg/kg) [26]. Given this lower cattle casein concentration, in order to claim whether ELISA result is reliable and obtain more information on dairying, an improved analysis strategy was adopted by using LC-MALDI-TOF/TOF-MS.

MATERIALS AND METHODS

Archaeological Samples

Ancient black food residues were taken from a pottery bowl (coded as No. M27: 9) in the tomb M27 (see Figure 1) at the Subeixi site located in the Turpan Basin, Shanshan County, Xinjiang Uighur Autonomous Region, China. The Turpan District is famous for its representative continental desert climate, with annual precipitation of 25.2 mm and an evaporation rate as high as 2500 mm. Site description and archaeological background were previously introduced in detail in other sources [27]–[29]. In brief, the food residues obtained are believed to have existed in cal. 500-300 BC. The tomb M27 in No.3 cemetery is an earth-pit burial consisting of three layers. The skeletons of an adult male, a female, and an infant were found in the upper layer; the skeleton of another adult female was buried in the middle layer; an elderly man was interred on the lower layer, and the black food residue which was analyzed was found in a pottery bowl to the left side of his waist.

Table 1. Results of MS/MS by LC-MALDI-TOF.

No.	Sequence	Protein	Species
A	FVVAPFPEVFR	α_{S1}-casein	sheep and goat
B	YLGYLEQLLR	α_{S1}-casein	sheep, and bovine
C	FFVAPFPEVFGK	α_{S1}-casein	bovine
D	DMPIQAFLLYQEPVLGPVR	β-casein	bovine, sheep, goat, water buffalo

doi:10.1371/journal.pone.0037053.t001

Protein extraction

The black food residue was firstly ground by hand with an agate mortar. Next, 0.1 g of ancient powder was extracted with 1 mL PBS at room temperature for 20 min. The supernatant was then collected and placed into a centrifuge vial (1.5 mL) without protein pollution, and clarified by centrifuging (2500 g for 10 min). Finally, protein extracts were diluted 10 times for later analysis.

Tryptic digestion

1 mL of protein extract was diluted with the addition of 20 μL of 50 mM DTT at 56°C for 30 min. The sample was then allowed to re-equilibrate to room temperature, alkylated with the addition of 20 μL of 100 mM iodoacetamide, and then incubated for 1 hour at 37°C under slight vortex. Trypsin was subsequently added in a 1:30 (trypsin:sample) ratio twice: once at the start and once after 2 hours. Finally, digestion was completed overnight at 37°C.

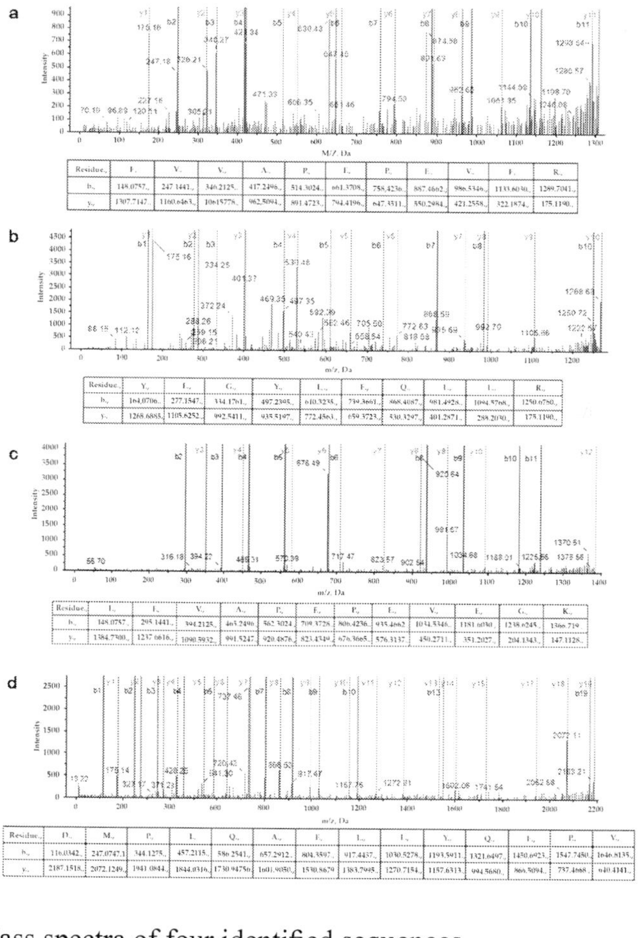

a

Residue	F	V	V	A	P	F	P	E	V	F	R
b_i	148.0757	247.1441	346.2125	417.2496	514.3024	661.3708	758,4236	887.4662	986.5346	1133.6030	1289.7041
y_i	1307.7147	1160.6463	10615778	962,5094	891.4723	794.4196	647.3511	550.2984	421.2558	322.1874	175.1190

b

Residue	Y	L	G	Y	L	E	Q	L	L	R
b_i	164,0706	277.1547	334.1761	497.2395	610.3235	739.3661	868.4087	981.4928	1094.5768	1250.6750
y_i	1268.6885	1105.6252	992.5411	935.5197	772.4563	659.3723	530.3297	401.2871	288.2030	175.1190

c

Residue	L	F	V	A	P	F	P	E	V	F	G	K
b_i	148.0757	295.1441	394.2125	465.2496	562.3024	709.3728	806.4236	935.4662	1034.5346	1181.6030	1238.6245	1366.719
y_i	1384.7300	1237.6616	1090.5932	991.5247	920.4876	823.4349	676.3665	576.3137	450.2711	351.2027	204.1343	147.1128

d

Residue	D	M	P	L	Q	A	F	L	L	Y	Q	E	P	V
b_i	116.0342	247.0747	344.1275	457.2115	586.2541	657.2912	804.3597	917.4437	1030.5278	1193.5911	1321.6497	1450.6923	1547.7450	1646.8135
y_i	2187.1518	2072.1249	1943.0844	1844.0316	1730.94756	1601.9050	1530.8679	1383.7995	1270.7154	1157.6313	994.5680	866.5094	737.4668	640.4141

Figure 2. Mass spectra of four identified sequences.

```
                            10        20        30        40        50        60        70        80
                    ....|....|....|....|....|....|....|....|....|....|....|....|....|....|....|....|...
bovine              MKLLILTCLVAVALARPKHPIKHQGLPQEVLNENLLRFFVAPFPEVFGKEKVNELS------KDIGSESTEDQAMEDIKQMEA
goat                .....................N....SP..P.......V........R..NI....------...............A...K.
sheep               ..........................SP..........V........R..NI....------.......I.......A...K.
water_buffalo       ..................Q.........G...........................------T....................
horse               ..................L.HRQPEII.NEQDSREKVLKERK..S-.AL.YI...NRQRELL.EKQKDEHKEYLI..PE.Q.S
Clustal Consensus   ****************** * .:   :     :..   :    **. *  * :***.      .:  .:. :: :** :* ::

                           110       120       130       140       150       160       170       180
                    ....|....|....|....|....|....|....|....|....|....|....|....|....|....|....|....|...
bovine              HIQKEDVPSERYLGYLEQLLRLKKYKVPQLEIVPNSAEERLHSMKEGIHAQQKEPMIGVNQELAYFYPELFRQFYQLDAYPSG
goat                Y....................K...N........K....Q.......NP.H..Q...A..........Q..............
sheep               Y........................N........K....Q.......NP.H..Q...A..........Q..............
water_buffalo       .........................N.........L...Q............................Q..............
horse               R.PR..ML---.QHT....R..S..NQL..QAIH--.Q.Q.IR...H---S.RK..RV....Q....L.P.QPS....V..YA
Clustal Consensus    * :**:     *   **** :*.**:  **: :    *:*:* ***.     *::**  **** **** : *:   ****.** .

                           210       220
                    ....|....|....|....|.
bovine              SDIPNPIGSENSEKT-TMPLW
goat                ............G.A-A....
sheep               ............G.I-.....
water_buffalo       ............G..-.....
horse               H.TAKL.A.......DII.E.
Clustal Consensus    * .: *.**** *   :* *
```

Figure 3. Multiple sequence alignment.

Multiple sequence alignment of alpha S1 casein from bovine (Accession: ACG63494.1 GI: 195973860), goat (Accession: CAA51022.1 GI: 311943), sheep (Accession: ACJ46472.1 GI: 213391435), water buffalo (Accession: AAX49507.1 GI: 61658244) and horse (Accession: NP_001075352.1 GI: 126352302). The detected peptides by LC-MALDI-TOF-MS/MS are light grey. "*" identical residues, ":" conserved residues, and "." semi-conserved residues.

Reversed-Phase Separation of Peptides

Peptides separation was performed on Tempo™ LC-MALDI spotting system (AB SCIEX, Foster City, CA). 10 μL of digested samples was injected and captured on a 0.3×10 mm ProteCol™ trap column (3 μm, C18, SGE Analytical Science, AU) and then eluted and separated from a trap column hooked in series with a 0.2×150 mm analytical column (3 μm, C18, Michrom Bioresource, Auburn, CA). Peptides were resolved by dual solvent gradient elution (A: 2% ACN, 0.1% TFA; B: 98% ACN, 0.1% TFA) of 5–35% B in 30 min, 35–90% B in 5 min at a flow rate of 2 μL/min. The column effluent was monitored at 214 nm using a 3-nL UV flow cell, and after UV detection, was mixed with the MALDI matrix in a 1:1 ratio through external syringe pump and spotted onto a MALDI plate in a 44×28 array at 8 s intervals. The matrix solution was 5 mg/mL α-cyano-4 hydroxycinnamic acid dissolved in 70:30 ACN-H_2O.

Table 2. The concentration of standard bovine casein and sample's OD absorbance.

No.	Solution added	Concentration of bovine casein (mg/kg)	OD value(450 nm)
1	Dilution buffer	0	0.073
2	bovine casein Standard 1	0.5	0.218
3	bovine casein Standard 2	1.5	0.366
4	bovine casein Standard 3	4.5	0.735
5	bovine casein Standard 4	13.5	1.507
6	ancient sample	0.432	0.180

doi:10.1371/journal.pone.0037053.t002

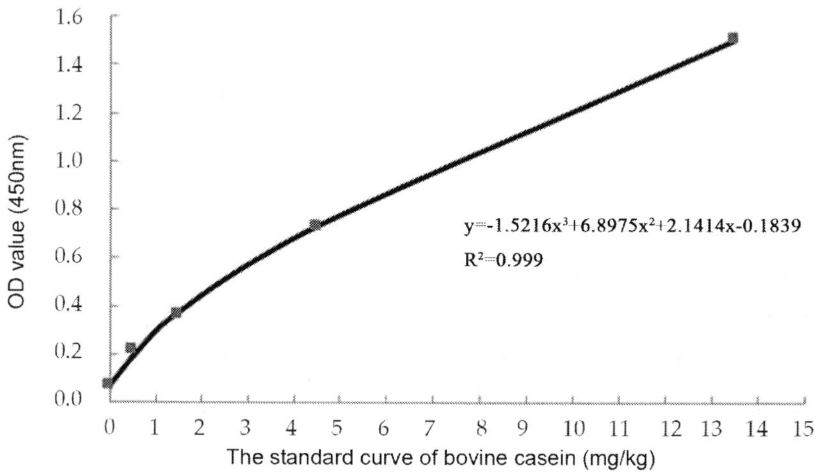

Figure 4. The standard curve of bovine casein.

According to the ELISA procedures and the manual of the kit (RIDASCREEN® FAST Casein (Art. No.: R4602), R-Biopharm AG, Darmstadt, Germany), about 0.1 g ancient sample was taken to do the test. The standard samples and the ancient sample's OD values can be seen in Table 2. Based on the data illustrated in Table 2, the bovine casein concentration is 0.432 mg/kg, as the standard curve showed.

MS/MS *analysis by* MALDI-TOF/TOF

Mass spectrometric analysis was performed on a 4800 MALDI-TOF/TOF Analyzer (AB SCIEX, Foster City, CA). MS-mode acquisitions consisted of 500 laser shots averaged from 10 sample positions ("search pattern positions"). After completing the MS analysis of an entire sample plate, precursors for MS/MS were selected by the instrument software. The software used signal-to-noise and mass tolerance thresholds to create a peak list for each 8-s RPHPLC fraction. Masses that recurred in several adjacent fractions were selected only from the fraction containing the peak apex based on intensity. The signal intensity for any given mass was required to be below the detection threshold for two fractions before that mass was eligible for reselection. The software was set to permit up to 30 MS/MS precursors to be selected for every 8-sec RPHPLC fraction. A mass range of 800 to 4000 Da. was considered for precursor selection. In the MS/MS acquisitions, 3000 laser shots were averaged from 60 sample positions. Two-keV MS/MS acquisition method (CID off) was used with a source voltage of 8 kV, a collision cell voltage of 6 kV, and a second accelerating voltage of 15 kV.

Data analysis

Database searching was performed with the ProteinPilot3.0 software (AB SCIEX, Foster City, CA) utilizing Paragon™ as the search engine. Each MS/MS spectrum was searched against a database of all species (SwissProt, released March 2009). The search parameters allowed for cysteine modification by iodoacetamide and biological modifications programmed in the algorithm (*i.e.*, phosphorylations, amidations, semitryptic fragments, etc.). The detected protein threshold (unused protscore (conf)) in the software was set to 1.3 to achieve 95% confidence level, and identified proteins were grouped by the Pro Group algorithm (AB SCIEX) to minimize redundancy.

RESULTS

At the onset, the digested protein extract was directly analyzed through MALDI-TOF/TOF-MS, but no specific peptide peak was obtained. In this case, another analytical strategy, which separated sample peptides first by means of a Tempo LC-MALDI spotting system, was applied to avoid other laborious procedures for the removal of impurities and greatly increased the purity of the peptides.

Four peptide sequences related to casein were identified in the ancient food residue and are listed in Table 1; their respective mass spectra are reported in Figure 2. Sequence No.a is a homology sequence of alpha S1 casein 38–48 of sheep (*Ovis aries*) and goat *(Capra hircus)*; sequence No.b is a homology sequence of alpha S1 casein 106–115 of sheep and cattle (*Bos taurus*); No.d is a homology sequence of beta-casein of sheep, goat, cattle and water buffalo (*Bubalus bubalis*). Of note, sequence No.c is a conserved amino acid sequence assigned as unique peptide of alpha S1 casein 38–49 of cattle. These four sequences are marked by shading as showed in Figure 3.

In previous ELISA analysis, about 0.1 g of ancient sample was used [26]. The optical density (OD) values of cattle casein standards and the ancient sample were obtained in terms of the instruction manual of ELISA test kit (RIDASCREEN® FAST Casein (Art. No.: R4602), R-Biopharm AG, Darmstadt, Germany). Based on the concentrations and OD values of cattle casein standards in Table 2, the standard curve was fitted in Figure 4 to compute the bovine casein concentration of the ancient sample.

The antibodies in this ELISA kit specifically detect α-, β- and κ-caseins of cow and have no cross-reactivity to casein from sheep and goat. The proteomic identified unique peptide of cattle alpha S1 casein, which confirmed the qualitative ELISA result.

DISCUSSION

The identified alpha S1 casein and beta-casein from cattle, sheep and goat were characteristic of milk. Given the low concentration of cattle casein, these foods, therefore, may have been once dipped in milk. Furthermore, the identification of the unique peptide of cattle alpha S1 casein and the homology sequence from sheep and goat demonstrated that cattle and sheep/goat milking activity in Subeixi region existed at least 2300 years ago, thus permitting speculation that goat/sheep and bovine were raised by ancient people for milk during that period. Therefore, ancient people could get more animal protein without slaughtering animals, thus helping them better adapt to the hostile desert environment.

In Subeixi cemeteries, sheep/goat bones were buried, but the presence of bovine bones in this site is inconclusive [27], [28]; the present work not only reflects sheep/goat exploitation, but also supports the presence of cattle, and can be considered as evidence of cattle raising. According to ancient literature and archaeological findings (houses, pottery, grinding stones, crops, noodles and cakes made of millet dough, bows, bridles, leather, wool clothing, sheep remains etc.), the inhabitants of Subeixi site had a semi-agricultural and semi-pastoral (stock raising) life style [29]; and this work, which implies milking activities, further demonstrates the nature of the pastoral life of ancient people.

As late as AD 1500, the Old World could be divided into regions of milking and non-milking, and one major area of non-milking was located in eastern Asia, which included Southeast Asia and most of the Far East. The boundary of milking and non-milking people in China was roughly between the areas of Tibet and Mongolia, and eastern China [30], [31]. In addition, the distribution of lactose malabsorption is reflected in the spread of milking in these areas [32]. Studies regarding the origin and spread of dairying in the Near East and Europe have made great progress [25]; however, the timing of the emergence of dairying and its spread across Eurasian steppe to East Asia still remains unclear. To date, the Subeixi site is the furthest eastern location in the Old World that prehistoric milking has been detected, and the present work provides BP2300 as the latest time for the eastern spread of milking into the West of China. During this period, milking was not part of the traditional means of subsistence in Chinese agricultural areas, including the Gansu corridor, located in the east of Turpan basin. Thus, this work lends support to the eastern limits of prehistoric milking in western China.

Among residue analysis methods for dairying research, lipid analysis, using isotope of individual fatty acids, played an important role; this method has been widely applied to demonstrate the presence of milk fats [6], [8], [13], [33]–[39] and has further revealed the events of horse domestication [7]. However, certain deficiencies in this method not being able to distinguish the milk fat of cattle, sheep, and goat still exist. Although immunological analysis is very specific for certain species, only cattle casein could be detected and identified until now. In contrast to lipid and immunological analysis, the proteomic method is able to identify a more precise origin of protein in one analysis. For example, a multiple alignment of alpha S1 casein (see Figure 3) from sheep, goat, cattle, water buffalo, and horse was performed through Bioedit software. Figure 3 clearly shows that these animals, possibly raised for milk use, have distinct differences in amino acid loci of alpha S1 casein.

This work also shows that the cattle casein with concentration as low as 0.432 mg/kg in ancient samples could be detected and identified by proteomic methods. In fact, the concentration of cattle casein absorbed in the pottery body may be as low as 0.3–0.5 mg/kg [40]. Although there are some difficulties in direct extraction of protein from pottery shards, some successful extraction methods have been developed [41], [42]. Therefore, when using these same methods, it would be possible to use proteomic methods to detect and identify milk protein absorbed in pottery. As a result, proteomic methods would seem to have great potential in dairy research related to the origin and spread of domesticated animals.

CONCLUSIONS

Although the focus on analysis of protein sequence has been one of regular approaches in modern biosciences for the past several years, similar methods have not been widely applied in ancient residue research. The present work proves the capability of LC-MALDI-TOF-MS/MS to identify the milk component preserved in ancient food remains, and clearly discerns the origin as well. Additionally, it is suggested that alpha S1 casein could be used for discrimination of the origin of milk in ancient residue. Moreover, considering both of the lower detection limit and higher accuracy, those involved in this study believe that not only the visible residue, but also the invisible residue, such as the protein absorbed in pottery, could be identified by proteomics methods. Furthermore, the results imply that animals such as goat/sheep and cattle were raised for milk, which enriches the understanding of Subeixi culture (cal. 500-300 BC), the furthest eastern location of prehistoric milking up to date.

ACKNOWLEDGMENTS

We would like to gratefully acknowledge the helpful comments provided by the anonymous reviewers and Dr. Dorian Q. Fuller.

Author Contributions

Conceived and designed the experiments: CH HJ CW YY. Performed the experiments: CH YW LG. Analyzed the data: CH LG YY. Contributed reagents/materials/analysis tools: EL HJ YW YX. Wrote the paper: CH CW YY.

REFERENCES

1. Greenfield HJ (1988) The origins of milk and wool production in the old world: A zooarchaeological perspective from the central Balkan. Current Anthropology 29: 573–593.HJ Greenfield1988The origins of milk and wool production in the old world: A zooarchaeological perspective from the central Balkan.Current Anthropology29573593
2. Sherratt A (1981) Plough and pastoralism: aspects of the secondary products revolution. Pattern of the past: Studies in honour of David Clarke (eds I Hodder, G Isaac and N Hammond). Cambridge: Cambridge University Press. pp. 261–305.A. Sherratt1981Plough and pastoralism: aspects of the secondary products revolution. Pattern of the past: Studies in honour of David Clarke (eds I Hodder, G Isaac and N Hammond)CambridgeCambridge University Press261305
3. Sherratt A (1983) The secondary exploitation of animals in the Old World. World archaeology 15(1): 90–104.A. Sherratt1983The secondary exploitation of animals in the Old World.World archaeology15190104
4. Vigne JD, Helmer D (2007) Was milk a "secondary product" in the Old World Neolithisation process? Its role in the domestication of cattle, sheep and goats. Anthropozoologica 42: 9–40.JD VigneD. Helmer2007Was milk a "secondary product" in the Old World Neolithisation process? Its role in the domestication of cattle, sheep and goats.Anthropozoologica42940
5. Krausmann F (2004) Milk, Manure, and Muscle Power. Livestock and the Transformation of Preindustrial Agriculture in Central Europe. Human Ecology 32: 735–772.F. Krausmann2004Milk, Manure, and Muscle Power. Livestock and the Transformation of Preindustrial Agriculture in Central Europe.Human Ecology32735772
6. Copley MS, Berstan R, Dudd SN, Docherty G, Mukherjee AJ, et al. (2003) Direct chemical evidence for widespread dairying in prehistoric Britain. Proceedings of the National Academy of Sciences 100: 1524–1529.MS CopleyR. BerstanSN DuddG. DochertyAJ Mukherjee2003Direct chemical evidence for widespread dairying in prehistoric Britain.Proceedings of the National Academy of

Sciences10015241529

7. Outram AK, Stear NA, Bendrey R, Olsen S, Kasparov A, et al. (2009) The earliest horse harnessing and milking. Science 323: 1332–1335. AK OutramNA StearR. BendreyS. OlsenA. Kasparov2009The earliest horse harnessing and milking.Science32313321335
8. Evershed RP, Payne S, Sherratt AG, Copley MS, Coolidge J, et al. (2008) Earliest date for milk use in the Near East and southeastern Europe linked to cattle herding. Nature 455: 528–531.RP EvershedS. PayneAG SherrattMS CopleyJ. Coolidge2008Earliest date for milk use in the Near East and southeastern Europe linked to cattle herding.Nature455528531
9. Heginbotham A, Millay V, Quick M (2006) The Use of Immunofluorescence Microscopy and Enzyme-Linked Immunosorbent Assay as Complementary Techniques for Protein Identification in Artists' Materials. Journal of the American Institute for Conservation 45: 89–105.A. HeginbothamV. MillayM. Quick2006The Use of Immunofluorescence Microscopy and Enzyme-Linked Immunosorbent Assay as Complementary Techniques for Protein Identification in Artists' Materials.Journal of the American Institute for Conservation4589105
10. Miller LJ (2004) Urban economies in early states: the secondary products revolution in the Indus Civilization. New York: New York University. (Ph. D dissertation). LJ Miller2004Urban economies in early states: the secondary products revolution in the Indus Civilization.New YorkNew York University. (Ph. D dissertation)
11. Enattah NS, Jensen TG, Nielsen M, Lewinski R, Kuokkanen M, et al. (2008) Independent introduction of two Lactase-Persistence alleles into human populations reflects different history of adaptation to Milk Culture. The American Journal of Human Genetics 82: 57–72.NS EnattahTG JensenM. NielsenR. LewinskiM. Kuokkanen2008Independent introduction of two Lactase-Persistence alleles into human populations reflects different history of adaptation to Milk Culture.The American Journal of Human Genetics825772
12. Balasse M, Bocherens H, Tresset A, Mariotti A, Vigne JD (1997) Emergence of dairy production in the Neolithic? Contribution of

isotopic analysis of cattle archaeological bones. Earth & Planetary Sciences 325: 1005–1010.M. BalasseH. BocherensA. TressetA. MariottiJD Vigne1997Emergence of dairy production in the Neolithic? Contribution of isotopic analysis of cattle archaeological bones.Earth & Planetary Sciences32510051010

13. Craig OE, Taylor G, Mulville J, Collins MJ, Parker Pearson M (2005) The identification of prehistoric dairying activities in the Western Isles of Scotland: an integrated biomolecular approach. Journal of Archaeological Science 32: 91–103.OE CraigG. TaylorJ. MulvilleMJ CollinsM. Parker Pearson2005The identification of prehistoric dairying activities in the Western Isles of Scotland: an integrated biomolecular approach.Journal of Archaeological Science3291103
14. Greenfield HJ (2005) A reconsideration of the secondary products revolution: 20 years of research in the central Balkans. In: The Zooarachaeology of Milk and Fats (Proceedings of the 9th ICAZ Conference, Durham 2002) (eds J. Mulville and A. Outram). Oxford: Oxbow, 14–31: HJ Greenfield2005A reconsideration of the secondary products revolution: 20 years of research in the central Balkans. In: The Zooarachaeology of Milk and Fats (Proceedings of the 9th ICAZ Conference, Durham 2002) (eds J. Mulville and A. Outram).Oxford: Oxbow,14–31
15. Tokarski C, Martin E, Rolando C, Cren-Olivé C (2006) Identification of Proteins in Renaissance Paintings by Proteomics. Anal Chem 78: 1494–1502.C. TokarskiE. MartinC. RolandoC. Cren-Olivé2006Identification of Proteins in Renaissance Paintings by Proteomics.Anal Chem7814941502
16. Fremout W, Kuckova S, Crhova M, Sanyova J, et al. (2011) Classification of protein binders in artist's paints bymatrix-assisted laser desorption/ionisation time-of-flight mass spectrometry: an evaluation of principal component analysis (PCA) and soft independent modelling of class analogy (SIMCA). Rapid Commun Mass Spectrom 25: 1631–1640.W. FremoutS. KuckovaM. CrhovaJ. Sanyova2011Classification of protein binders in artist's paints bymatrix-assisted laser desorption/ionisation time-of-flight mass spectrometry: an evaluation of principal component analysis (PCA) and soft independent modelling of class analogy (SIMCA).Rapid

Commun Mass Spectrom2516311640

17. Solazzo C, Fitzhugh WW, Rolando C, Tokarski C (2008) Identification of protein remains in archaeological potsherds by proteomics. Anal Chem 80: 4590–4597.C. SolazzoWW FitzhughC. RolandoC. Tokarski2008Identification of protein remains in archaeological potsherds by proteomics.Anal Chem8045904597
18. Chambery A, Maro AD, Sanges C, Severino V, Tarantino M, et al. (2009) Improved procedure for protein binder analysis in mural painting by LC-ESI/Q-q-TOF mass spectrometry: detection of different milk species by casein proteotypic peptides. Analytical and Bioanalytical Chemistry 395: 2281–2291.A. ChamberyAD MaroC. SangesV. SeverinoM. Tarantino2009Improved procedure for protein binder analysis in mural painting by LC-ESI/Q-q-TOF mass spectrometry: detection of different milk species by casein proteotypic peptides.Analytical and Bioanalytical Chemistry39522812291
19. Kuckova S, Crhova M, Vankova L, Hnizda A, Hynek R, et al. (2009) Towards proteomic analysis of milk proteins in historical building materials. International Journal of Mass Spectrometry 284: 42–46.S. KuckovaM. CrhovaL. VankovaA. HnizdaR. Hynek2009Towards proteomic analysis of milk proteins in historical building materials. International Journal of Mass Spectrometry2844246
20. Fremout W, Dhaenens M, Saverwyns S, Sanyova J, Vandenabeele P, et al. (2010) Tryptic peptide analysis of protein binders in works of art by liquid chromatography–tandem mass spectrometry. Analytica Chimica Acta 658: 156–162.W. FremoutM. DhaenensS. SaverwynsJ. SanyovaP. Vandenabeele2010Tryptic peptide analysis of protein binders in works of art by liquid chromatography–tandem mass spectrometry.Analytica Chimica Acta658156162
21. Buckley M, Kansa SW, Howard S, Campbell S, Thomas-Oates J, et al. (2010) Distinguishing between archaeological sheep and goat bones using a single collagen peptide. Journal of Archaeological Science 37: 13–20.M. BuckleySW KansaS. HowardS. CampbellJ. Thomas-Oates2010Distinguishing between archaeological sheep and goat bones using a single collagen peptide.Journal of Archaeological Science371320

22. Stevens SM , Wolverton S, Venables B, Barker A, Seeley KW, et al. (2010) Evaluation of microwave-assisted enzymatic digestion and tandem mass spectrometry for the identification of protein residues from an inorganic solid matrix: implications in archaeological research. Anal Bioanal Chem 396: 1491–1499.Stevens SM JrS. WolvertonB. VenablesA. BarkerKW Seeley2010Evaluation of microwave-assisted enzymatic digestion and tandem mass spectrometry for the identification of protein residues from an inorganic solid matrix: implications in archaeological research.Anal Bioanal Chem39614911499
23. Boros-Major A, Bona A, Lovasz G, Molnar E, Marcsik A, et al. (2011) New perspectives in biomolecular paleopathology of ancient tuberculosis: a proteomic approach. Journal of Archaeological Science 38: 197–201.A. Boros-MajorA. BonaG. LovaszE. MolnarA. Marcsik2011New perspectives in biomolecular paleopathology of ancient tuberculosis: a proteomic approach.Journal ofArchaeological Science38197201
24. Yuan J (2010) Zooarchaeological study on the domestic animals in ancient China. Quaternary Sciences 30(2): 298–306. (in Chinese). J. Yuan2010Zooarchaeological study on the domestic animals in ancient China.Quaternary Sciences 30(2)298–306. (in Chinese)
25. Greenfield HJ (2010) The Secondary Products Revolution: the past, the present and the future, World Archaeology 42(1): 29–54.HJ Greenfield2010The Secondary Products Revolution: the past, the present and the future, World Archaeology4212954
26. Hong C, Jiang HE, Yang YM, Lü EG, Wang CS (2011) Application of Enzyme-linked immunosorbent assay (ELISA) method in the detection of ancient milk residues. Sciences of Conservation and Archaeology 23: 25–28 (in Chinese). C. HongHE JiangYM YangEG LüCS Wang2011Application of Enzyme-linked immunosorbent assay (ELISA) method in the detection of ancient milk residues. Sciences of Conservation and Archaeology 2325–28 (in Chinese)
27. XIA (Xinjiang Institute of Archaeology), MT (Museum of Turpan) (2002) Subeixi site and cemeteries in Shanshan County, Xinjiang. Kaogu 6: 42–57. (in Chinese). XIA (Xinjiang Institute of Archaeology), MT (Museum of Turpan)2002Subeixi site and

cemeteries in Shanshan County, Xinjiang.Kaogu 642–57. (in Chinese)

28. Lü EG (2006) Main achievements of the excavation of Subeixi Site. In: Yin Q, editor. New Discussions on Turpan. Urumchi: Xinjiang People's Publishing House. pp. 239–252. (in Chinese). EG Lü2006Main achievements of the excavation of Subeixi Site.Q. YinNew Discussions on TurpanUrumchiXinjiang People's Publishing House. pp. 239–252. (in Chinese)
29. Gong YW, Yang YM, Ferguson DK, Tao DW, Li WY, et al. (2011) Investigation of ancient noodles, cakes, and millet at the Subeixi Site, Xinjiang, China. Journal of Archaeological Science 38: 470–479. YW GongYM YangDK FergusonDW TaoWY Li2011Investigation of ancient noodles, cakes, and millet at the Subeixi Site, Xinjiang, China.Journal of Archaeological Science38470479
30. Simoons FJ (1970) The traditional limits of milking and milk use in southern Asia. Anthropos 65: 547–593.FJ Simoons1970The traditional limits of milking and milk use in southern Asia. Anthropos65547593
31. Simoons FJ (1979) Dairying, milk use, and lactose malabsorption in Eurasia: A problem in culture history. Anthropos 74: 61–80. FJ Simoons1979Dairying, milk use, and lactose malabsorption in Eurasia: A problem in culture history.Anthropos746180
32. Simoons FJ (1978) The geographic hypothesis and lactose malabsorption. American Journal of Digestive Diseases 23: 963–980.FJ Simoons1978The geographic hypothesis and lactose malabsorption.American Journal of Digestive Diseases23963980
33. Copley MS, Berstan R, Dudd SN, Aillaud S, Mukherjee AJ, et al. (2005) Processing of milk products in pottery vessels through British prehistory. Antiquity 79: 895–908.MS CopleyR. BerstanSN DuddS. AillaudAJ Mukherjee2005Processing of milk products in pottery vessels through British prehistory.Antiquity79895908
34. Copley MS, Berstan R, Dudd SN, Straker V, Payne S, et al. (2005) Dairying in antiquity. I. Evidence from absorbed lipid residues dating to the British Iron Age. Journal of Archaeological Science 32: 485–503.MS CopleyR. BerstanSN DuddV. StrakerS. Payne2005Dairying in antiquity. I. Evidence from absorbed lipid residues dating to the

British Iron Age.Journal of Archaeological Science32485503

35. Copley MS, Berstan R, Straker V, Payne S, Evershed RP (2005) Dairying in antiquity. II. Evidence from absorbed lipid residues dating to the British Bronze Age. Journal of Archaeological Science 32: 505–521.MS CopleyR. BerstanV. StrakerS. PayneRP Evershed2005Dairying in antiquity. II. Evidence from absorbed lipid residues dating to the British Bronze Age.Journal of Archaeological Science32505521
36. Copley MS, Berstan R, Dudd SN, Straker V, Payne S, et al. (2005) Dairying in antiquity. III. Evidence from absorbed lipid residues dating to the British Neolithic. Journal of Archaeological Science 32: 523–546.MS CopleyR. BerstanSN DuddV. StrakerS. Payne2005Dairying in antiquity. III. Evidence from absorbed lipid residues dating to the British Neolithic.Journal of Archaeological Science32523546
37. Copley MS, Clark K, Evershed RP (2005) Organic-residue Analysis of Pottery Vessels and Clay Balls. In: Hodder I, editor. Changing Materialities at Çatalhoyuk: Reports from the 1995–99 Seasons. Cambridge: McDonald Institute for Archaeological Research. pp. 169–174.MS CopleyK. ClarkRP Evershed2005Organic-residue Analysis of Pottery Vessels and Clay Balls.I. HodderChanging Materialities at Çatalhoyuk: Reports from the 1995–99 SeasonsCambridgeMcDonald Institute for Archaeological Research169174
38. Craig OE, Chapman J, Heron C, Willis LH, Bartosiewicz L, et al. (2005) Did the first farmers of central and eastern Europe produce dairy foods? Antiquity 79: 882–894.OE CraigJ. ChapmanC. HeronLH WillisL. Bartosiewicz2005Did the first farmers of central and eastern Europe produce dairy foods?Antiquity79882894
39. Spangenberg JE, Jacomet S, Schibler J (2006) Chemical analyses of organic residues in archaeological pottery from Arbon Bleiche 3, Switzerland- evidence for dairying in the late Neolithic. Journal of Archaeological Science 33: 1–13.JE SpangenbergS. JacometJ. Schibler2006Chemical analyses of organic residues in archaeological pottery from Arbon Bleiche 3, Switzerland- evidence for dairying in the late Neolithic.Journal of Archaeological Science33113

40. Craig OE, Mulville J, Pearson MP, Sokol R, Gelsthorpe K, et al. (2000) Detecting milk proteins in ancient pots. Nature 408: 312.OE CraigJ. MulvilleMP PearsonR. SokolK. Gelsthorpe2000Detecting milk proteins in ancient pots.Nature408312
41. Craig OE, Collins MJ (2000) An improved method for the immunological detection of mineral bound protein using hydrofluoric acid and direct capture. Journal of Immunological Methods. 236(1–2): 89–97.OE CraigMJ Collins2000An improved method for the immunological detection of mineral bound protein using hydrofluoric acid and direct capture. Journal of Immunological Methods.2361–28997
42. Craig OE (2002) The Removal of Protein from Mineral Surfaces Implications for Residue Analysis of Archaeological Materials. Journal of Archaeological Science 29: 1077–1082.OE Craig2002The Removal of Protein from Mineral Surfaces Implications for Residue Analysis of Archaeological Materials.Journal of Archaeological Science2910771082ss

CHAPTER

8

A SENSITIVE AND EFFECTIVE PROTEOMIC APPROACH TO IDENTIFY SHE-DONKEY'S AND GOAT'S MILK ADULTERATIONS BY MALDI-TOF MS FINGERPRINTING

Francesco Di Girolamo [1], Andrea Masotti [2], Guglielmo Salvatori [3], Margherita Scapaticci [4], Maurizio Muraca [1] and Lorenza Putignani [5,6]

[1]Department of Laboratory Medicine, Bambino Gesù Children's Hospital, IRCCS, Piazza Sant'Onofrio 4, Rome 00165, Italy

[2]Gene Expression-Microarrays Laboratory, Bambino Gesù Children's Hospital, IRCCS, Piazza Sant'Onofrio 4, Rome 00165, Italy

[3]Neonatal Intensive Care Unit, Department of Medical and Surgical Neonatology, Bambino Gesù Children's Hospital, IRCCS, Piazza Sant'Onofrio 4, Rome 00165, Italy

[4]Department of Laboratory Medicine, San Camillo Hospital, Viale Vittorio Veneto 18, Treviso 31100, Italy

[5]Parasitology Unit, Bambino Gesù Children's Hospital, IRCCS, Piazza Sant'Onofrio 4, Rome 00165, Italy

[6]Metagenomics Unit, Bambino Gesù Children's Hospital, IRCCS, Piazza Sant'Onofrio 4, Rome 00165, Italy

ABSTRACT

She-donkey's milk (DM) and goat's milk (GM) are commonly used in newborn and infant feeding because they are less allergenic than other milk types. It is, therefore, mandatory to avoid adulteration and contamination by other milk allergens, developing fast and efficient analytical methods to assess the authenticity of these precious nutrients. In this experimental work, a sensitive and robust matrix-assisted laser desorption/ionization time-of-flight mass spectrometry (MALDI-TOF MS) profiling was designed to assess the genuineness of DM and GM milks. This workflow allows the identification of DM and GM adulteration at levels of 0.5%, thus, representing a sensitive tool for milk adulteration analysis, if compared with other laborious and time-consuming analytical procedures.

Keywords: milk; adulteration; MALDI-TOF MS

INTRODUCTION

Milk is the most important nutrient for young mammals and is traditionally considered an ideal source of proteins and microelements for adult people. Confirming the popular tradition, recent clinical studies have clearly demonstrated that the she-donkey's milk (DM) represents the best mother's milk substitute for newborns, which are allergic to cow's milk (CM) proteins, because of its low allergenic properties together with its high nutritional value [1].

Goat's milk (GM) has also become popular in infant feeding because it is less allergenic than CM [2]; indeed, for infants suffering of CM-induced gastrointestinal allergy and chronic enteropathy, GM feeding has been reported [3].Unfortunately, the intolerance based on the immunological response to CM proteins is the most common form of food allergy in the pediatric population (from 0.3% to 7.5% of affected children) [4].For this reason, revealing adulterations of DM and GM is of fundamental importance for both economic and clinical reasons. To this aim, effective analytical methods to assess the genuineness of these milks are desirable. Currently, both immunological and chemical methods are used to detect milk and dairy product adulteration. The immunological methods are based on the recognition of antigens (*i.e.*, whole casein, β-casein and β-lactoglobulin) by specific antibodies [5,6,7]. However, even well-validated analytical techniques, such as the enzyme-linked immunosorbent assay (ELISA) and immunoblotting, are affected by false-positive results. On the other hand, chemical methods that are based on the detection of distinctive fatty acid and proteins, are very laborious and time-consuming. In fact, they rely on gas chromatography (GC) analyses for fatty acid characterization [8] and on gel or capillary electrophoresis, high-performance liquid chromatography (HPLC), and mass spectrometry (MS) for protein separation, identification and quantification [9,10]. Many MS methods have been developed for investigating protein profiles and structures in milk [11,12,13,14,15,16,17,18]. Matrix-assisted laser desorption/ionization time-of-flight MS (MALDI-TOF MS), owing to its simplicity of use and to the high reproducibility of the mass spectra, has become a powerful technique to obtain informative fingerprints of milk proteins. Moreover, as the MALDI-TOF MS analysis of a milk sample is able to reveal the characteristic peaks of the most abundant proteins, milk adulteration can be, therefore, assessed by identifying the additional protein peaks in the mass spectrum of the analyzed milk. In the last few years, the identification of whey proteins, such as α-lactalbumin and β-lactoglobulins, by MALDI-TOF MS have been employed by: (i) Cozzolino *et al.* [8] to reveal CM and ewe milk (EM) addition to fresh water buffalo mozzarella cheese with a limit of detection (LOD) below 5% and up to 2% respectively; (ii) Cozzolino *et al.* [19] to discover CM in EM and buffalo milk (BM) with a LOD below 5% and Consulo *et al.* [20] to identify CM or GM in DM with a LOD up to 2% and 0.5% respectively. Moreover, with the aim to identify possible fraudulent addition of CM to EM, employed for the production of commercial ewe cheese (Pecorino), Fanton *et al.* [21] described a MALDI-TOF MS method, able to assess the adulteration of ewe cheese by CM (LOD 10%), by detecting the presence of the γ2-casein mass peak. Following a bottom-up proteomic strategy (MS analysis of enzymatic digests of milk proteins), Chen *et al.* [10], and Calvano *et al.* [22], analyzed CM adulteration in GM by assessing β-lactoglobulin whey protein as the molecular marker, employing on-line HPLC electrospray MS (HPLC/ESI-MS) and MALDI-TOF MS, respectively (LOD 5%).

In this experimental work, we evaluated the performance of the total MALDI-TOF MS profiles, in the mass range 2000–25,000 Da, for the identification of DM and GM adulteration by CM, EM and BM. We have demonstrated that MALDI-TOF MS coupled to unsupervised hierarchical clustering, principal component (PCA) and Pearson's correlation analyses represent a rapid, robust and very sensitive method, to reveal CM, EM, and BM adulteration in DM and GM at very low levels (down to 0.5%). Considering the high sensitivity that we obtained following this approach (more sensitive than the analytical method proposed by Consulo *et al.* [20]), we envisage that our analytical method could be successfully applied also for the evaluation of water buffalo mozzarella cheese, buffalo milk and ewe cheese (Pecorino) adulterations with a higher precision than that reported by Cozzolino *et al.* [8,19] and Fanton *et al.* [21].

RESULTS AND DISCUSSION

Assessing the Performance of the MALDI-TOF MS Profiles in Discriminating Milk Species

After acquisition of six independent MALDI-TOF MS samples from she-donkey's, goat's, cow's, ewe's, and buffalo's crude milks, spectra were visually inspected (Figure 1) and the resulting flattened profiles were compared by gel-like representations, analyzed by unsupervised hierarchical clustering (Figure 2a) and PCA (Figure 2b) using the integrated software Biotyper 3.1 (Bruker Daltonics, Bremen, Germany), and finally by bootstrapped (n = 1000) clustering (Figure 2c) and correlation matrices (Figure 2d) with external statistical software (R Bioconductor, Fred Hutchinson Cancer Research Center, Seattle, WA, USA).

According to the dendrogram (Figure 2a) obtained by clustering the replicated profiles of the five milks and the PCA representation (Figure 2b), we found that the six mass spectra for each milk closely clustered together, whereas the different types of milk resulted well separated from each other. GM and EM displayed the lowest distance (D = 0.39) and clustered in the same clade, proximate to CM. BM and DM (D = 1.18) generated a second clade which in turn clustered with the larger clade formed by GM, EM and CM (D = 1.16). The bootstrapping procedure performed on external software (Figure 2c) confirmed the reliability of the Biotyper elaboration. The Pearson's correlation matrix obtained from all the profiles displayed a high intra-group similarity and a low inter-group correlation (Figure 2d). GM and EM samples correlated quite well, and this result is in agreement also with clustering analysis that locates these two milks in the same clade.

All together these data indicate that the high reproducibility and discriminating power of MALDI-TOF MS analysis, coupled to a set of robust data analysis tools, allows a straightforward identification of the five types of milk and that the different types can be easily distinguished from each other.

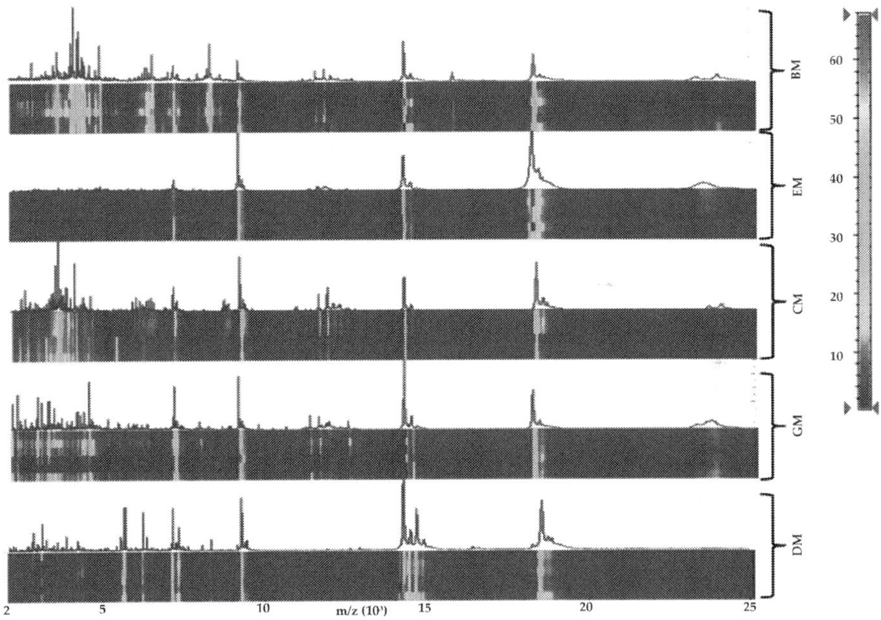

Figure 1. Pseudo-gel like and mass spectrometry (MS) proteomic profiling of crude milk from six she-donkey, goat, cow, ewe and buffalo animals. The mass values (m/z) are reported on the X axis, while the color bar on the Y axis indicates the peak intensity.

Assessing the Performance of the MALDI-TOF MS Profiles in Discriminating She-Donkey's Milk (DM) and Goat's Milk (GM) Adulteration with Cow's Milk (CM)

To track milk adulteration, eight mixtures were prepared by adding CM at different percentages (0.2%, 0.5%, 1%, 2%, 5%, 10%, 30%, and 50% *v/v*) to DM (Figure 3) and GM (Figure 4). Six replicas for each sample were collected and analyzed.

In this set-up the discriminant region of the MALDI-TOF MS profiles between 2000 and 25,000 kDa (Figure 3 and Figure 4) was very reproducible among the technical replicates and to gain further information about the correlation between the various mass profiles, we decided to perform a correlation analysis (Pearson's coefficients) and represent the obtained results in a graphical representation. The Pearson's correlation matrix representation for DM and GM adulteration with CM (Figure 5 and Table 1) displayed a high intra-class similarity as shown by the replicas in the diagonal, whereas the inter-class correlation appeared variable, as expected, owing to the different percentages of adulteration.

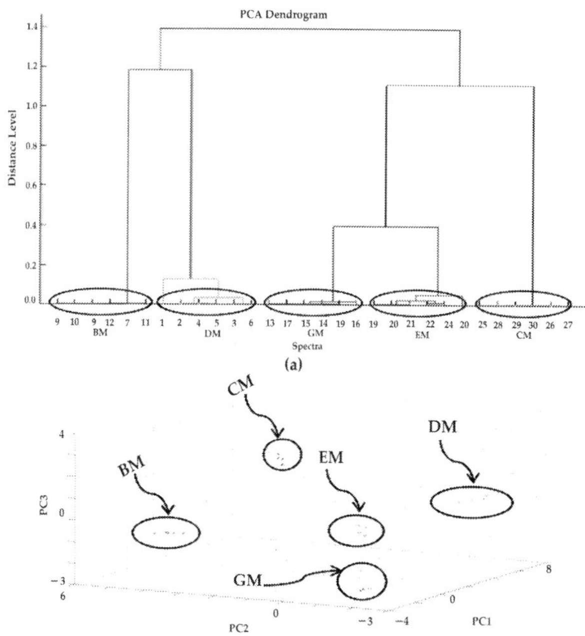

(b)
Cluster dendrogram with AU/BP values (%)

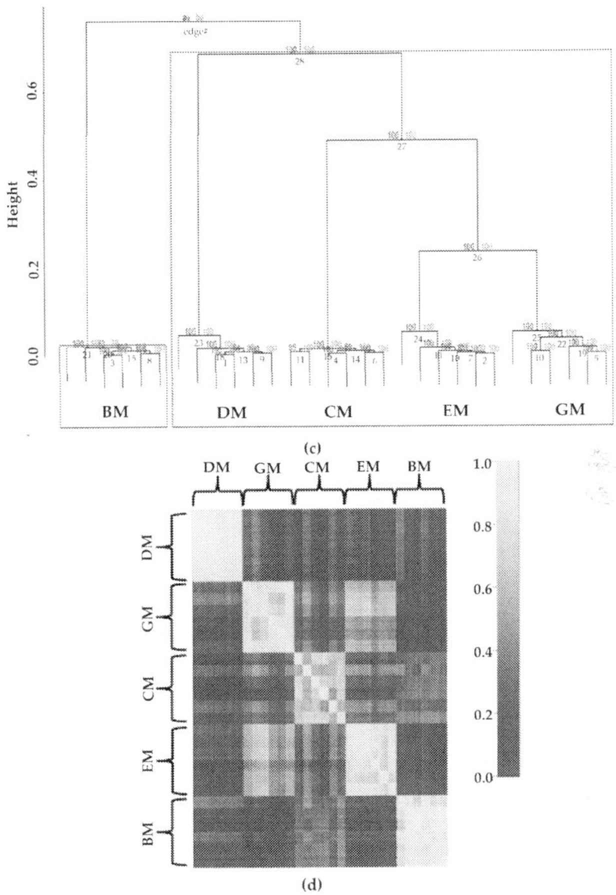

Figure 2. Data analysis of six independent matrix-assisted laser desorption/ionization time-of-flight mass spectrometry (MALDI-TOF MS)profiles of crude she-donkey's milk (DM), goat's milk (GM), cow's milk (CM), ewe milk (EM) and buffalo milk (BM). (a) Hierarchical clustering dendrogram; (b) Principal component (PCA) analysis with corresponding 3D scatter plot image obtained by Biotyper; (c) Bootstrapped (n = 1000) hierarchical clustering tree generated via pvclust. Values on the edges of the clustering are p-values (%). Red values are AU (Approximately Unbiased) p-values, and green values are BP (Bootstrap Probability) values as explained in methods section. Clusters with AU larger than 95% are highlighted by red boxes and indicate reliable clades; (d) Pearson's correlation coefficients represented as a correlation matrix. The correlation coefficients have been colored according to a scale ranging from 0 to 1, where blue corresponds to 0 and yellow to 1.

In particular, the native and the 50%-adulterated milks displayed a modest mean correlation value (0.720 for CM/DM and 0.782 for CM/GM), which gradually increased as the percentage of adulteration decreased (up to 0.2%–0.5%) (Table 1). The mean correlation coefficients that we obtained indicate that this simple analysis is not feasible to discern milk adulterations even at high percentages (10%–30%) as the correlation is rather good (0.961 for DM adulterated with CM at 10% *v*/*v*).

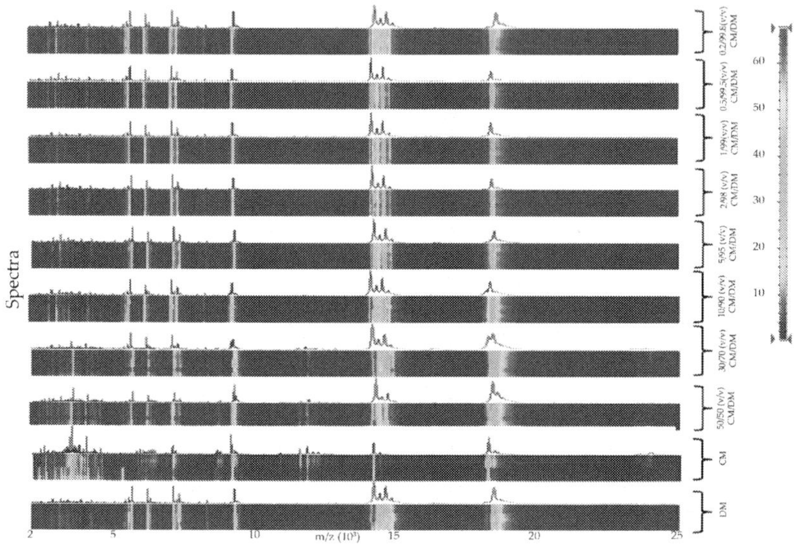

Figure 3. Pseudo-gel like image and MALDI-TOF MS spectra of the simulated adulteration of DM by CM.

Table 1. The identification ability (expressed by the symbol √ = Yes, × = No) of different percentages of DM and GM adulteration obtained by our combined analytical and statistical approach; the correspondent correlation coefficients are also reported.

Adulteration	50/50 (*v*/*v*)	30/70 (*v*/*v*)	10/90 (*v*/*v*)	5/95 (*v*/*v*)	2/98 (*v*/*v*)	1/99 (*v*/*v*)	0.5/99.5 (*v*/*v*)	0.2/99.8 (*v*/*v*)
CM/DM	√ 0.720	√ 0.907	√ 0.961	√ 0.921	√ 0.902	√ 0.954	√ 0.885	× 0.946
CM/GM	√ 0.782	√ 0.854	√ 0.908	√ 0.967	√ 0.961	√ 0.968	√ 0.974	× 0.961

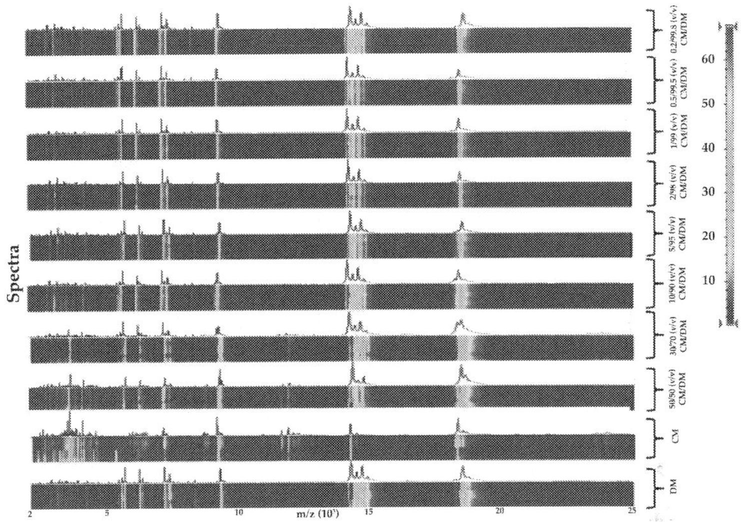

Figure 4. Pseudo-gel like image and MALDI-TOF MS spectra of the simulated adulteration of GM by CM.

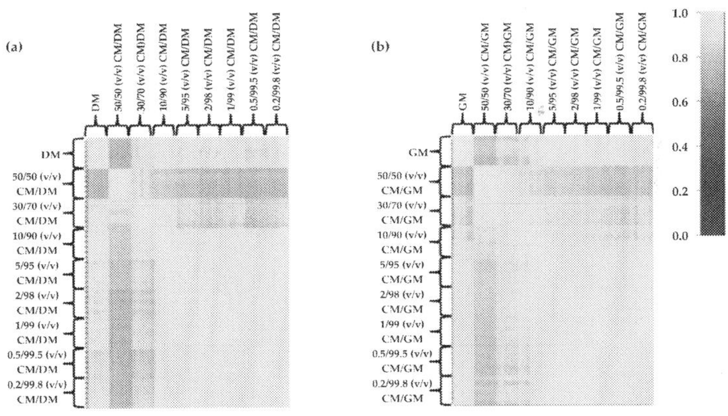

Figure 5. Pearson's correlation matrices of all spectral replicates for the eight mixtures (from 50% to 0.2%) of simulated adulterations of (a) DM with CM and (b) GM with CM. Correlation coefficients are represented with decreasing blue and yellow colors according to a scale ranging from 0 to 1, respectively.

In order to increase the ability to discern milk adulterations up to a very low limit (0.2% of CM addition), we grouped all spectra within consistent clades and PCA clusters (Figure 6 and Figure 7).

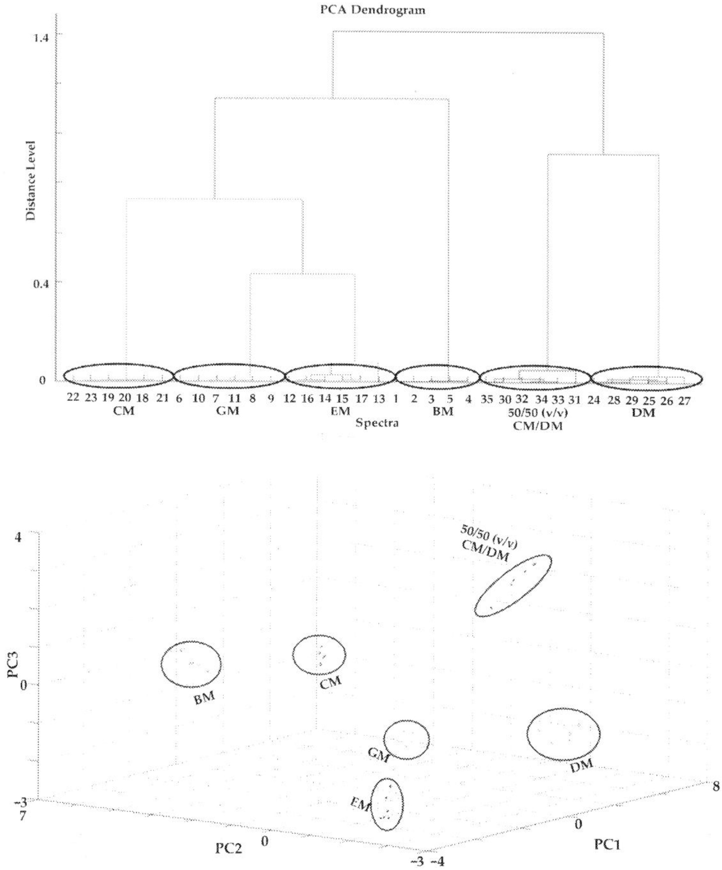
PCA Dendrogram
1.4
0.4
0
Distance Level
22 23 19 20 18 21 6 10 7 11 8 9 12 16 14 15 17 13 1 2 3 5 4 35 30 32 34 33 31 24 28 29 25 26 27
CM
GM
EM
BM
50/50 (v/v) CM/DM
DM
Spectra
50/50 (v/v) CM/DM
BM
CM
GM
DM
EM
PC3
PC2
PC1
4
0
-3
7
0
-3 -4
0
8

(a)

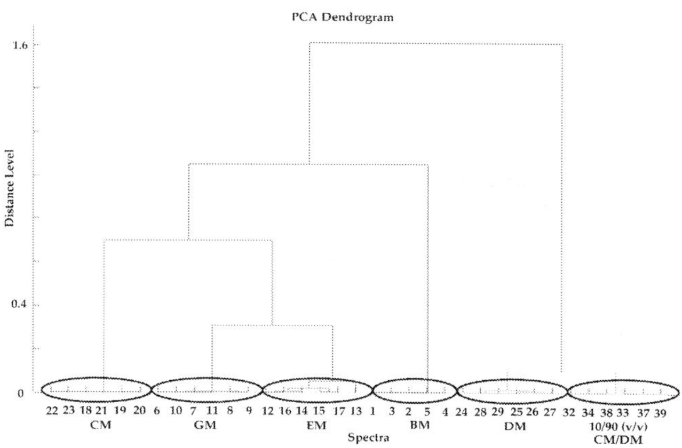
PCA Dendrogram
1.6
0.4
0
Distance Level
22 23 18 21 19 20 6 10 7 11 8 9 12 16 14 15 17 13 1 3 2 5 4 24 28 29 25 26 27 32 34 38 33 37 39
CM
GM
EM
BM
DM
10/90 (v/v) CM/DM
Spectra

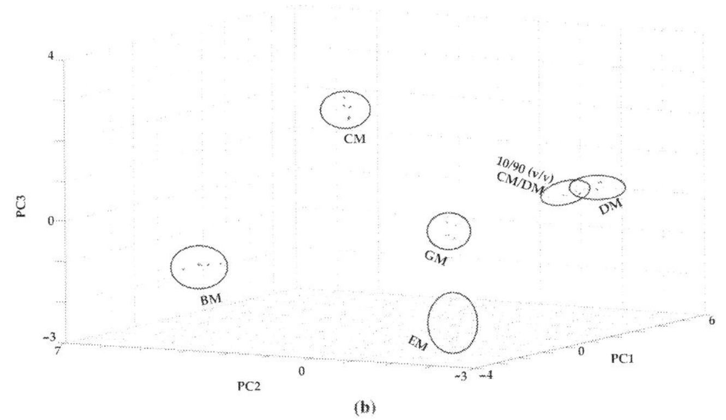

4
0
–3
PC3
CM
10/90 (v/v)
CM/DM
DM
GM
BM
EM
7
0
PC2
–3
–4
0
6
PC1

(b)

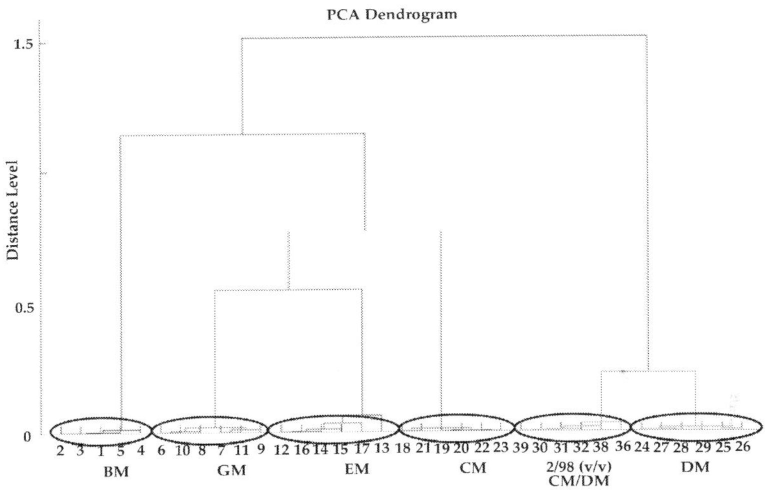

PCA Dendrogram
1.5
0.5
0
Distance Level
2 3 1 5 4 6 10 8 7 11 9 12 16 14 15 17 13 18 21 19 20 22 23 39 30 31 32 38 36 24 27 28 29 25 26
BM
GM
EM
CM
2/98 (v/v)
CM/DM
DM

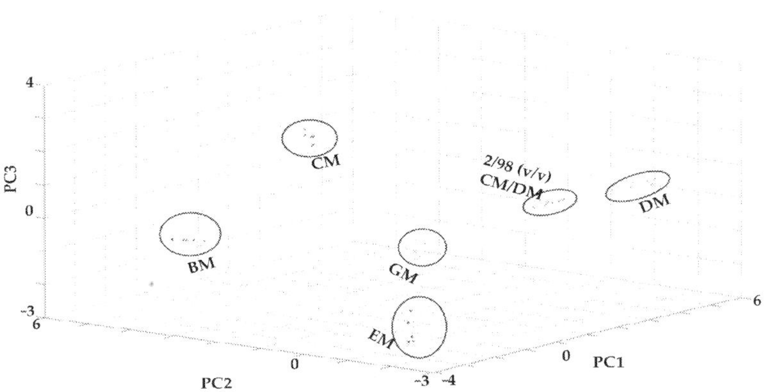

4
0
–3
PC3
CM
2/98 (v/v)
CM/DM
DM
BM
GM
EM
6
0
PC2
–3
–4
0
PC1
6

(c)

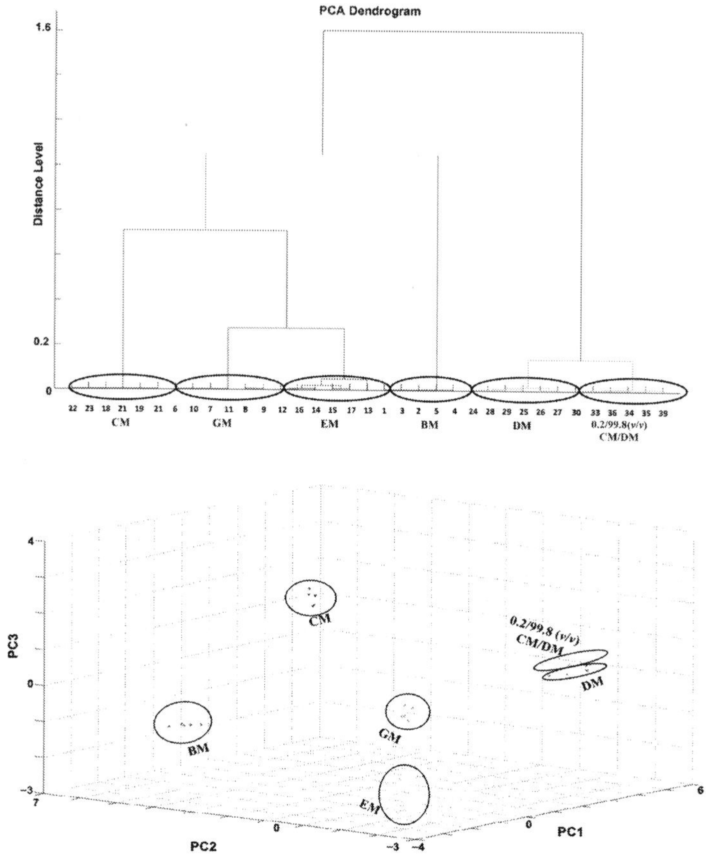

(d)

Figure 6. Hierarchical clustering dendrograms and principal component (PCA) 3D scatter plots of the first three principle components for the four mixtures (50%, 10%, 2%, and 0.2%) of the simulated adulteration of DM by CM (Panels a–d). All milk adulteration grouped inside the correct clade and cluster.

Both hierarchical clustering and PCA analysis obtained with Biotyper outlined the good separation between the spectra belonging to the different percentages of adulteration in DM and GM, although with some differences.

In particular, hierarchical clustering is a 2D representation of the euclidean distance between the various spectra, whereas PCA is a 3D representation where suitable principal components can be properly selected to group similar samples. The advantage of PCA is the possibility to visually rotate the principal components graph for a better inspection of the acquired samples.

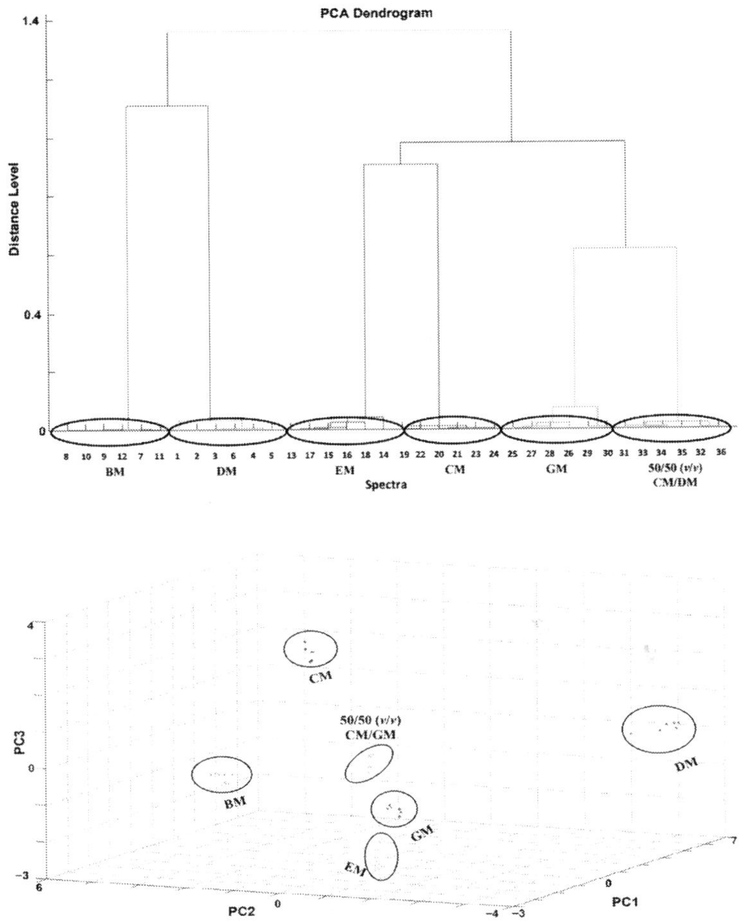

(a)

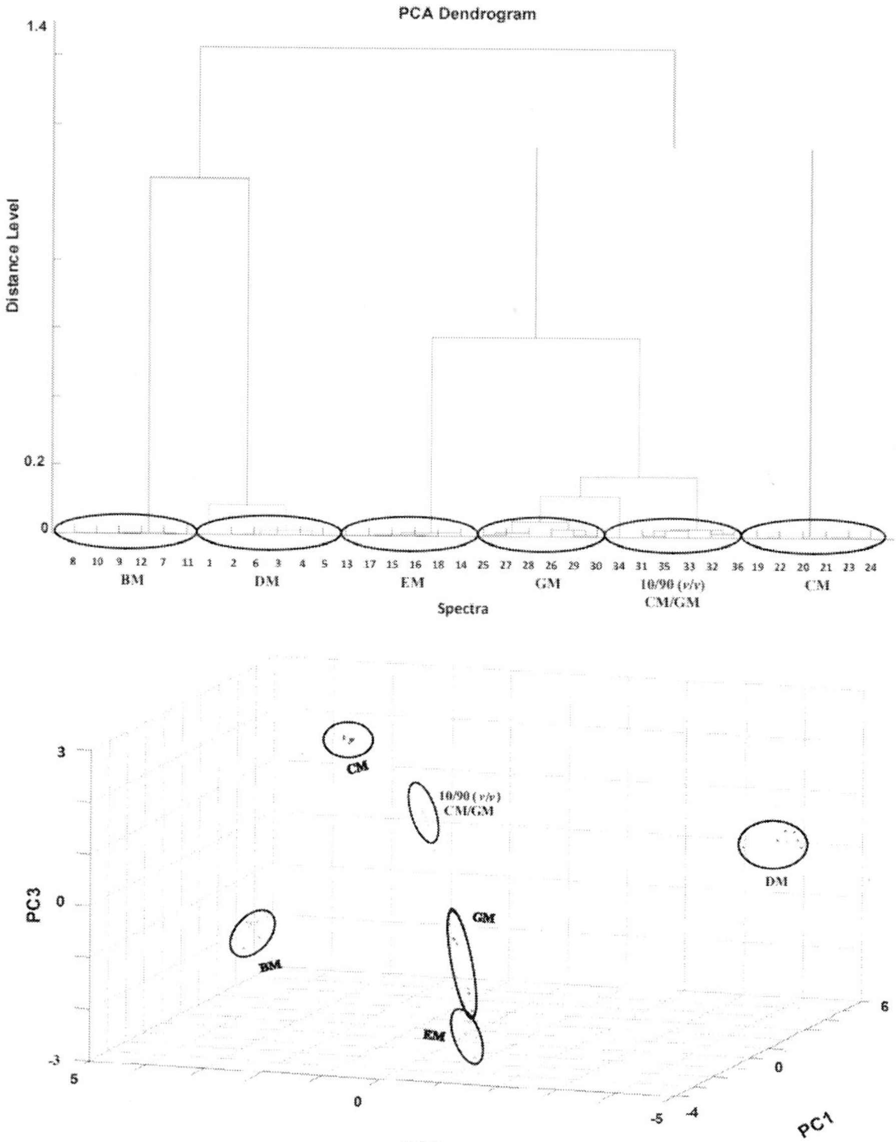
PCA Dendrogram
Distance Level
1.4
0.2
0
8 10 9 12 7 11 1 2 6 3 4 5 13 17 15 16 18 14 25 27 28 26 29 30 34 31 35 33 32 36 19 22 20 21 23 24
BM
DM
EM
GM
10/90 (v/v) CM/GM
CM
Spectra
CM
10/90 (v/v) CM/GM
DM
GM
BM
EM
PC3
3
0
-3
5
0
-5
-4
0
6
PC2
PC1

(b)

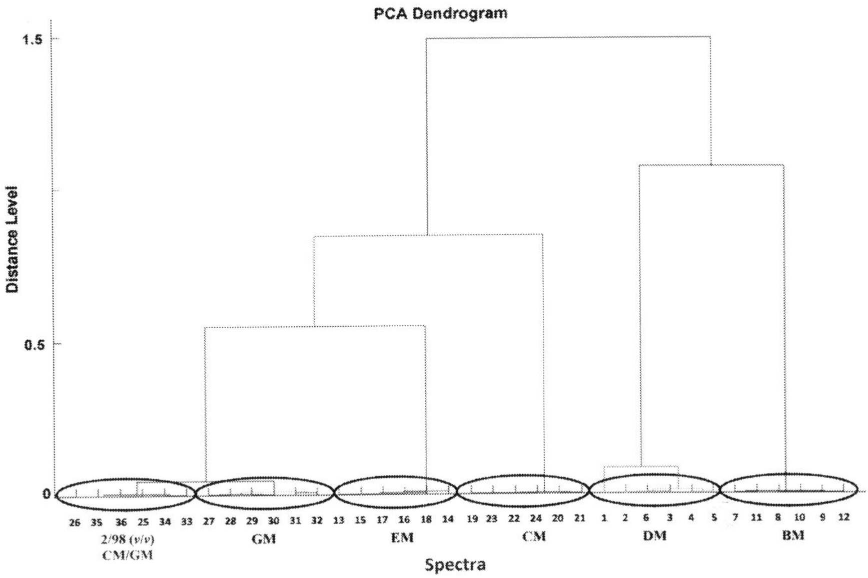
PCA Dendrogram
Distance Level
1.5
0.5
0
26 35 36 25 34 33 27 28 29 30 31 32 13 15 17 16 18 14 19 23 22 24 20 21 1 2 6 3 4 5 7 11 8 10 9 12
2/98 (v/v)
CM/GM
GM
EM
CM
DM
BM
Spectra

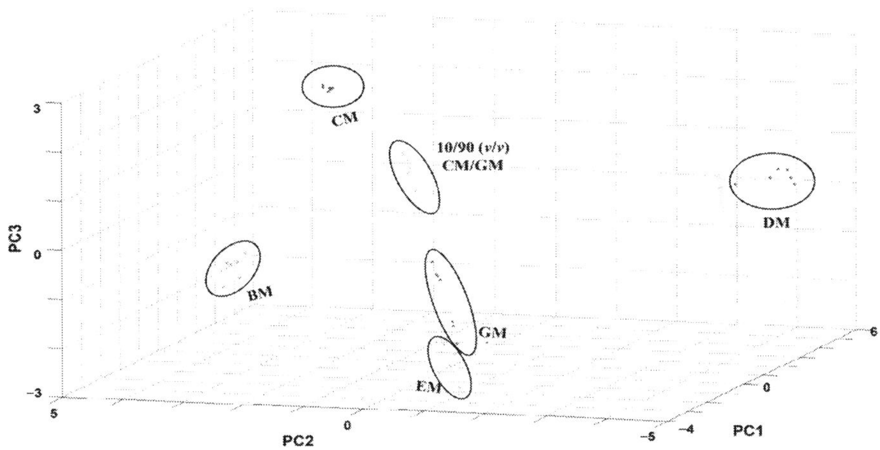
CM
10/90 (v/v)
CM/GM
DM
BM
GM
EM
PC3
3
0
−3
5
0
PC2
−5
−4
0
6
PC1

(c)

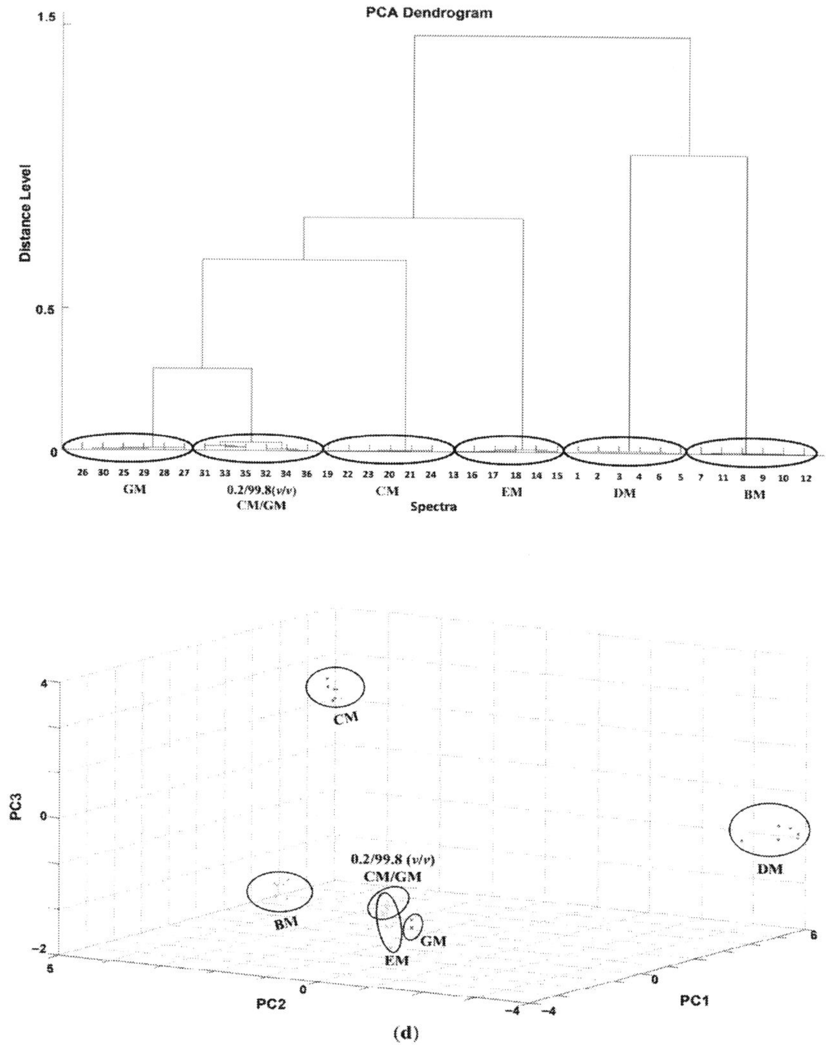

Figure 7. Hierarchical clustering dendrograms and PCA 3D scatter plots of the first three principle components for the four mixtures (50%, 10%, 2%, and0.2%) of the simulated adulteration of GM by CM (**Panels a–d**). All milk adulteration grouped inside the correct clade and cluster.

However, although Biotyper was able to discriminate adulterations up to a limit of 0.2%, the external clustering validation limited the recognition ability of our systems to a percentage of 0.5% for both DM (Figure 8a) and GM (Figure 8b).

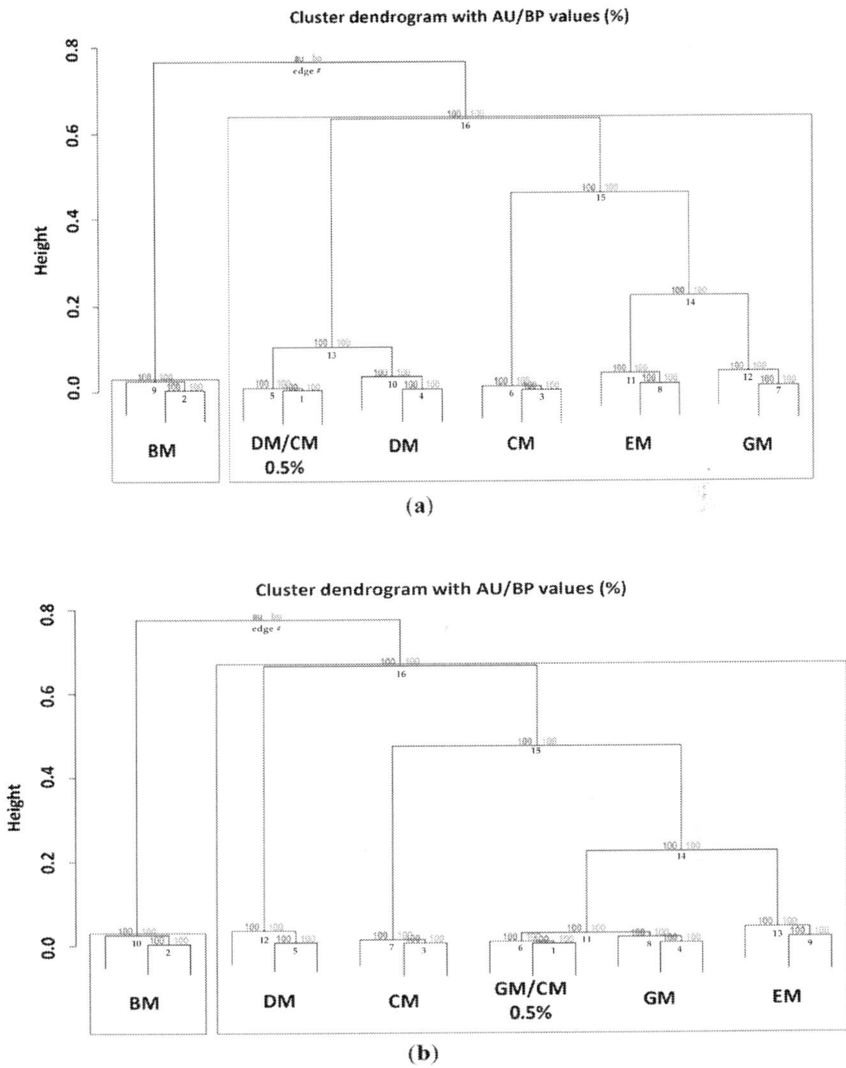

Figure 8. Bootstrapped ($n = 1000$) hierarchical clustering tree of (**a**) adulterated DM and (**b**) GM with CM (0.5% adulteration) generated via pvclust. Values on the edges of the clustering are p-values (%). Red values are AU (Approximately Unbiased) p-values, and green values are BP (Bootstrap Probability) values as explained in methods section. Clusters with AU larger than 95% are highlighted by red boxes and indicate reliable clades.

To validate further the robustness of our analytical method, we prepared other seven DM samples added with different amounts (*v*/*v*) of CM (CM/DM mixtures: 0.3/99.7, 0.4/99.6, 0.7/99.3, 1.5/98.5, 9/91, 15/85, 40/60) and another investigator analyzed them (independent blind analysis). Again, although the researcher was able to identify all the adulteration levels by means of Biotyper software the in-depth statistical analysis revealed that the limit of detection is a little bit higher, as the 0.3% adulterated samples were not recognized correctly (Figure 9a) compared to 0.4% adulterated samples (Figure 9b).

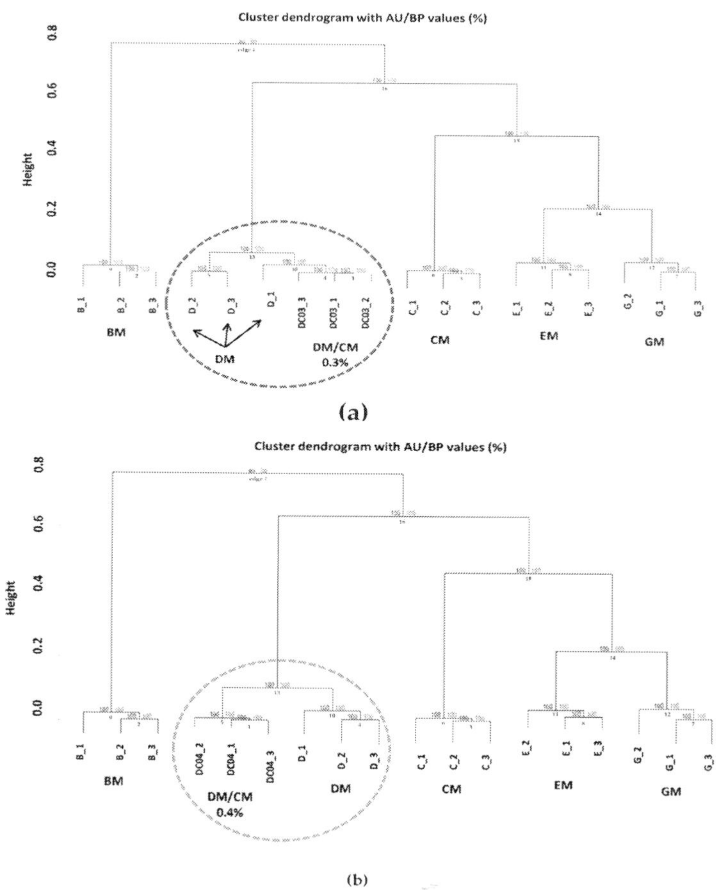

Figure 9. Hierarchical clustering tree of (a) adulterated DM with CM (0.3% *v*/*v*) and (b) DM with CM (0.4% *v*/*v*). Red circle indicates that the classification did not allow to obtain separate clades, whereas the green circle indicates that a good separation of the clades is obtained at an adulteration level of 0.4%.

This analytical method (using Biotyper 3.1 software) was also tested by using DM and GM samples treated by high (pasteurization) and ultra-high temperature (UHT) method. CM/DM and CM/GM mixtures (0.2/99.8, 10/90, and 50/50 volume ratios) were prepared and subsequently analyzed. Supplementary Figure S4 and S5 show that the method allowed the identification of adulterated milk as previously done with raw milk, suggesting that the proposed workflow is well-suited also for the analysis of commercial/market milks.

Assessing the Performance of the MALDI-TOF MS Profiles in Discriminating DM and GM Adulteration with EM and BM

We performed the same analysis also for DM and GM adulterated with other types of milk (EM and BM) obtaining similar results, briefly described here and illustrated in more details in the Supplementary section (Supplementary Figure S6–S8).

Briefly, correlation analysis (Pearson's coefficients) emphasized once more that this kind of analysis is not useful to discriminate low percentage of adulteration. In fact, similarly to what obtained for DM and GM adulterated with CM, adulteration of DM and GM with EM and BM follows a similar behavior.

In particular, the native and the 50%-adulterated milks displayed a modest correlation value (from 0.583 for BM/GM to 0.846 for EM/GM), which gradually increased as the percentage of adulteration decreased (up to 0.2%–0.5%) (Table 2). Again, at adulteration percentages of about 10%–30% the values of the correlation coefficients are quite high (up to 0.959 for a 10% adulteration of DM with EM) indicating that the spectra are overall very well correlated. These results confirm that correlation analysis alone is not sufficient to discriminate efficiently milk adulteration.

Of note, from hierarchical clustering analysis (Biotyper software) of GM adulteration by EM at a percentage of 0.2% we found that the MALDI-TOF MS profiles grouped outside the correct clade and the PCA cluster, limiting the correct identification of adulteration of GM with EM at 0.5% mixtures. This is however a good results if we consider that the limit of detection generally obtained with other electrophoretic analyses is about 10%.

Table 2. The identification ability (expressed by the symbol √ = Yes, × = No) of different percentages of DM and GM adulteration obtained by the Biotyper analysis; the correspondent correlation coefficients are also reported.

Adulteration	50/50 (*v/v*)	30/70 (*v/v*)	10/90 (*v/v*)	5/95 (*v/v*)	2/98 (*v/v*)	1/99 (*v/v*)	0.5/99.5 (*v/v*)	0.2/99.8 (*v/v*)
EM/DM	√ 0.664	√ 0.804	√ 0.959	√ 0.974	√ 0.984	√ 0.967	√ 0.888	√ 0.895
EM/GM	√ 0.846	√ 0.902	√ 0.891	√ 0.823	√ 0.871	√ 0.790	√ 0.878	× 0.964
BM/DM	√ 0.644	√ 0.813	√ 0.948	√ 0.969	√ 0.954	√ 0.983	√ 0.978	√ 0.957
BM/GM	√ 0.583	√ 0.754	√ 0.822	√ 0.876	√ 0.873	√ 0.805	√ 0.828	√ 0.778

EXPERIMENTAL SECTION

Milk Sampling and Preparation

The following samples were analyzed: (i) raw GM, DM, CM, EM, and BM collected from Italian farms (Lazio and Puglia); and (ii) adulterated samples made of CM/GM, EM/GM, BM/GM, CM/DM, EM/DM, and BM/DM mixtures, in volume ratios 0.2/99.8, 0.5/99.5, 1/99, 2/98, 5/95, 10/90, 30/70, and 50/50. Raw GM (from six goats belonging to Maltese breed), DM (from six she-donkeys belonging to Amiata, Viterbese and Martina Franca breeds), CM (from six cows belonging to Frisona breed), EM (from six ewes belonging to Tuscolania breed), and BM (from six buffalos belonging to Mediterranean Italian breed) samples were collected at middle lactation stage (125 ± 20 days), from two Italian herds located in Lazio and Puglia. Raw milk samples were mechanically collected into sterile polystyrene containers, immediately frozen, and stored at −80 °C until use, to prevent undesired proteolysis. After thawing, raw milk samples were defatted by centrifugation (using the Eppendorf Centrifuge 5417 R) and analyzed by linear MALDI-TOF MS for the generation of proteomic phenotyping profiles. The milk was defatted by two-step centrifugation (3000× *g*, 10 min) at 4 °C and the skimmed milk was centrifuged (20,000× *g*, 20 min) at 4 °C to remove bacteria and cell debris. The skimmed milk's fractions were subsequently diluted 1:100 with ultrapure water (Milli-Q Millipore) and analyzed by linear MALDI-TOF MS as reported in the next paragraph.

MALDI-TOF MS Spectra Acquisition

One microliter of diluted skim milk was placed onto an MSP 96 polished steel target (Bruker Daltonics, Bremen, Germany) and allowed to dry at room temperature. Each sample was overlaid with 1 μL of matrix, which consisted of a solution of 10 mg/mL of sinapinic acid (Sigma–Aldrich, St. Louis, MO, USA) in 50% acetonitrile-0.1% trifluoroacetic acid (Sigma–Aldrich). Measurements were performed with a Microflex LT linear mass spectrometer (Bruker Daltonics), using FlexControl software package (version 3.0 Bruker Daltonics) [23]. Spectra were recorded in the positive linear mode (laser frequency, 20 Hz; ion source 1 voltage, 20 kV; ion source 2 voltage, 18.4 kV; lens voltage, 9.1 kV; mass range, 2000 to 25,000 Da). Six independent spectra (500 shots one step from different positions of the target spot, for spectrum) for each skimmed milk's fraction were manually collected, externally calibrated by using Bacterial Test Standard (Bruker Daltonics) and subsequently analyzed.

MALDI-TOF MS Spectra Analysis

The entire set of data considered in this work consists in a collection of 468 MS spectra (six animals for 5 milk species for six spectra replicas and, 8 different percentage of adulteration for each adulterated sample's mixture for six spectra replicas). MS spectra were manually acquired, and visually inspected before statistical analysis.

Spectra were loaded into FlexAnalysis software, version 3.0, (Bruker Daltonics) and preprocessed for: (i) mass adjustment, spectra were compressed by a factor of 10 in the total mass range; (ii) smoothing, mass data were adjusted by the Savitsky–Golay algorithm with a frame size of 25 Da; (iii) baseline subtraction, was applied the minimum value for finding the baseline; (iv) normalization, was applied the maximum morm to normalize the baseline subtracted data; and (v) peak picking, was applied spectra differentiation algorithm for finding the peaks, maximum peaks 100, threshold 0.1, method peak fitting. The entire preprocessed raw datasets of the 180 spectra replicas from the milk samples and 288 spectra replicas from the simulated fraudulent addition of DM and GM by CM, EM and BM were imported in R Bioconductor [24] for correlation matrices calculation (Pearson's correlation) and hierarchical clustering. To gain confidence on the structure of the tree, we used the resampling package pvclust. For each cluster in hierarchical clustering, p-values (between 0 and 1) are calculated via multiscale bootstrap resampling. The package *pvclust* provides two types of p-values: AU (Approximately Unbiased) and BP (Bootstrap Probability). AU, which is computed by multiscale bootstrap resampling, is a better approximation to unbiased p-value than BP value computed by normal bootstrap resampling. The same set of spectra, were imported in Biotyper 3.1 software and converted into a virtual gel (pseudo-gel-like) format. The mass values (m/z) were reported on the X axis, while the color bar, reported on the Y axis, showed the relationship between the color and the pick intensity.

To visualize differences between milk and milk-adulterated populations, hierarchical clustering analysis was performed by principal component analysis (PCA) and dendrogram creation, via the external MATLAB (Matrix Laboratory) tool integrated in the MALDI Biotyper (Bruker Daltonik GmbH, Bremen, Germany).

In order to the PCA clustering method, we used the following parameters: Method, hierarchical; distance measure, correlation; linkage algorithm, average.

CONCLUSIONS

The detection of fraudulent adulterations and unintended contaminations of DM and GM by fast and efficient analytical methods is a matter of fundamental importance because many allergic children are fed with these fundamental nutrients. To this purpose, in this work we presented a suitable analytical method based on MALDI-TOF MS coupled to a robust statistical analysis (hierarchical clustering and PCA analysis) that provides substantial advances in the typing of milk and allows the investigation of the commonest types of adulterations in a straightforward, accurate and sensitive way.

This approach does not involve laborious pre-analytical sample separation steps and represents a fast reliable and robust method for routine analyses also for dairy industry. We would like to emphasize that although the Biotyper analysis allows to recognize adulterated milks (when compared with other types of milk) made by low percentage (even below 0.5%) of adulterant, a *post hoc* refinement of the analysis (bootstrapping analysis) indicated that the method is reliable for recognition of samples adulterated up to 0.5%. Overall, these results suggest that MALDI-TOF spectra of milk samples preliminarily analyzed by the integrated Biotyper software, allows a rapid determination of specimens adulterated even at low percentage of adulterant. Of note, the suggested workflow is far from being a quantitative analytical method, as the discrimination of classes of samples at different percentages of adulteration is still not possible with the analytical and statistical methods employed in this work.

Finally, the analytical strategy herein described is highly sensitive and robust compared to the techniques employed and reported so far in the literature and allows the qualitative identification of adulterated DM and GM milks obtained by the fraudulent addition of other types of milk up to a limit of 0.5%.

Acknowledgments

This work was supported by San Camillo Hospital, Treviso, Italy Fellowship Program to Maurizio Muraca; Ministry of Health, Ricerca Corrente (RC201402G003251) to Lorenza Putignani.

AUTHOR CONTRIBUTIONS

Francesco Di Girolamo conceived and designed the study, recruited samples, acquired mass spectra, interpreted the data and drafted the manuscript; Andrea Masotti performed the statistical analysis of the data and participated in drafting the manuscript; Margherita Scapaticci performed all pre-analytical procedures and contributed to data generation; Maurizio Muraca made substantial contributions to the interpretation of data and gave his valuable support in the critical revision of the manuscript and conceptual content; Lorenza Putignani contributed to conceive and designed the study and to draft the manuscript. All the authors read and approved the final manuscript.

REFERENCES

1. Salimei, E.; Fantuz, F. Equid milk for human consumption. *Int. Dairy J.* 2012, *24*, 130–142. .

2. Roncada, P.; Gaviraghi, A.; Liberatori, S.; Canas, B.; Bini, L.; Greppi, G.F. Identification of caseins in goat milk. *Proteomics* 2002, *2*, 723–726. .
3. Park, Y.W. Hypo-allergenic & therapeutic significance of goat milk. *Small Rumin. Res.* 1994, *14*, 151–159. .
4. Iacono, G.; Carroccio, A.; Cavataio, F.; Montalto, G.; Soresi, M.; Balsamo, V. Use of ass's milk in multiple food allergy. *JPGN* 1992, *14*, 177–181. .
5. Addeo, F.; Pizzano, R.; Nicolai, M.A.; Caira, S.; Chianese, L. Fast isoelectric focusing and antipeptide antibodies for detecting bovine casein in adulterated water buffalo milk and derived mozzarella cheese. *J. Agric. Food Chem.*2009, *57*, 10063–10066. .
6. Enne, G.; Elez, D.; Fondrini, F.; Bonizzi, I.; Feligini, M.; Aleandri, R. High performance liquid chromatography of governing liquid to detect illegal bovine milk's addition in water buffalo mozzarella: Comparison with results from raw milk and cheese matrix. *J. Chromatogr. A* 2005, *1094*, 169–174. .
7. Cunsolo, V.; Muccilli, V.; Saletti, R.; Foti, S. Applications of mass spectrometry techniques in the investigation of milk proteome. *Eur. J. Mass Spectrom.* 2011, *17*, 305–320. .
8. Cozzolino, R.; Passalacqua, S.; Salemi, S.; Garozzo, D. Identification of adulteration in water buffalo mozzarella and in ewe cheese by using whey proteins as biomarkers and matrix-assisted laser desorption/ionization mass spectrometry. *J. Mass Spectrom.* 2002, *37*, 985–991. .
9. Czerwenka, C.; Muller, L.; Lindner, W. Detection of the adulteration of water buffalo milk and mozzarella with cow's milk by liquid chromatography–mass spectrometry analysis of β-lactoglobulin variants. *Food Chem.* 2010, *122*, 901–908. .
10. Chen, R.K.; Chang, L.W.; Chung, Y.Y.; Lee, M.H.; Ling, Y.C. Quantification of cow milk adulteration in goat milk using high-performance liquid chromatography with electrospray ionization mass spectrometry. *Rapid Commun. Mass Spectrom.* 2004, *18*, 1167–1171. .
11. Cunsolo, V.; Cairone, E.; Muccilli, V.; Saletti, R.; Foti, S. Sequence and phosphorylation level determination of two donkey β-caseins

by mass spectrometry. *Rapid Commun. Mass Spectrom.* 2009, *13*, 1907–1916. .

12. Cunsolo, V.; Cairone, E.; Fontanini, D.; Criscione, A.; Muccilli, V.; Saletti, R.; Foti, S. Sequence determination of α_{s1}-casein isoforms from donkey by mass spectrometry methods. *J. Mass Spectrom.* 2009, *44*, 1742–1753. .
13. Cunsolo, V.; Muccilli, V.; Fasoli, E.; Saletti, R.; Righetti, P.G.; Foti, S. Poppea's bath liquor: The secret proteome of she-donkey's milk. *J. Proteomics* 2011, *74*, 2083–2099. .
14. Saletti, R.; Muccilli, V.; Cunsolo, V.; Fontanini, D.; Capocchi, A.; Foti, S. MS-based characterization of α_{s2}-casein isoforms in donkey's milk. *J. Mass Spectrom.* 2012, *47*, 1150. .
15. Giuffrida, M.G.; Cantisani, A.; Napolitano, L.; Conti, A.; Godovac-Zimmermann, J. The amino acid sequence of two isoforms of α-lactalbumin from donkey (*Equus asinus*) milk is identical. *Biol. Chem. Hoppe-Seyler* 1992, *373*, 931–935. .
16. Godovac-Zimmermann, J.; Conti, A.; Napolitano, L. The primary structure of donkey (*Equus asinus*) lysozyme contains the Ca(II) binding site of α-lactalbumin. *Biol. Chem.* 1988, *369*, 1109–1115. .
17. Herrouin, M.; Mollé, D.; Fauquant, J.; Ballestra, F.; Maubois, J.L.; Leonil, J. New genetic variants identified in donkey's milk whey proteins. *J. Protein Chem.* 2000, *19*, 105–115. .
18. Godovac-Zimmermann, J.; Conti, A.; James, L.; Napolitano, L. Microanalysis of the amino-acid sequence of monomeric β-lactoglobulin I from Donkey (*Equus asinus*) milk. *Biol. Chem.* 1988, *369*, 171–179. .
19. Cozzolino, R.; Passalacqua, S.; Salemi, S.; Malvagna, P.; Spina, E.; Garozzo, D. Identification of adulteration in milk by matrix-assisted laser desorption/ionization time-of-flight mass spectrometry. *J. Mass Spectrom.* 2001, *36*, 1031–1037. .
20. Cunsolo, V.; Muccilli, V.; Saletti, R.; Foti, S. MALDI-TOF mass spectrometry for the monitoring of she-donkey's milk contamination or adulteration. *J. Mass Spectrom.* 2013, *48*, 148–153. .
21. Fanton, C.; Delogu, G.; Maccioni, E.; Podda, G.; Seraglia, R.; Traldi, P. Matrix-assisted laser desorption/ionization mass spectrometry in the dairy industry 2. The protein fingerprint of ewe cheese and

its application to detection of adulteration by bovine milk. *Rapid Commun. Mass Spectrom.* 1998, *12*, 1569–1573. .

22. Calvano, C.D.; de Ceglie, C.; Monopolia, A.; Zambonina, C.G. Detection of sheep and goat milk adulterations by direct MALDI-TOF MS analysis of milk tryptic digests. *J. Mass Spectrom.* 2012, *47*, 1141–1149. .
23. Putignani, L.; del Chierico, F.; Onori, M.; Mancinelli, L.; Argentieri, M.; Bernaschi, P.; Coltella, L.; Lucignano, B.; Pansani, L.; Ranno, S.; *et al.* MALDI-TOF mass spectrometry proteomic phenotyping of clinically relevant fungi. *Mol. Biosyst.* 2011, *7*, 620–629. .
24. Gentleman, R.C.; Carey, V.J.; Bates, D.M.; Bolstad, B.; Dettling, M.; Dudoit, S.; Ellis, B.; Gautier, L.; Ge, Y.; Gentry, J.; *et al.* Bioconductor: Open software development for computational biology and bioinformatics. *Genome Biol.*2004, *5*, R80. .

SECTION III

PROTEOMICS AND FOOD ALLERGIES

CHAPTER

9

CURRENT CHALLENGES IN DETECTING FOOD ALLERGENS BY SHOTGUN AND TARGETED PROTEOMIC APPROACHES: A CASE STUDY ON TRACES OF PEANUT ALLERGENS IN BAKED COOKIES

Romina Pedreschi, Jørgen Nørgaard and Alain Maquet

European Commission (EC), DG Joint Research Centre (JRC), Institute for Reference Materials and Measurements (IRMM), Geel 2440, Belgium

ABSTRACT

There is a need for selective and sensitive methods to detect the presence of food allergens at trace levels in highly processed food products. In this work, a combination of non-targeted and targeted proteomics approaches are used to illustrate the difficulties encountered in the detection of the major peanut allergens Ara h 1, Ara h 2 and Ara h 3 from a representative processed food matrix. Shotgun proteomics was employed for selection of the proteotypic peptides for targeted approaches via selective reaction monitoring. Peanut presence through detection of the proteotypic Ara h 3/4 peptides AHVQVVDSNGNR (*m/z* 432.5, 3+) and SPDIYNPQAGSLK (*m/z*695.4, 2+) was confirmed and the developed method was able to detect peanut presence at trace levels (≥10 μg peanut g^{-1} matrix) in baked cookies.

Keywords:data dependent acquisition; peanut; food allergens; proteomics; shotgun; selective reaction monitoring; targeted

INTRODUCTION

An increasing incidence of food allergies in Europe and USA [1,2] is being reflected in clinical studies. European legislation recognizes so far 14 major allergenic foods [3] and requires a mandatory declaration when they are part of the ingredients; in cases where the manufacturer cannot exclude their presence as a result of accidental contamination in foodstuffs the label often contains the phrase "may contain". But so far, there are no established threshold limits below which an allergen poses only a small risk of causing harm to an allergic consumer. Even though commonly accepted trigger levels have not been established yet, there is consensus among the scientific community, that allergen detection methods should be capable of covering the low ppm range (1–10 mg allergenic ingredient kg^{-1} food product).

Currently, to detect allergens in food products, enzyme-linked immunosorbent assays (ELISA) and PCR analyses were adopted as methods of choice by the food industry and official food control agencies [2,4]. However, food matrix and processing effects can result in large numbers of false positives and negatives using these methods [5], which lack the precision and rigor needed in cases of liability issues [6]. Mass spectrometry driven approaches on the other hand can be successfully used as a confirmatory platform given the specificity and sensitivity that can be achieved. In addition, multi-allergen detection and quantification is feasible after some stringent considerations are fulfilled [6].

The bottom up mass spectrometry approach (reconstruction of the protein based on peptides) either for characterization of the allergenic protein or by a targeted approach for sensitive detection/quantification are being used as platforms for allergen detection with promising outcomes [7,8].

The lack of established guidelines addressing different issues during sample preparation for analysis of food allergens makes it difficult to have a generic platform for multi-allergen detection and quantification. Published investigations in many cases lack experimental details and pose question marks relating to critical points of the food allergen detection workflows. It is of urgent necessity that certain aspects of the protocol used are stated: target analyte, source of allergenic food used (e.g., raw, roasted, defatted, *etc.*), incurred or spiked allergenic food (before or after processing); clear statement of reporting units (mass ratio of allergenic food, total proteins, allergenic protein target), digestion conditions (possible peptide modifications), *etc.*, just to cite a few aspects of the sample preparation phase.

In this manuscript, we would like to illustrate with peanut serving as an example: (i) the challenges to be faced to detect food allergens at trace levels in complex food matrices (1–10 mg peanuts kg^{-1} cookie); (ii) the need for improvements in sample preparation and mass spectrometry analytical tools to achieve low levels of detection; and (iii) the awareness on key issues related to the development of a robust multi-allergen and quantitative method for trace analysis of food allergens.

MATERIALS AND METHODS

Materials

Peanut included in the IRMM-481f peanut test material (mixture of five varieties, five different processing conditions) were utilized (Table 1). Wheat flour based cookie containing different amounts of IRMM peanut mixture (0, 10, 100, 1000 and 10,000 $\mu g \cdot g^{-1}$ matrix) were prepared in house. Incurred cookies were baked at 180 °C for 16 min. The wheat flour based cookie recipe used consisted of: wheat flour 49.0%, butter 19.6%, dust sugar 18.4%, skimmed milk powder 5.9%, water 6.6%, sodium chloride 0.3%, sodium hydrogen carbonate 0.1%, ammonium bisulfate 0.1%. Baked cookies were grinded to a particle size <250 μm in liquid nitrogen and stored at −20 °C until use. Cookie dough was incurred with different amounts of peanuts before baking. Homogeneity of the cookie material was tested using ELISA kits.

Table 1. Detailed information of the IRMM-481 peanut test material used to prepare the incurred cookies.

Peanut variety, origin	Peanut treatment
Runners Argentina	Blanched air-roasted at 140 °C for 20 min
Common Natal, South-Africa	Raw, air-roasted at 160 °C for 13 min
Virginia, USA	Blanched, oil roasted at 145 °C for 25 min
Virginia, China	Blanched, oil roasted at 140 °C for 9 min
Jumbo Runners, USA	Blanched only

Chemicals

All chemicals used for sample preparation were purchased from VWR International (West Chester, PA, USA) and were at least analytical reagent grade. PlusOne chemicals for gel electrophoresis (Tris, glycine, 3-[(3-cholamidopropyl)dimethylammonio]-1-propanesulfonate (CHAPS), urea, thiourea, dithiothreitol (DTT), dimethylformamide), 2D clean up and 2D quantification kits were purchased from GE Healthcare (Uppsala, Sweden). Mini gels NUPAGE® 12% Bis-Tris (1.0 mm) were obtained from Invitrogen (Carlsbad, CA, USA). Water from a Milli-Q water system (Millipore, Bedford, MA, USA) was used. Trypsin mass spectrometry grade (Cat. # 786-578) was obtained from G Biosciences (St. Louis, MO, USA).

Protein Extraction and Quantification

Tris buffered saline (TBS) extraction as previously described [5] was used. One to five g of cookie sample were extracted in 10 to 20 mL of TBS buffer (20 mM Tris, 150 mM NaCl, pH 7.4) at 4 °C in an ultrasonic bath for 20 min. After centrifugation at 3500 g for 30 min, proteins in the supernatant were quantified with the 2D Quant kit and further purified and precipitated with the 2D clean up kit following the instruction manuals of GE Healthcare (Uppsala, Sweden).

Protein Enrichment

A large and highly diverse bead-based library of combinatorial peptide ligands known as ProteoMiner™ (Biorad, MO, USA) was used for protein enrichment. The large sample loading capacity ProteoMiner™ protein enrichment kit was employed. One to five g of cookie samples incurred with peanuts (0, 10, 100, 1000 and 10,000 $\mu g \cdot g^{-1}$ matrix) were extracted with TBS buffer. After equilibration of the columns as described in the kit, proteins were bound to the ligands for 3 h at room temperature under agitation. Samples were washed with phosphate buffered saline solution (PBS) composed of 150 mM NaCl, 10 mM NaH_2PO_4 at pH 7.4 following the instructions of the manufacturer. The unbound fraction was collected in a separate tube. The enriched sample was eluted with 8 M urea and 2% CHAPS according to the manufacturer's instructions. For all cookie samples incurred with peanuts, SDS-PAGE was used to analyze the protein profiles of the initial extract, enriched and unbound fractions. Protein content was normalized to 5 µg per lane to allow comparisons among the different fractions. For mass spectrometry analysis, samples (initial extract, enriched and unbound fractions) were cleaned up and precipitated with the GE Healthcare kit described in the material and method section and further trypsin digested.

SDS-PAGE

Fifteen well mini gels NUPAGE® 12% Bis-Tris (1.0 mm) from Invitrogen were loaded with 5 µg protein sample and run in a NUPAGE® electrophoresis system (Invitrogen). Two gels were run at 200 V for approximately 50 min. Silver stained gels [9] were scanned and evaluated with Labscan software (GE Healthcare). Molecular weight standards of 3.5–260 kDa from Invitrogen and of 8–220 kDa from Sigma-Aldrich (USA) were run in parallel.

Gel Free and Gel Based Trypsin Digestion

Gel free trypsin digestion was performed as follows: Fifty micrograms of protein were dissolved in 50 µL of Rapigest™ (Waters, Manchester, UK) in a 50 mM ammonium bicarbonate solution. Two and a half microlitres of a 50 mM DTT solution in 50 mM ammonium bicarbonate was added and let to stand for 30 min at 60 °C under agitation. Samples were cooled down and 5 µL of a 100 mM iodoacetamide in 50 mM ammonium bicarbonate was added and let to stand for 30 min in darkness. One microlitre of a sequencing grade 1 $\mu g \cdot \mu L^{-1}$ trypsin solution (1:50, enzyme:protein ratio) was added and let to incubate for 5 h at 37 °C. After incubation,

5 μL of 500 mM HCl solution was added and transferred to a molecular mass cut-off filtration device (3000 MWCO) and centrifuged at 14,000 g for 10 min. The filtrate was recovered for nano UPLC-Q-TOF-MS/MS analysis.

Gel bands were destained with 15 mM potassium ferrocyanide per 50 mM sodium thiosulfate solution and then reduced with 10 mM DTT in 50 mM ammonium bicarbonate for 30 min. Alkylation was done with 55 mM iodoacetamide in 50 mM ammonium bicarbonate for 30 min. Following, gel plugs were dehydrated with ACN in a vacuum concentrator until dryness and then trypsin (20 ng·μL^{-1}) was added and let to stand overnight at 37 °C. Peptides were extracted once with 1% formic acid and 2% ACN and twice with 50% ACN. After evaporation until dryness in a vacuum concentrator, peptides were re-suspended in 0.1% formic acid.

Nano LC Separation of Peptides

Separation of peptides was performed on a nano LC reversed phase chromatography system. A 180 μm × 20 mm Symmmetry C_{18} (5 μm) nano Acquity™ was used as trap column. Separation of peptides was carried out on a 75 μm × 100 mm (1.7 μm) BEH 130 C_{18} nano Acquity™ column. Eluent A was composed of 0.1% formic acid in water and eluent B was 0.1% formic acid in ACN. A flow rate of 0.4 μL·min^{-1} was used. The linear gradient used to achieve separation of peptides was as follows: (initial) 97% A, 3% B; (0.5–80 min) 60% A, 40% B; (80–85 min) 15% A, 85% B; (85–90 min) 15% A, 85% B; (90–95 min) 97% A and 3% B and (95–100 min) 97% A and 3% B. One microlitre of a sample mixture of ~0.8 μg·μL^{-1} prepared in Rapigest™ was injected for in liquid digested samples and 5 μL of the same sample mixture for gel digested samples.

Nano Electrospray Q-TOF Tandem Mass Spectrometry (Nano-ESI Q-TOF MS/MS)

MS/MS experiments were carried out in a quadrupole time of flight mass spectrometer (Q-TOF Ultima Global, Waters, Manchester, U.K) equipped with a nano-electrospray Z spray source. The operating conditions of the Q-TOF-MS were: Capillary voltage 3.0 kV, sample cone voltage, 100 V; source temperature, 80 °C. The instrument was operated in positive ion mode. Time of flight was operated in a continuous extraction mode. In the positive linear mode (V mode) an accelerating voltage of 9.10 kV was used for the TOF and the MCP value of the detector was set to 2200 V. A mass calibrant of Glu-Fibrinopeptide B (Sigma Aldrich) at 500 *f*mol·μL^{-1} was utilized in MS/MS mode. Lock spray was utilized during the whole acquisition. Full scans (MS mode) were performed over the 350–1900 *m/z* range with scan time of 0.9 s and interscan time of 0.1 s. For peptide fragmentation, the instrument was used in a data dependent acquisition MS survey mode (MS/MS). Thus, the fragmentation of an ion is achieved when a minimum intensity (specified value) is detected (simultaneous fragmentation of the three most abundant ions). The collision energy was varied between 5 and 55 V according to the mass and charge state of the respective peptides.

Database Searching

The fragmentation ion spectra obtained from the MS survey mode were processed using Mass Lynx version 4.0 (Waters), a software that converts MS/MS raw data to peak lists. After centroiding and background subtraction, the generated PKL files obtained were used for Mascot (Matrix of science, London, UK) database searching against a customized peanut allergen and Swiss-Prot database. This customized allergen database consisted of Swiss-Prot sequences of: allergen Ara h 1 (P43237, clone P17 precursor), allergen Ara h 1 (P43238, clone P41B precursor), allergen Ara h 2 isoform (Fragment, Q7Y1C0), allergen Ara h 2.02 (Q8GV20), allergen II (Fragment, Q941R0), allergen Ara h 3/Ara h 4 (Q8LKN1), Ara h 3 Glycinin (Fragment, O82580), Ara h 4 Glycinin (Q9SQH7), allergen Ara h 5 Profilin (Q9SQI9), allergen Ara h 6 (Fragment, Q9SQG5), allergen Ara h 7 (Q9SQH1), allergen Ara h 8 (Q6VT83). A maximum of one missed cleavages were allowed. Peptide tolerance and MS/MS tolerance were set to 100 ppm and 0.1 Da respectively. Modifications on cysteine residues by carboxyamidation were set as fixed and a possible modification of methionine by oxidation was set as variable modification.

Selective Reaction Monitoring (SRM)

SRM data were acquired on a Waters Quattro Premier triple quadrupole coupled to a Waters nanoAcquity Ultra Performance system fitted with a Waters Symmetry 5 µm particle diameter C_{18} 180 µm × 20 mm trap column and a 75 µm × 150 mm (1.7 µm) BEH 130 C_{18} column. Peptides were eluted using a linear gradient of 1–35% B over 50 min at a flow rate of 0.3 µL·min^{-1}. Solvent A corresponded to 0.1% formic acid in milli-Q water and solvent B to 0.1% formic acid in ACN.

The mass spectrometer was operated in SRM mode. For ionization, 2.75 kV of capillary voltage, 35 V of cone voltage and 100 °C capillary temperature were used. The collision energy was calculated using the following formulas: CE = 0.044 * (*m/z*) + 5.5 for doubly charged ions and CE = 0.051 * (*m/z*) + 0.5 for triply charged ions. Selection of transitions (precursor/fragment combinations) was based on the DDA experimental data generated on an Ultima Global Q-TOF instrument (Waters). The SRM transitions monitored for Ara h 1, Ara h 2 and Ara h 3 are presented in Table 2. A Blast similarity search using the tool [10] was performed confirming the specificity of these peptides. The dwell time (DT) for each transition was 0.05 s.

Table 2. Selective Reaction Monitoring (SRM) transitions monitored for detection of Ara h 1, Ara h 2 and Ara h 3.

Peptides	**Parention (*m/z*) (+)**	**Fragmention(s) (*m/z*)**	**CE** [a]
Ara h 1			
VLLEENAGGEQEER	786.9 (+2)	989.5 (y9)/875.4 (y8)/804.4 (y7)/747.3 (y6)/561.3 (y4)/304.2 (y2)	40
DLAFPGSGEQVEK	688.8 (+2)	1077.5 (y10)/930.5 (y9)/833.4 (y8)/447.2 (b4)/300.2 (b3)/229.1 (b2)	36
Ara h 2			
CCNELNEFENNQR	863.8 (+2)	1050.5 (y8)/807.4 (y6)/660.3 (y5)/531.3 (y4)	43
NLPQQCGLR	543.3 (+2)	858.4 (y7)/761.4 (y6)/633.3 (y5)/200.1 (a2)	29
CDLEVESGGR	561.2 (+2)	846.4 (y8)/604.3 (y6)/505.2 (y5)/376.2 (y4)	30
CMCELQQIMENQSDR	1006.9 (+2)	1721.8 (y14)/1361.7 (y11)/1248.6 (y10)/992.5 (y8)/879.4 (y7)/748.3 (y6)/619.3 (y5)/292.1 (b2)	49
Ara h 3			
LNAQRPDNR	361.9 (+3)	1083.6 (ymax)/970.5 (y8)/856.4 (y7)/657.4 (y5)/228.1 (b2)	19

SPDIYNPQAGSLK	695.4 (+2)	1389.7 (ymax)/1302.7 (y12)/977.5 (y9)/814.5 (y8)/700.4 (y7)/475.3 (y5)/300.1 (b3)	36
AHVQVVDSNGNR	432.5 (+3)	749.5 (b7)/663.3 (y6)/535.4 (b5)	

RESULTS AND DISCUSSION

Allergen detection methods must be sensitive and selective enough to detect the target proteins at levels as low as 1–10 mg allergenic food kg^{-1} food matrix. When investigating peanuts as the allergenic food source, given that they are largely incorporated in processed foods and thus a potential cross contact of intended peanut free products cannot be ruled out representing a constant risk to allergenic consumers [11]. In addition, detection of traces of peanut allergens in food products is difficult because they are obscured by the food matrix and processing hinders detection.

The workflow of Figure 1 is a typical approach followed to detect food allergens. It includes basic steps such as protein extraction, enrichment (if necessary), protein digestion and detection of peptides from the allergenic proteins through mass spectrometry platforms. To date there are eleven allergenic proteins in peanuts identified of which Ara h 1, Ara h 2 and Ara h 3 are recognized as the most abundant and major peanut allergens [12,13,14]. Ara h 1, Ara h 2 and Ara h 3 represent 12–16%, 5.9–9.3% and 21.8–38.5% of the total protein content determined by SDS-PAGE quantitative studies [12]. If the target is to detect 10 ppm of allergenic food kg^{-1} cookies the question arises, *how much is the quantity of the major allergens reaching the detector in the mass spectrometer? Since Ara h 3 is the most abundant peanut allergen across different varieties, all model calculations were done with this target.* It can be assumed that the average protein content of peanuts is 20%, and that Ara h 3 represents 30% of total peanut proteins. Five grams of cookies contaminated with 10 μg peanuts g^{-1} matrix are extracted with 20 mL of TBS buffer giving an average content of Ara h 3 of 2.45 *f*mol μL^{-1}solution. For protein digestion, 50 μg of total proteins (from the cookie matrix plus peanuts) are treated and approximately 0.8 μg total protein is loaded on the column which results in an ideal and theoretical content of 480 *a*mol Ara h 3. This is an optimistic scenario, since 100% recovery during protein extraction, complete digestion and 100% ionization efficiency of peptides is assumed. Therefore, *detection methods should be capable to detect targeted peptides in the amol-low femtomol range.*

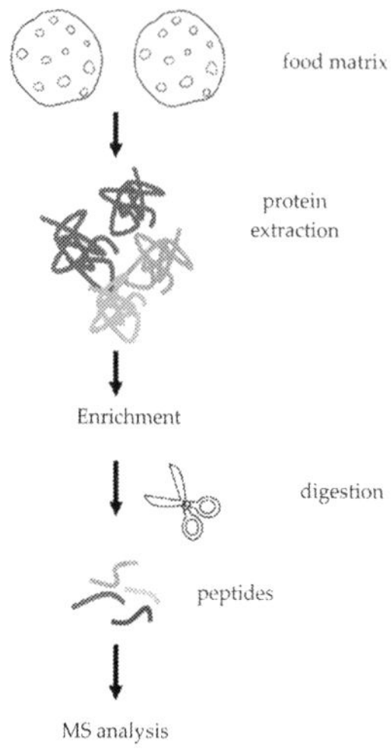

Figure 1. Workflow for sample preparation for mass spectrometry analysis. The different steps involve: protein extraction from the matrix, protein enrichment (optional), enzymatic (e.g., trypsin) digestion and MS analysis of the peptide mix either through shotgun (DDA) or targeted (SRM) proteomic approaches.

To complicate this scenario, the presence of other highly abundant proteins from the matrix (cookie) may hamper detection of the low abundant ones (trace amounts of target allergenic foods) by ion suppression of the target peptides. In this particular case, wheat flour and skimmed milk powder make up 50% and 6% of the cookie composition. Assuming total protein contents of 10% and 36% respectively for wheat flour and skimmed milk powder, we are talking about differences in protein or peptide levels of at least 6 orders of magnitude. *A reasonable question would be if current mass spectrometry technology is able to cope with such huge differences?* This will be discussed in the coming sections.

Up to now processing effects are not sufficiently being taken into account. Food processing may induce some unknown protein modifications in the allergen targets [15]. Figure 2 displays a SDS-PAGE protein profile for the IRMM-481f peanut mix used to prepare the incurred cookies and cookies incurred with high amounts of peanuts (100,000 $\mu g \cdot g^{-1}$ matrix). The allergome of peanut is rather complex. Ara h 1 is a 63–68 kDa glycoprotein assembled in di and trimeric complexes. Of the allergen Ara h 2 two isoforms with masses of *ca.* 16 and 17 KDa have been isolated. Ara h 3 and Ara h 4 are isoallergens and can be designated as the allergen Ara h 3/4. The allergenic protein Ara h 3/4 consists of an acidic and basic subunit which remain covalently linked by an intermolecular disulfide bridge and associate into a very stable hexameric structure. The acidic subunit has a molecular mass in the range of 40–45 kDa, whereas the basic subunit has a mass of *ca.* 25 kDa. Ara h 3/4 is mainly proteolytically modified (truncation at multiple sites) with possible glycosylation. Proteolytic truncation was observed for the acidic subunit but not for the basic resulting in a series of polypeptides ranging from 13–45 kDa [16]. Ara h 1 and Ara h 3 were clearly identified through mass spectrometry in both samples (bands 1–4; 11 and 14; Table 3). Band 11 identified Ara h 1 but also the presence of alpha-casein which was detected and confirmed by mass spectrometry (Table 3). Co-migration of major allergens with matrix proteins was observed and thus interference with their detection at trace levels is expected. Band 13 was negative for Ara h 3 and identified casein instead, while Ara h 2 (bands 5 and 7) could not be detected in the processed food product even when the concentration of added peanut was very high. Ara h 1 and Ara h 3 are present in higher concentrations than Ara h 2 in the different peanut varieties [12]. IRMM-481f peanut mix presents a low amount of Ara h 2 which limited its qualitative detection by SDS-PAGE and could not be detected even when the matrix contained very high concentrations of peanuts. *Is it because of the low abundance of this protein and low sensitivity of the gel based approach or are there any interactions with other matrix components that resulted in protein modification and thus protein migrate differently and consequently was not identified by MS analysis?* These are certainly issues that need to be answered.

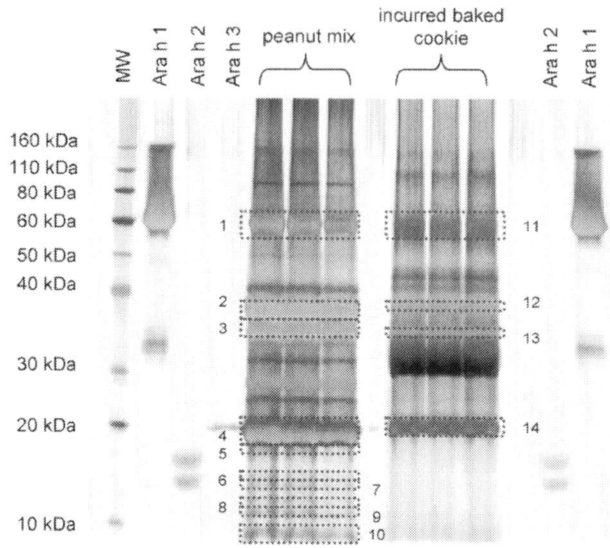

Figure 2. SDS-PAGE protein profiles for IRMM-481f peanut mix and 16 min baked cookie incurred with 100,000 μg IRMM-481f peanut mix g^{-1} matrix. The different band numbers were submitted to trypsin digestion and peptides submitted to nano LC-Q-TOF-MS/MS for protein identification. The list of peptides and identified proteins are presented in Table 3. Five micrograms protein was loaded per lane. A molecular weight standard (MW) of 3.5–260 kDa was run in parallel.

Table 3. Tryptic peptides identified from IRMM peanut mix and 16 min baked cookies incurred with 100,000 μg peanuts g^{-1} matrix of IRMM peanut mix.

Protein identification	Band # [a]	Peptides matched	Mascot Score
IRMM peanut mix			
Ara h 1 clone P41B precursor	1	18, 15, 26	756, 744, 1473
Ara h 1 clone P17 precursor		16, 15, 29	704, 726, 1619
Allergen Ara h 3 Ara h 4		1, 3, 9	81, 76, 509
Allergen Ara h 3 Ara h 4	2	13, 7, 12	637, 369, 728
Ara h 3 Glycinin	3	1, -, 1	90, -, 70
Allergen Ara h 3 Ara h 4	4	6, 11, 9	365, 771, 576
Allergen Ara h 2	5	3, 3	140, 114
Ara h 3 Glycinin	6	2	98
Allergen Ara h 3 Ara h 4	7	2	96

Ara h 1 clone P17 precursor		2	67
Ara h 1 clone P41B precursor		2	67
Allergen Ara h 2 isoform		1	51
Allergen Ara h 6	8	6	209
Allergen Ara h 3 Ara h 4	9	8	431
Allergen Ara h 6		10	369
Allergen Ara h 3 Ara h 4	10	10	558
Cookie incurred at 100,000 µg·g^{-1}			
Ara h 1 clone P41B precursor			
Ara h 1 clone P17 precursor	11	14, 10, 8	766, 582, 481
Allergen Ara h 3 Ara h 4		13, 11, 10	686, 569, 552
		4, 4, 2	264, 190, 134
Allergen Ara h 3 Ara h 4			
	12	3, 1, 1	103, 20, 50
Alpha casein S2			
	13	3, 3	102, 117
Allergen Ara h 3 Ara h 4			
Alpha casein S2	14	6, 7	340, 524
		(−), 2	(−), 103

[a] Band numbers correspond to those reported in Figure 2. When sufficient amount of sample, three independent replicates were submitted to DDA MS/MS protein identification (three corresponding values for peptides matched, Mascot score and expect value) otherwise one or two samples were used. (−) Stands for a negative hit. The expect value is the probability to obtain a random protein identification. Expect values were in all cases <0.0001 peanuts g^{-1} matrix.

Shotgun Proteomics Approach: Nano LC-Q-TOF-MS/MS

Protein extracts from incurred cookies with 0, 10, 100, 1000 and 100,000 µg peanuts g^{-1} matrix were digested with trypsin and the peptide mixture was submitted to data dependent acquisition (DDA) with a Q-TOF instrument. Results are presented in Table 4. Ara h 1 was detected in samples containing at least 10,000 µg peanuts g^{-1} matrix. Ara h 3 was detected in samples containing ≥1000 µg·g^{-1}.

Table 4. DDA MS/MS results for 16 min baked cookies incurred with different amounts of peanuts (0–100,000 μg g^{-1} matrix).

Amount of peanut μg·g^{-1} matrix	**Protein Identification**	**Peptides**	***m/z***	**charge**
1	Negative			
10	Negative			
100	Negative			
1000	Ara h 3/Ara h 4	LNAQRPDNR	362	3+
	Ara h 4 Glycinin	AHVQVVDSNGNR	433	3+
	Ara h 3 Glycinin			
10,000	Ara h 3/Ara h 4	LNAQRPDNR	362	3+
	Ara h 4 Glycinin	AHVQVVDSNGNR	433	3+
	Ara h 3 Glycinin	SPDIYNPQAGSLK	695	2+
	Ara h 1 clone P17	VLLEENAGGEQEER	787	2+
	Ara h 1 clone P41B	DLAFPGSGEQVEK	689	2+
		DQSSYLQGFSR	644	2+
		GTGNLELVAVR	565	2+
100,000	Ara h 3/Ara h 4	LNAQRPDNR	362	3+
	Ara h 4 Glycinin	AHVQVVDSNGDNR	433	3+
	Ara h 3 Glycinin	SPDIYNPQAGSLK	695	2+
		FNLAGNHEQEFLR	788	2+
		GENESDEQGAIVTVR	802	2+
		FFVPPSEQSLR	654	2+
		TANDLNLLILR	628	2+
		RPFYSNAPQEIFIQQGR	684	3+
	Ara h 1 clone P17	VLLEENAGGEQEER	787	2+
	Ara h 1 clone P41B	DLAFPGSGEQVEK	689	2+
		DQSSYLQGFSR	644	2+
		GTGNLELVAVR	565	2+
		WGPAEPR	407	2+
		QFQNLQNHR	593	2+
		SSDNEGVIVK	524	2+
		GSEEEDITNPINLR	794	2+
		DGEPDLSNNFGR	661	2+
		IFLAGDKDNVIDQIEK	607	3+
		EGEQEWGTPGSHVR	524	2+
		SSENNEGVIVK	588	2+

		LFEVKPDK	488	2+
		EGALMLPHFNSK	672	2+

Peptides were selected based on E-values lower than 0.05. *E*-value is the probability of a random peptide identification. Results are based on three independent replicates.

Ara h 2 could not be detected even in samples containing 100,000 μg peanuts g^{-1} matrix. The differences in detectability of the three major allergens in the matrix seem to be related to their initial content in the peanut source. Ara h 3 is the most abundant protein followed by Ara h 1 and Ara h 2 [12]. Baked cookies incurred with trace amount of peanuts are an excellent representation of a processed complex food matrix where huge differences in the content of individual proteins (food allergens *vs.* bulk proteins from other ingredients) can be expected. In addition, not all proteins generate a sufficient amount of detectable peptides: poor ionization and/or fragmentation behaviour are main issues that prevent peptide detection with the state-of-the-art mass spectrometry technology, which is even more accentuated when very complex matrices are used [15]. A way to circumvent the huge differences in concentration of proteins/peptides in the sample and ion suppression is to enrich the targeted analytes [15,17,18]. There is resilience for the use of enrichment techniques in trace analysis not only because of the expense of time and work but because they might complicate quantitative analysis. However it was the strategy taken in this work, otherwise it was not feasible to attain the required detection limits.

In the present study, the shotgun approach (data dependent acquisition—DDA) was limited by the required rate to carry out a precursor scan and fragmentation event for a single peptide, which was not fast enough to analyze all peptides. Consequently, bias towards highly abundant peptides was observed. Thus, a simple protein enrichment protocol based on a peptide ligand library known as ProteoMiner™ was evaluated. First, a possible enrichment of the major peanut allergens was confirmed in the peanut material used to prepare the incurred cookies (IRMM-481f peanut mix). Ara h 1 and Ara h 3 were satisfactorily enriched in the IRMM peanut mix sample (Figure 3A). Ara h 2, however, was not satisfactorily enriched and seemed to be washed away (present in the unbound fraction). A different situation was observed when the matrix was incorporated (baked cookie), only enrichment of Ara h 3 could be evidenced through SDS-PAGE and was dependent on the amount of the extracted material (Figure 3B). At least a 10× enrichment for Ara h 3 from cookies was achieved. In cookies incurred with 100 μg peanuts g^{-1} matrix, Ara h 3 could be detected after enrichment through LC-Q-TOF MS/MS (Table 5). Previous works in allergen detection have made used of enrichment techniques such as ultrafiltration, precipitation, and peptide ligand libraries [12,19,20]. Enrichment can increase the identification rate of the low abundant fraction proteins/peptides but the intrinsic limitation of a shotgun approach will remain [21].

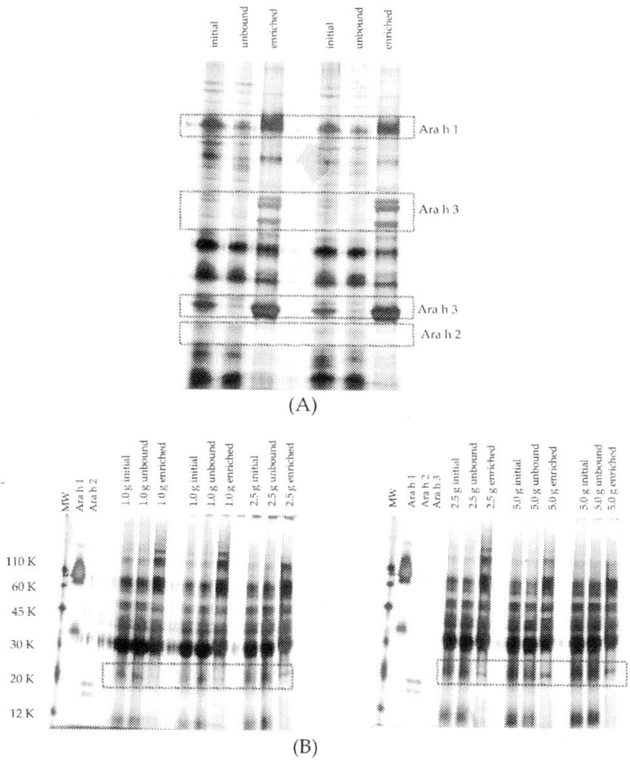

Figure 3. Enrichment of major peanut allergens through peptide ligand libraries (ProteoMiner™). (A) IRMM-481f peanut mix sample; (B) 16 min baked cookies incurred with 1000 μg IRMM-481f peanut mix g^{-1} matrix. Different amounts of starting material were extracted. Protein loaded was normalized to 5 μg per lane to allow comparisons. Initial corresponds to whole protein extract.

Table 5. DDA MS/MS results for 16 min baked cookies incurred with trace amounts of peanuts (0–1000 μg·g^{-1} matrix) after protein enrichment via peptide ligand library—ProteoMiner™.

Amount of peanut μg·g^{-1} matrix	Protein identification	Peptides (*m/z*, charge)	MS/MS
10	Negative		
100	Ara h 3/Ara h 4	LNAQRPDNR (361.9, 3+)	1, 1, 1
	Ara h 4 Glycinin		
	Ara h 3 Glycinin		
1000	Ara h 3/Ara h 4	LNAQRPDNR (361.9, 3+)	2, 1, 1
	Ara h 4 Glycinin	AHVQVVDSNGDNR (432.5, 3+)	1, 1, 1
	Ara h 3 Glycinin	SPDIYNPQAGSLK (695.4, 2+)	1, 1, 1
		QIVQNLR (435.8, 2+)	1, 1, 1
		GENESDEQGAIVTVR (802.4, 2+)	1, 1, 0

Peptides were selected based on *E*-values lower than 0.05. *E*-value is the probability of a random peptide identification. Results are based on three independent extractions.

It is worth noting the importance of a shotgun approach. Indeed a targeted approach such as selective reaction monitoring (SRM) is developed based on experimental data obtained through a shotgun approach (e.g., data dependent acquisition) for selection of the proteotypic peptides. In addition, a shotgun approach is useful to assess unexpected or known protein/peptide modifications which might be frequently encountered when working with processed foods.

Targeted Proteomics Approach: SRM

SRM is a highly sensitive approach to selectively detect and quantify peptides previously selected and relying on the monitoring of specific ion transitions [22].

Selection of SRM transitions to be monitored was based on experimentally obtained data with a DDA approach in an ESI-Q-TOF instrument (Table 2). Peptides highly observable and detectable for Ara h 1, Ara h 2 and Ara h 3 were selected. Fragment ions were selected based on MS/MS spectra of these highly observable precursor ions (Table 2). At least three transitions per peptide were monitored to facilitate reliable identification of the peptide. Previously enriched incurred cookie samples (1, 10 and 100 μg peanuts g^{-1} matrix) were submitted to SRM. Two proteotypic peptides from Ara h 3 were detected in enriched cookie samples containing ≥10 μg peanuts g^{-1} matrix. These peptides corresponded to: AHVQVVDSNGNR (*m/z* 432.5, triply charge ion) from which three out of the three fragment ions were detected and SPDIYNPQAGSLK (*m/z* 695.4, doubly charged ion, Figure 4) from which 4 out of 7 fragment ions were detected. However, Ara h 1 was only detected in samples containing 100 μg peanuts g^{-1} matrix (DLAFPGSGEQVEK, *m/z* 688.8 doubly charged ion) and Ara h 2 peptides were not detected in the enriched samples tested. In the present study, a gain of a factor of ten in the average signal was achieved by fractionation/enrichment techniques previous to SRM analysis as already reported by Picotti *et al.* analysing yeast samples [23].

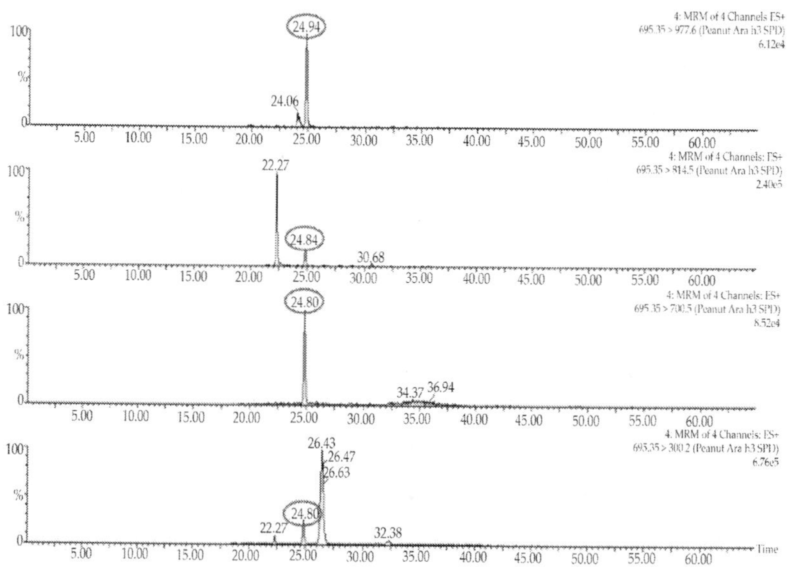

Figure 4. Selective Reaction Monitoring (SRM) of the Ara h 3 peptide SPDI-YNPQAGSLK (*m/z* 695.4, 2+) in enriched cookie samples incurred with 10 μg peanuts g^{-1} matrix. SRM analysis was carried out on a Waters Quatro Premier triple quadrupole.

Previous SRM investigations focused on detection of peanut allergens in different food matrices have been reported [7,19,20,24]. However, one should be critical with issues concerning sample preparation and analysis that pose some reasonable questions on achieved detection limits. In the study of Careri *et al.* [19], for instance, the different food matrices tested (rice crispy and cacao based snacks) were fortified with different amounts of raw peanuts and it was not clearly reported if the different amount of peanut proteins were added before or after processing. Thus, they were dealing with spiked samples and raw peanuts. The effect of food processing on the extractability of Ara h 2 and Ara h 3/4 was not considered. The detection limits of 5 μg protein g^{-1} matrix for Ara h 2 and 1 μg protein g^{-1} matrix for Ara h 3/4 might be too optimistic. In addition, limits of detection were expressed as μg protein g^{-1} matrix which is different than μg peanuts g^{-1} matrix. In our study, we achieved detection of Ara h 3/4 in incurred cookies with 10 μg peanuts g^{-1}matrix targeting the same Ara h 3 peptides (*i.e.*, AHVQVVDSNGNR and SPDIYNPQAGSLK) reported by Careri *et al.* [19]. For Ara h 2, the same peptides reported by Careri *et al.* [19] CCNELNEFENNQR (*m/z* 863.8, doubly charged ion) and CMCELQQIMENQSDR (*m/z* 1006.9, doubly charged ion) were also targeted but unsuccessfully in our study even though an alkylation step was introduced. In our case, to confirm detection of a certain peptide in SRM mode at least three fragment ions needed to be detected as to be considered a true hit while in the case of Careri *et al.* [19] it was based on one fragment ion which could render confirmation of certain peptide doubtful. At least two fragment ions should be used: one as quantifier and the other as qualifier fragment ions. The ratio quantifier/qualifier must remain constant throughout samples and it is considered as another quality control step in SRM [7,8]. In the study of Shefcheck *et al.* [20], detection levels of 2 μg protein g^{-1} matrix (Ara h 1) from dark chocolate with SRM have been reported. The Ara h 1 peptides VLLEENAGGEQEER (*m/z* 786.9, doubly charged ion) and DLAFPGSGEQVEK (*m/z* 688.9, doubly charged ion) were targeted detecting three fragment ions per peptide. With this particular matrix, the most important factor affecting the extraction of the target analyte(s) seemed to be the interaction protein-tannins. The range in terms of concentration between the target analyte(s) and matrix proteins do not seem to be the major problem for detection since the sample was dark chocolate (proteins present belong to the cacao matrix). In cookie as food matrix, not only the huge range in terms of protein concentrations but the interaction and processing effects seem to be the limiting factors for detecting at least Ara h 1 and Ara h 2 at lower levels.

In a recent study conducted by Heick *et al.* [7,24] targeting seven different allergens from milk, egg, soy, hazelnut, peanut, walnut and almond in flour and bread as food matrices, detection limits of 11 $\mu g \cdot g^{-1}$ matrix have been reported for the allergens from peanuts (Ara h 1 and Ara h 3/4, respectively). However, this detection limit is based on spiking different amounts of semi purified peanut proteins in blank matrix. Unfortunately, the source of peanut material was not described (e.g., raw or processed peanut, one variety or a mix or varieties representative of what is commercially available and used by the food industry) which might explain the differences in the selected proteotypic peptides. With the exception of DLAFPGSGEQVEK (*m/z* 688.8, doubly charged ion) from Ara h 1, the other three peptides largely differed from the ones selected in the present study. For example, the proteotypic peptide RPFYSNAPQEIFIQQGR (*m/z*684.5, triply charged ion) from Ara h 3/4 was reported as the most intense marker [7]. However, in our study, we could only detect this peptide when the amount of peanuts was very high (>10,000 $\mu g\ g^{-1}$ matrix) in the cookie matrix. The reported LOD value of 11 $\mu g \cdot g^{-1}$ matrix might also be very optimistic since the starting material was defatted peanut, and peanut contains approximately 50% fat and the calculated LOD was based on matrix spiked with total peanut proteins.

There is consensus in the community that SRM is the approach to take for trace level detection of major food allergens, especially as confirmatory method when liability issues are raised. However, still aspects remain that need to be common practice and improved before a solid detection method is developed. Recommendations of the scientific community [6] working on the subject includes but are not limited to: history and description of the allergen food source, nature of the reporting units, quality control of MS data, *etc.*

Perspectives

Previous evaluations of the reproducibilty of ProteoMiner enrichment opens perspectives for its use in food allergen quantitation [25,26]. The quantitative reproducibility of cytokines per trial and not only in terms of enrichment factor was confirmed by immunodetection and nano LC-MS/MS [25]. A method that incorporated ProteoMiner enrichment has been successfully used for label free quantification of series of cerebrospiral fluid samples processed in parallel [26]. The signal intensity of peptides coming from growing amounts of exogenous spiked proteins in the matrix was evaluated.

Data independent acquisition (DIA) known as MS^E presents itself as a promising approach to detect trace levels of food allergens in complex food matrices due to the higher sensitivity that can be achieved compared to the classical DDA approach. An advantage of MS^E is the possibility to account for certain modifications simultaneously, higher sequence coverage, and label free quantification [27]. Wei *et al.* [28] performed data independent acquisition (DIA) or MS^E to analyze Ara h 1 detectable peptides in raw and roasted peanuts. The potential of certain Ara h 1 peptides as markers was based on their presence in both raw and roasted peanuts at relatively high intensities. We detected the same Ara h 1 peptides: DLAFPGSGEQVEK and VLLEENAGGEQEER as the most abundant ones through DDA (Table 3). Wei and co-workers also introduced a matrix (*E. coli* tryptic digest) in a 1:200 v/v (Ara h 1 matrix) proportion and were able to perform label free Ara h 1 quantification with the introduction of an ADH tryptic digest as internal standard. However, this material does not represent a processed food product. In the present work, a much higher dynamic range was encountered in terms of concentration of proteins besides that a representative matrix (cookie) was incorporated. This MS^E approach has been explored by our team, but so far no comparable or better sensitivity has been achieved than the ones achieved with SRM.

Multi allergen detection and quantification are hot topics nowadays. An LC-MS based method claiming detection of seven different food allergens at trace levels in bread has been recently reported [7]. The LOD values reported need to be considered with caution since the allergen food sources are not fully described and might not be representative of real scenarios, being by far too optimistic. However, it is a first attempt that brings new insights into this possible multi allergen detection. For absolute allergen quantification, the topic is much more sensitive and we should be cautious on how to interpret quantification. Ideally, an isotopically labeled target protein that can be incorporated during the sample preparation workflow or even before processing simulating real processing and extraction conditions should be used. However, so far it is more realistic to use isotopically labeled peptides [6]. The digestion step for quantification of proteins/peptides is a key aspect and requires being reproducible and complete [29]. Assessment of the stability of target peptides (e.g., possible deaminations) during the proteolysis and afterwards is also of key importance for absolute protein quantification. Absolute quantification of milk allergens in a variety of food matrices has been recently reported [8]. Many of the target peptides present Q and N in their sequence. We have experimentally observed with our target peptides that when these amino acids are present they can undergo deamination to a certain degrees. Of course, this issue might be circumvented by accounting for the amidated and de-aminated forms in the SRM method. Given the different physico-chemical properties of the different allergenic proteins to target, the complexity of the different potential food matrices encountered and the diversity of food processing conditions, a generic sample preparation platform might not be realistic. However, it might be feasible to rely on a few number of sample preparation workflows grouping food products sharing similar characteristics (e.g., high protein content, high polyphenol content, *etc.*). This would certainly help for validation and standardization of food allergen detection methods through SRM targeted approaches.

CONCLUDING REMARKS

A well described and representative food allergen source (IRMM-481f), which is a mix of five commercially available different peanut varieties and five different processing conditions, was used contrary to previous reported investigations in which the food allergen source is vague or not representative enough. A representative and highly processed food matrix (cookies) with a high protein content of non target proteins (milk and wheat) was employed. This matrix is representative of what is to be expected when analyzing food samples contaminated with allergens: a wide range in protein concentration and a highly processed sample. This work mainly focused on method development through shotgun proteomics and sample preparation for future targeted confirmatory approaches such as SRM. The SRM approach followed, allowed us to detect Ara h 3 peptides at levels as low as 10 μg peanuts g^{-1} cookies. The two proteotypic Ara h 3/4 peptides AHVQVVDSNGNR and SPDIYNPQAGSLK were confirmed as markers of peanut presence in cookies containing as low as 10 μg peanuts g^{-1} cookies. This targeted SRM is being further optimized by our group to allow detection, confirmation and potential quantification of Ara h 3 and Ara h 1 peanut allergens in different food matrices.

ACKNOWLEDGEMENTS

Arnoud Groen and Kathryn Lilley from the Cambridge Centre for Proteomics, University of Cambridge are kindly acknowledged for their collaboration with the SRM work.

REFERENCES

1. Poms, R.E.; Klein, C.L.; Anklam, E. Methods for allergen analysis in food: A review. *Food Addit. Contam.* 2004, *21*, 1–31.
2. Rona, R.J.; Keil, T.; Summers, C.; Gislason, D.; Zuidmeer, L.; Sodergren, E.; Sigurdardottir, S.T.; Lindner, T.; Goldhahn, K.; Dahlstrom, J.; *et al.* The prevalence of food allergy: A meta-analysis. *J. Allergy Clin. Immunol.* 2007, *120*, 638–646.
3. European Commission, Directive 2000/13/ec of the European Parliament and of the Council of 20 March 2000 on the Approximation of the Laws of the Member States Relating to the Labelling, Presentation and Advertising of Foodstuffs. In *OJ L 109, 6.5.2000*; European Commission: Bruxelles, Belgium, 2009; p. 29.
4. Kerbach, S.; Alldrick, A.J.; Crevel, R.W.; Dömötör, L.; DunnGalvin, A.; Clare Mills, E.; Pfaff, S.; Poms, R.E.; Popping, B.; Tömösközi, S. Managing food allergens in the food supply chain-viewed from different stakeholder perspectives. *Qual. Assur. Saf. Crops Foods* 2009, *1*, 50–60.
5. Chassaigne, H.; Nørgaard, J.V.; van Hengel, A.J. Proteomics-based approach to detect and identify major allergens in processed peanuts by capillary LC-Q-TOF (MS/MS). *J. Agric. Food Chem.* 2007, *55*, 4461–4473.

6. Johnson, P.E.; Baumgartner, S.; Aldick, T.; Bessant, C.; Giosafatto, V.; Heick, J.; Mamone, G.; O'Connor, G.; Poms, R.E.; Popping, B.; *et al.* Current perspectives and recommendations for the development of mass spectrometry methods for the determination of allergens in foods. *J. AOAC Int.* 2011, *94*, 1026–1033.
7. Heick, J.; Fischer, M.; Kerbach, S.; Tamm, U.; Popping, B. Application of a liquid chromatography tandem mass spectrometry method for the simultaneous detection of seven allergenic foods in flour and bread and comparison of the method with commercially available Elisa test kits. *J. AOAC Int.* 2011, *94*, 1060–1068.
8. Lutter, P.; Parisod, V.; Weymuth, H. Development and validation of a method for the quantification of milk proteins in food products based on liquid chromatography with mass spectrometric detection. *J. AOAC Int.* 2011, *94*, 1043–1059.
9. Blum, H.; Beier, H.; Gross, H.J. Improved silver staining of plant proteins, RNA and DNA in polyacrylamide gels. *Electrophoresis* 1987, *8*, 93–99.
10. Uniprot. Blast tool. Available online: http://www.uniprot.org (accessed on 15 July 2011).
11. Schmitt, D.A.; Nesbit, J.B.; Hurlburt, B.K.; Cheng, H.; Maleki, S.J. Processing can alter the properties of peanut extract preparations. *J. Agric. Food Chem.* 2010, *58*, 1138–1143.
12. Kang, I.-H.; Gallo, M.; Tillman, B.L. Distribution of allergen composition in peanut (*Arachis hypogaea L.*) and wild progenitor (*Arachis*) species. *Crop Sci.* 2007, *47*, 997–1003.
13. Li, J.; Shefcheck, K.; Callahan, J.; Fenselau, C. Primary sequence and site-selective hydroxylation of prolines in isoforms of a major peanut allergen protein Ara h 2. *Protein Sci.* 2010, *19*, 174–182.
14. Pele, M. Peanut allergens. *Rom. Biotechnol. Lett.* 2010, *15*, 5204–5212.
15. Beck, M.; Claassen, M.; Aebersold, R. Comprehensive proteomics. *Curr. Opin. Biotechnol.* 2011, *22*, 3–8.
16. Chassaigne, H.; Trègoat, V.; Nørgaard, J.V.; Maleki, S.J.; van Hengel, A.J. Resolution and identification of major peanut allergens using a combination of fluorescence two-dimensional differential gel electrophoresis, western blotting and q-tof mass spectrometry. *J.*

Proteomics 2009, *72*, 511–526.

17. Gosetti, F.; Mazzucco, E.; Zampieri, D.; Gennaro, M.C. Signal suppression/enhancement in high-performance liquid chromatography tandem mass spectrometry. *J. Chromatogr. A* 2010, *1217*, 3929–3937.
18. Ly, L.; Wasinger, V.C. Protein and peptide fractionation, enrichment and depletion: Tools for the complex proteome. *Proteomics* 2011, *11*, 513–534.
19. Careri, M.; Costa, A.; Elviri, L.; Lagos, J.B.; Mangia, A.; Terenghi, M.; Cereti, A.; Garoffo, L. Use of specific peptide biomarkers for quantitative confirmation of hidden allergenic peanut proteins Ara h 2 and Ara h 3/4 for food control by liquid chromatography-tandem mass spectrometry. *Anal. Bioanal. Chem.* 2007, *389*, 1901–1907.
20. Shefcheck, K.J.; Callahan, J.H.; Musser, S.M. Confirmation of peanut protein using peptide markers in dark chocolate using liquid chromatography-tandem mass spectrometry (LC-MS/MS). *J. Agric. Food Chem.* 2006, *54*, 7953–7959.
21. Picotti, P.; Aebersold, R.; Domon, B. The implications of proteolytic background for shotgun proteomics. *Mol. Cell. Proteomics* 2007, *6*, 1589–1598.
22. Sherwood, C.A.; Eastham, A.; Lee, L.W.; Risler, J.; Mirzaei, H.; Falkner, J.A.; Martin, D.B. Rapid optimization of MRM-MS instrument parameters by subtle alteration of precursor and product *m/z* targets. *J. Proteome Res.* 2009, *8*, 3746–3751.
23. Picotti, P.; Bodenmiller, B.; Mueller, L.N.; Domon, B.; Aebersold, R. Full dynamic range proteome analysis of *S. cerevisiae* by targeted proteomics. *Cell* 2009, *138*, 795–806.
24. Heick, J.; Fischer, M.; Pöpping, B. First screening method for the simultaneous detection of seven allergens by liquid chromatography mass spectrometry. *J. Chromatogr. A* 2011, *1218*, 938–943.
25. Li, L.; Sun, C.; Freeby, S.; Yee, D.; Kieffer-Jaquinod, S.; Guerrier, L.; Boschetti, E.; Lomas, L. Protein sample treatment with peptide ligand library: Coverage and consistency. *J. Proteomics Bioinform.* 2009, *2*, 485–494.
26. Mouton-Barbosa, E.; Roux-Dalvai, F.; Bouyssié, D.; Berger, F.; Schmidt, E.; Righetti, P.; Guerrier, L.; Boschetti, E.; Burlet-Schiltz,

O.; Monsarrat, B.; Gonzalez de Peredo, A. In depth exploration of cerebrospinal fluid by combining peptide ligand library treatment and label free protein quantification. *Mol. Cell. Proteomics* 2010, *9*, 1006–1021.

27. Levin, Y.; Hradetzky, E.; Bahn, S. Quantification of proteins using data-independent analysis (MS^E) in simple and complex samples: A systematic evaluation. *Proteomics* 2011, *11*, 3273–3287.
28. Wei, H.; Gledhill, A.; Maleki, S. The Analysis of Allergens in Raw and Roasted Peanuts Using Nanoacquity UPLC and Xevo Q-TOF MS. In *Application Note*; Waters: Beverley, MA, USA, 2010; p. 8.
29. Brownridge, P.; Beynon, R.J. The importance of the digest: Proteolysis and absolute quantification in proteomics. *Methods* 2011, *54*, 351–360.

CHAPTER

10

PROTEOMIC APPROACH TO EVALUATE MECHANISMS THAT CONTRIBUTE TO FOOD ALLERGENICITY: COMPARATIVE 2D-DIGE ANALYSIS OF RADIOALLERGOSORBENT TEST POSITIVE AND NEGATIVE PATIENTS

Bindukumar Nair, John C. Wheeler, Donald E. Sykes, Paula Brown, Jessica L. Reynolds, Ravikumar Aalinkeel, Supriya D. Mahajan, and Stanley A. Schwartz

Division of Allergy, Immunology, and Rheumatology, Department of Medicine, University at Buffalo, 640 Ellicott Street, Room 444 Innovation Center, Buffalo Niagara Medical Campus, 640 Ellicott Street, Buffalo, NY 14203, USA

ABSTRACT

Proteomic profiles of RAST$^+$ subjects with severe food allergies and RAST$^-$ subjects were compared using 2D-DIGE analysis to obtain candidate biomarkers specific to food allergies. Our analysis highlighted 52 proteins that were differentially expressed between the RAST$^+$and RAST$^-$ groups of which 37 were successfully identified that include chondroitin sulfates, zinc finger proteins, C-type lectins, retinoic acid binding proteins, heat shock proteins, myosin, cytokines, mast cell expressed proteins, and MAP kinases. Biological network analysis tool Metacore revealed that most of these regulated proteins play a role in immune tolerance, hypersensitivity and modulate cytokine patterns inducing a Th2 response that typically results in IgE-mediated allergic response which has a direct or indirect biological link to food allergy. Identifying unique biomarkers associated with certain allergic phenotypes and potentially cross-reactive proteins through bioinformatics analyses will provide enormous insight into the mechanisms that underlie allergic response in patients with food allergies.

INTRODUCTION

The increased prevalence of food allergies is a significant clinical and public health problem in the USA. It is estimated that in the USA alone, food-allergic reactions accounted for over 20,000 emergency department visits, 2,300 episodes of anaphylaxis, and over 500 hospitalizations over just a 2-month period in 2003 [1, 2]. Despite the prevalence of food allergy, diagnosis and treatment are still far from optimal. Eight types of food account for over 90% of allergic reactions in affected individuals: milk, eggs, peanuts, tree nuts, fish, shellfish, soy, and wheat [3, 4]. Reactions to these foods by an allergic person can range from a tingling sensation around the mouth and lips and hives to death, depending on the severity of the allergy. Food allergy is more prevalent in children than adults. Additionally, allergic reactions to various foods often develop into allergy-related respiratory diseases such as asthma, allergic rhinitis, and atopic dermatitis.

Several studies provide strong evidence that both host genetic susceptibility and environmental factors determine the complex regulation of IgE-mediated food allergies, however, the mechanisms by which a person develops an allergy to specific foods are largely unknown. As the IgE-mediated, "immediate type" food hypersensitivity reactions account for significant morbidity and mortality, further understanding of these mechanisms that contribute to the allergenicity could improve the prediction, diagnosis, and management of severity food allergy in individuals.

Food allergies are caused by IgE-dependent or IgE-independent immunologic reactions, which then lead to an inflammatory reaction, in which mast cells, eosinophilic granulocytes, and other cells are involved. Genetic and environmental causes predominantly underlie food allergies. Food intolerances or nonimmunologic food incompatibilities are often caused by specific enzyme deficiencies and must be diagnostically differentiated from food allergies. Constant allergen exposure and other environmental factors determine whether a sensitised individual will become chronically allergic and experience persistent symptoms. Therapeutics for the treatment of allergy are focused on prevention or diminishing the specific serum immunoglobulin E (IgE) responsible for the appearance of the allergic reaction [5, 6].

Proteomic technologies play an important role in drug discovery, diagnostics, and molecular medicine because they are the link between genes, proteins, and disease states. Identifying unique patterns of protein expression or biomarkers associated with IgE-mediated food allergy is a rapidly emerging area of clinical proteomics. Genomics provides only a partial picture, while proteomics identifies specific proteins responding to gene expression. Progress in protein annotation and in our understanding of protein-protein interactions will undoubtedly lead to diagnostic and therapeutic advances in the treatment of food allergies. Proteomics research technologies are rapidly changing our understanding of complex and dynamic biological systems by providing information relevant to functionally associated changes in protein abundances, protein–protein interactions, and posttranslational modifications [7–10]. 2-dimensional fluorescence difference gel electrophoresis (2D-DIGE) has emerged as a robust method to study protein expression profiling in clinical samples. 2D gel electrophoresis can simultaneously separate and display hundreds to thousands of different proteins. This method separates proteins in 2 dimensions according to their isoelectric point and their molecular size. Fluorescent, 2-D DIGE [11–13] allows the multiplex analysis of 3 sample proteomes on the same gel. Data from proteomic analysis provide enormous amount of biological information and can help establish multimodal markers for early diagnosis and prognosis. Additionally, these proteomic data can provide an opportunity for identifying molecular pathways that underlie various clinical allergic phenotypes. Identifying unique biomarkers associated with certain allergic phenotypes, and potentially cross-reactive proteins that can be identified through bioinformatics analyses will provide enormous insight into the mechanisms that underlie allergic response in patients with food allergies.

The current study examines the proteomic profile of a group of RAST$^+$ subjects with severe food allergies and compares their proteomic profile with RAST$^-$ subjects. Our goal is to not only identify biomarkers that are specific to food allergies that result in IgE-mediated atopy but also to identify molecular mechanism that underlie the allergic response in patients with food allergies.

MATERIALS AND METHODS

Study Design

Both pediatric and adult patients with mild to severe food allergies who were evaluated in our allergy clinic were informed about the research study and some patients volunteered to participate. Institutional IRB approval was obtained prior to commencement of the study. Patients were consented based on the HSIRB guidelines; in case of minors, parental consent was obtained as per the IRB guidelines. A total of 10 mL of blood sample was drawn from the patient, of which a portion of the sample approx 3 mL was used for a RAST (Radioallergosorbent test) as a routine clinical procedure and the remaining sample was used for isolation of peripheral blood mononuclear cells (PBMC) which were used for the proteomics analysis. Further, gene expression studies were done using RNA which was extracted from these cells. RAST is a blood test used to detect specific IgE antibodies to suspected or known allergens, in this case, specific to food allergens. The IgE antibody is associated with Type I allergic response. Based on the results of the RAST test, the patients were classified into RAST positive (RAST$^+$) or RAST negative (RAST$^-$). A total of RAST$^-$ and RAST$^+$ patients were recruited for the study. Table 1 provides demographic and clinical information for the food RAST$^+$ and RAST$^-$ subjects. Comparative analysis was done between RAST$^+$ and RAST$^-$ patient groups using 2D-DIGE proteomic analysis. Three independent sample pools from RAST$^+$ versus RAST$^-$ patient samples were analyzed by 2D-DIGE for identifying the differentially expressed proteins. Each pooled sample consisted of 3 RAST$^+$ and 3 RAST$^-$ patient samples, respectively.

Table 1: Lists of the clinical and demographic information for subjects enrolled in the study. * Denotes in mm the measure of wheal/flare in a antigen-specific skin test done relative to a saline control.

(a) Food rast negative patients.

Pt number	Age	Gender	Ethnicity	Clinical H/O	Skin test	RAST Antigen-specific IgE (ng/mL)
1	3	F	White	Peanut: urticaria and rash after peanut food challenge	Negative: peanut 2/4*	Negative: peanut and tree nuts
2	15	F	Black	Tree nut: anaphylaxis (angioedema, shortness of breath and abdominal pain)	Not done	Negative: peanut and tree nuts
3	1	M	White	Eczema workup	Positive: milk 12/20*	Negative: egg, milk and wheat
4	2	M	Hispanic	Eczema workup	Negative: egg 5/6*	Negative: egg
5	3	M	Black	Fish: urticaria	Positive: mixed Whitefish	Negative: tuna, salmon, cod
6	8	F	White	Cocoa: headache, strawberry: rash	Positive: cocoa 5/45* Negative: fresh strawberry	Negative: cocal and strawberry
7	7	M	Not specified	Peanut: hives, egg: hives	Not done	Negative: egg, peanut and tree nuts
8	7	F	White	Tuna: hives	Negative: tuna 4/5* (negative)	Negative: tuna
9	9	M	Not specified	Tomato: facial edema	Not done	Negative: tomato

(b) Food rast positive patients.

Pt number	Age	Gender	Ethnicity	Clinical H/O	Skin test	RAST Antigen-specific IgE (ng/mL)
1	6	M	Black	Peanut: angioedema, shortness of breath and itchy eyes	Not done	Peanut: >100
2	4	M	Black	Peanut: rash	Positive: peanut (5/20)*	Peanut: 0.86
3	5	F	Black	Peanut: hives, soy: eczema flair	Positive: soy (5/10) and egg (3/7)	Peanut: 7.32, soy: 5.15 and tree nuts: 4.14–>100
4	3	M	Black	Peanut: hives	Positive: peanut (12/20)*	Peanut: 72.8
5	2	F	White	Peanut: eczema flare	Positive: peanut (15/40)*	Peanut: 4.07
6	6	M	White	Tree nut: difficulty breathing	Not done	Tree nuts: 15–>100
7	16	M	Black	Crab: angioedema	Positive: crab (10/12)*, lobster (10/13)* and clam (4/6)*	Crab: 14.70, Clam: 1.06 and lobster: 9
8	4	M	Not specified	Peanut: anaphylaxis	Not done	Peanut: 9.39
9	11	M	Black	Cashew: throat itching	Positive: cashew (9/9)* and walnut (18/20)*	Tree nuts: 0.51–9.24
10	16	F	Black	Peanut: itchy throat	Not done	Peanut: 20
11	1.5	F	Black	Egg: anaphylaxis	Not done	Egg: 62.80
12	9	F	White	Egg: urticaria	Negative	Egg: 2.78

PBMC Isolation

Peripheral blood leukocytes were isolated from whole blood using Ficoll-Hpaque gradient (Amersham-Pharmacia, Piscataway, NJ, Cat number 17-1440-03). Samples were centrifuged for 30 min at 700 ×g and 20°C without applying a brake. The PBMC interface was carefully removed by pipetting and was washed twice with PBS/EDTA and resuspended in

2 mL of complete RPMI media and total number of cells counted using a hemocytometer.

RNA Extraction

Cytoplasmic RNA is extracted using Trizol reagent (Invitrogen- Life Technologies, Carlsbad, Calif) [14]. The amount of RNA is quantitated using a Nanodrop ND-1000 spectrophotometer (Nanodrop Wilmington, Del) and isolated RNA is stored at −80°C until used.

Real Time Quantitative PCR (Q-PCR)

Q-PCR was used to quantitate gene expression. RNA is reverse transcribed to cDNA using the reverse transcriptase kit from Promega (Promega Inc, Madison, Wis, USA; Cat number A3500). Relative abundance of each mRNA species is quantitated by real time quantitative PCR for specific primers using the Brilliant SYBR green QPCR master mix from Stratagene (Stratagene Inc, La Jolla, Calif, USA; Cat number 600548-51). Relative expression of mRNA species is calculated using the comparative C_T method [15]. All data are controlled for quantity of RNA input by performing measurements on an endogenous reference gene, β-actin. In addition, results on expression levels of different genes from Rast$^+$ samples are normalized to results obtained from Rast$^-$ samples. Results are expressed as transcript accumulation index (TAI) as described earlier [16]. This calculation assumes that all PCR reactions are working with 100% efficiency.

Protein Extraction

PBMC isolated from RAST$^+$ and RAST$^-$ subjects were washed twice with 1X PBS (Invitrogen, Grand Island, NY, USA). Total protein was extracted using standard cell lysis buffer (30 mM TrisCl; 8 M urea; 4% (w/v) CHAPS, adjusted to pH 8.5) for 10 min on ice. The cell lysate was centrifuged at 4°C for 10 min at 12000 ×g and was further purified by precipitation with chloroform/methanol as described [17]. Samples were resuspended in standard cell lysis buffer. Protein concentrations were determined using the Coomassie Protein Reagent (Bio-Rad, Hercules, Calif, USA) prior to DIGE analysis.

Two-Dimensional Difference Gel Electrophoresis (2D-DIGE)

The Ettan DIGE technique developed by GE Healthcare (Piscataway, NJ, USA) was used to detect differences in protein abundance between the $RAST^+$ and $RAST^-$ samples. The Ettan DIGE system uses 3 CyDye DIGE fluors (Cy2, Cy3, Cy5), each with a unique fluorescent wavelength, matched for mass and charge. CyDyes form a covalent bond with the free epsilon amino group on lysine residues of the sample proteins. CyDyes label approximately 2% of the lysine residues. This system allows in each set of gel 2 experimental samples and an internal standard to be simultaneously separated on the same gel. The internal standard is comprised a pool of an equal amount of all the experimental samples. The use of an internal standard facilitates accurate inter-gel matching of spots, and allows for data normalization between gels to minimize gel to gel experimental variability [12]. Cell lysates were labeled with CyDye per the manufacturer. All reagents used were from GE Healthcare (Amersham Biosciences, Piscataway, NJ, USA).

Three independent (n=3) 2D-DIGE experiments were done comparing $RAST^+$ and $RAST^-$patient samples. Briefly, six samples (3 $RAST^+$ and 3 $RAST^-$), each of 50 µg cell lysate were labeled with 400 pmoL of either Cy3 or Cy5, kept on ice for 30 min and then quenched with a 50-fold molar excess of free lysine. 50 µg of protein pooled from equal amounts of protein from all the six experimental samples were labeled with Cy2 and used as an internal standard in each of the three sets of Cy3/Cy5 gels. Cy3-, Cy5-, and Cy2-labeled samples and unlabelled protein (500–1000 µg) were pooled. Unlabeled protein was added to enhance the protein staining and hence a better image quality of the SYPRO ruby staining. An equal volume of 2X sample buffer (8 M urea; 2% (v/v) Pharmalytes 3–10; 2% (w/v) dithiothreitol (DTT); 4% (w/v) CHAPS) was added and incubated on ice for 10 min. The total volume of sample was adjusted to 450 µL with rehydration buffer (4% (w/v) CHAPS; 8 M urea; 1% (v/v) Pharmalytes 3–10 nonlinear (NL); 13 mM DTT). Samples were applied to immobilized pH gradient (IPG) strips (24 cm, pH 3–10 NL) and absorbed by active rehydration at 30 V for 13 hr. Isoelectric focusing was carried out using an IPGphor IEF system with a 3-phase program; first phase at 500 V for 1 hr, second phase at 1000 V for 1 hr, and third phase (linear gradient) 8000 V to 64000 V for 2 hr (50 uA maximum per strip). Prior to separation in the second dimension, strips were equilibrated for 15 min in equilibration buffer I (50 mM Tris-HCl, 6 M urea, 30% (v/v) glycerol, 2% (w/v) SDS, 0.5% (w/v) DTT). The strips were again equilibrated for 15 min in equilibration buffer II (50 mM Tris-HCl, 6 M urea, 30% (v/v) glycerol, 2% (w/v) SDS, 4.5% (w/v) iodoacetamide) and the equilibrated IPG strips were transferred onto 18×20 cm, 12.5% uniform polyacrylamide gels poured between low fluorescence glass plates. Gels were bonded to inner plates using Bind-Silane solution (Promega, Madison, Wis, USA) according to the manufacturer. Strips were overlaid with 0.9% agarose in 1X running buffer containing bromophenol blue and were run for 16 hr (1.8 W/gel overnight) at 15°C in an Ettan DALT electrophoresis system. After the run was completed, the 2D gels were scanned 3 times with a Typhoon 9410 imager, each time at different excitation wavelengths (Cy3, 532 nm; Cy5, 633 nm; Cy2, 488 nm). Images were cropped with ImageQuant v5.2 software and then imported into DeCyder differential in-gel analysis (DIA) v5.0 software from GE Healthcare for spot identification and normalization of spot intensities within each gel.

Gels were fixed in 30% (v/v) methanol, 7.5% (v/v) acetic acid for 3 hr and stained with SYPRO-ruby dye (Molecular Probes, Eugene, Ore, USA) overnight at room temperature. Gels were destained in water and then scanned using the Typhoon 9410 scanner. Spots of interest were excised from the gel using the Ettan Spot Picker. DeCyder software (GE Healthcare) was specifically developed for use with the Ettan DIGE system. DeCyder software allows for automatic detection of spots, background subtraction, quantitation, normalization, internal standardization, and integral matching. The differential in-gel analysis (DIA) component of DeCyder software draws boundaries around spots in a composite gel image obtained from the intragel overlap of the Cy2-, Cy3-, and Cy5-scanned images and normalizes the data from each CyDye to account for differences in dye fluorescence intensity and scanner sensitivity. For all the three sets of experiments (six images), the abundance difference between samples ($RAST^{+}$ and $RAST^{-}$) run on the same gel was then analyzed. The biological variation analysis (BVA) component of DeCyder software was then used to match all image comparisons from individual Cy3/Cy5 gel sets for a cross-gel statistical analysis. DeCyder BVA initially calculates normalized intensities (standard abundance) for all spots by comparison to the internal standard, and from this, an average volume ratio and a Student's paired t-test derived p value were calculated for each spot. A paired t-test derived p value of ≤ 0.05 was considered statistically significant [12].

Mass Spectrometry and Protein Identification

Spots were excised from 2D-gels loaded with 500 to 1000 μg of total protein and stained with Sypro ruby (Molecular Probes). The gel plugs were incubated in 200 μL of 200 mM ammonium bicarbonate with 50% acetonitrile for 30 min at 37°C and then dried with a Speed Vac. Samples were then incubated for 1 h at room temperature in 3.5 μL of 20 μg/mL trypsin (Promega) in 40 mM ammonium bicarbonate. An additional 30 μL of 40 mM ammonium bicarbonate with 10% acetonitrile was added and samples were incubated overnight at 37°C. The samples were desalted using μ C_{18} ZipTips (Millipore, Bedford, Mass, USA) and eluted directly onto a MTP ground steel 384-target (Bruker Daltonics, Billerica, Mass, USA) with the matrix solution (a saturated solution of α-cyano-4-hydroxy-cinnamic acid [CHCA: Bruker Daltonics] dissolved in 0.1% trifluoroacetic acid:75% acetonitrile). MALDI spectra in positive ion mode were obtained using a Bruker Daltonics Biflex IV MALDI-TOF mass spectrometer operating in the reflectron mode. The spectra in the range of 500–3200 Da were obtained by the summation of 50–200 laser shots. Identification and labeling of the peaks were done manually. External calibration was carried out using the peptide calibration standard II (700–4000 Da) (Bruker Daltonics, number 222570). Data acquisition was executed using the Flex Control program (Build 2.4.0.0) and data processing was performed using the Bruker XTOF 5 software (version 5.1.5). The signatures of proteins of interest formed by the masses of the products of tryptic digestion were analyzed using an on-line search engine MASCOT (http://www.matrixscience.com/) to compare the calculated masses of a theoretical tryptic digestion. Searches were conducted for matching masses from theoretical tryptic peptides from the NCBI protein databases. We used the following search parameters: monoisotopic masses, with a tolerance of 200 ppm or less (typically 75–100 ppm or 0.1–0.5 Da), one missed tryptic cleavage, a carbamidomethyl modification of cysteine (to account for iodoacetamide treatment) and variable modifications of oxidation of methionine and phosphorylation of serine, threonine or tyrosine. A protein identification was considered significant for a Mowse score with $p < 0.05$.

RESULTS

2D-DIGE Analysis and Protein Identification

Protein profiles of RAST$^-$ and RAST$^+$ PBMCs were analyzed by DIGE to examine the differential expression of proteins among the two groups. Protein samples from RAST$^+$ and RAST$^-$ PBMC cell lysates were labeled with fluorescent CyDyes, pooled and analyzed in triplicate by 2D-DIGE. The pooled mixture of equal amounts of protein from each of the six experimental samples labeled with Cy2 and run on every gel provided an internal standard to ensure that every protein in each sample appears on all of the gels. Each Cy3- or Cy5-labeled sample was then compared to the same internal standard and measurements were taken relative to the internal Cy2 standard, thereby reducing gel-to-gel variation and increasing statistical confidence. Dye-swapping was performed to minimize labeling-dependent bias.

The DIA (differential in-gel analysis) module of the differential analysis software DeCyder (GE Healthcare) was used to calculate protein spot volumes and the normalized volume ratio for each differentially labeled comigrated protein. On an average of about 1000 spots were detected and matched successfully in three RAST$^+$ and RAST$^-$ CyDye-labeled proteome gels (Figure 1). The ratio of spot intensity of the two groups of protein spots is then calculated. Gel to gel matching of the standard spot maps from each gel was performed using the DeCyder BVA software module. This allowed for the statistical analysis of changes in protein abundance between samples. Statistical analysis (paired t-test) was performed for the difference between the abundance of proteins from RAST$^+$ cells versus proteins from and RAST$^-$ cells. A total of 52 protein spots showed a significant change in abundance with. Proteins from 37 gel spots were successfully identified. Figure 2 is the SYPRO ruby-stained 2D gel image showing all protein spots. Differentially expressed protein spots showing significant modulation were selected for subsequent identification and are shown as numbered outlines. The maximum increase in protein expression was 1.65-fold (spot number

1302 in Figure 2) and the maximum decrease observed was 10.7-fold (spot #242, Figure 2).

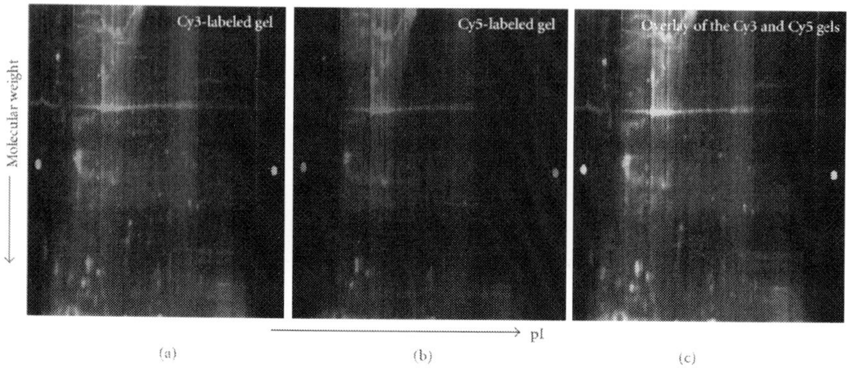

Figure 1: Representative 2D-DIGE gel images of Cy3 and Cy5 Dye-labeled RAST$^+$ and RAST$^-$ samples. Protein samples were labeled with Cy Dyes and analyzed by 2D-DIGE. (a) RAST$^-$ (Cy3); (b) RAST $^+$(Cy5); (c) overlay of Cy3 and Cy5 gel images.

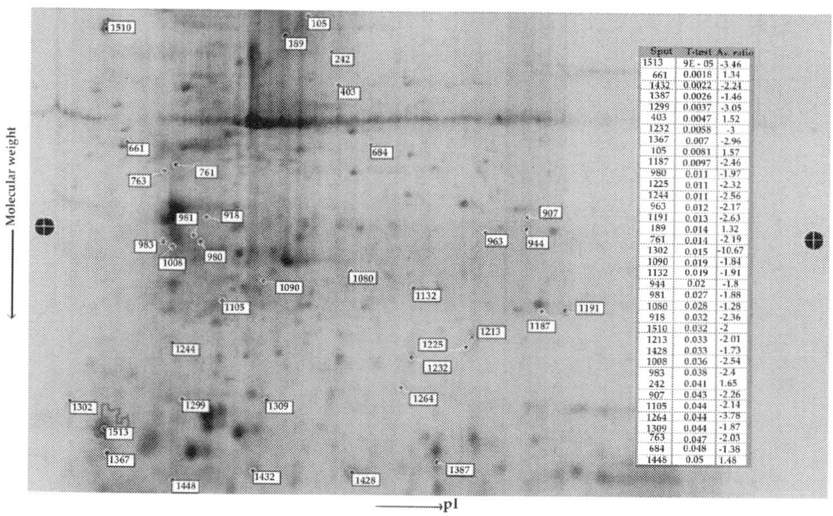

Spot	T-test	Av. ratio
1513	9E - 05	-3.46
661	0.0018	1.34
1432	0.0022	-2.24
1387	0.0026	-1.46
1299	0.0037	-3.05
403	0.0047	1.52
1232	0.0058	-3
1367	0.007	-2.96
105	0.0081	1.57
1187	0.0097	-2.46
980	0.011	-1.97
1225	0.011	-2.32
1244	0.011	-2.56
963	0.012	-2.17
1191	0.013	-2.63
189	0.014	1.32
761	0.014	-2.19
1302	0.015	-10.67
1090	0.019	-1.84
1132	0.019	-1.91
944	0.02	-1.8
981	0.027	-1.88
1080	0.028	-1.28
918	0.032	-2.36
1510	0.032	-2
1213	0.033	-2.01
1428	0.033	-1.73
1008	0.036	-2.54
983	0.038	-2.4
242	0.041	1.65
907	0.043	-2.26
1105	0.044	-2.14
1264	0.044	-3.78
1309	0.044	-1.87
763	0.047	-2.03
684	0.048	-1.38
1448	0.05	1.48

Figure 2: SYPRO ruby-stained 2-D gel image of a 2D-DIGE separation of RAST$^+$ and RAST$^-$ samples. 2-D electrophoresis was performed using 24 cm, pH 3–10 linear IPG strips and 10% SDS-PAGE gels. Spots that showed statistically significant differences in intensity between the RAST$^-$ and RAST$^+$ are marked. Differences in Cye Dye-labeled sample abundances were analyzed using DeCy-

der Software. Differentially expressed protein spots with their respective fold differences and p values are given as a table. A paired t-test derived p value of ≤0.05 was considered statistically significant. Protein spots were excised and identified using MALDI-TOF and the identified proteins are listed in Table 2.

MALDI_TOF Analysis for Protein Identification

Proteomic evaluation of differential expression of proteins in PBMCs of RAST$^+$ and RAST$^-$subjects by 2D-DIGE and MALDI-TOF resulted in a short list of 37 proteins that were found to be significantly modulated (Table 2). The table lists these proteins and includes their symbols, % sequence coverage of the identified peptide, number of matching peptides, theoretical mass, and calculated pI of the identified protein.

Table 2: List of the up/down regulated proteins from RAST$^+$ and RAST$^-$samples analyzed by 2D-DIGE and further identified by MALDI-TOF. Data shown in the table include spot number, calculated mass, theoretical pI, number of matching peptides, % sequence coverage, name and symbol of identified proteins.

Spot number	Mass	PI value	Peptides matches	Sequence coverage (%)	Protein name	Symbol
189	14505	6.81	7	59	40S ribosomal protein S12	RS12_HUMAN
1244	54005	6.21	11	28	6-phosphofructo-2-kinase/fructose-2,6-biphosphatase 4	F264_HUMAN
1008	35830	6.51	7	32	Aldose reductase	ALDR_HUMAN
1213	15683	5.42	6	27	Cellular retinoic acid-binding protein 2	RABP2_HUMAN
1302	61312	8.63	8	17	Chondroitin sulfate N-acetylgalactosaminyl-transferase 1	CGAT1_HUMAN
1387	91751	5.98	15	19	Coiled-coil domain-containing protein	CF204_HUMAN
1299	83214	7.23	8	17	Complement C2	CO2_HUMAN
1428	36702	8.45	10	39	C-type lectin domain family 4 member K	CLC4K_HUMAN
963	38901	5.27	9	26	DDB1- and CUL4-associated factor 7	DCAF7_HUMAN

1105	40517	9.00	10	28	Developmentally regulated GTP-binding protein 1	DRG1_HUMAN
1225	30836	8.71	6	18	DnaJ homolog subfamily C member 27	DJC27_HUMAN
416	16373	10.08	7	59	DnaJ(HSP40) homolog	DJC15_HUMAN
944	77221	5.36	11	21	Elongation factor G 2	EFG2_RALME
1448	104682	6.08	12	12	Exostosin-like 3	EXTL3_HUMAN
105	81024	6.45	13	20	FACT complex subunit SSRP1	SSRP1_HUMAN
661	16459	8.67	6	51	Globin-1 OS=Petromyzon marinus PE=1 SV=2	GLB1_PETMA
1309	37162	10.80	8	31	Heparan sulfate glucos-amine 3-O-sulfotransfer-ase6	HS3S6_HUMAN
1080	57528	6.12	14	22	IKAROS zinc finger protein 1	IKZF1_HUMAN
1513	18033	9.15	6	49	Interleukin-17F	IL17F_HUMAN
980	129489	7.14	11	12	Laminin subunit beta-3	LAMB3_HU-MAN
403	36288	7.08	8	24	L-lactate dehydrogenase C chain	LDHC_HUMAN
242	47983	9.64	9	23	MAPK/MAK/MRK over-lapping kinase	MOK_HUMAN
1132	21215	9.03	7	24	Mast cell expressed mem-brane protein 1	MCEM1_HU-MAN
684	181570	5.94	13	17	Mitogen-activated protein kinase kinase kinase 4	M3K4_HUMAN
907	121648	9.46	11	14	Myosin-Ic	MYO1C_HU-MAN
1191	60539	9.62	19	28	Parafibromin	CDC73_HUMAN
1187	43939	9.33	10	24	Pentatricopeptide repeat-containing protein 2	PTCD2_HUMAN
1090	56852	8.52	10	21	Probable histidyl-tRNA synthetase, mitochondrial	SYHM_HUMAN
761	50550	5.00	10	34	Rab GDP dissociation inhibitor alpha	GDIA_HUMAN
983	79767	5.54	13	18	RUN and FYVE domain-containing protein 1	RUFY1_HUMAN
981	12439	10.54	7	73	Somatoliberin	SLIB_HUMAN
763	33047	6.64	7	31	Tissue factor	TF_HUMAN
918	23294	11.33	7	44	Transcription factor 23	TCF23_HUMAN

1264	64218	8.56	9	23	Transmembrane anterior posterior transformation protein 1 homolog	TAPT1_HUMAN
1510	121711	5.95	16	22	Transmembrane protein 132C OS	T132C_HUMAN
1432	23069	9.51	6	38	UPF0684 protein C5orf30	CE030_HUMAN
1367	57986	6.11	8	19	Zinc finger protein Aiolos	IKZF3_HUMAN

Validation of the Differentially Expressed Proteins

Having shortlisted these key proteins that are differentially expressed in the RAST$^+$ and RAST$^-$ subjects, we evaluated if the gene expression levels of these proteins were significantly different in these two groups. Based on an exhaustive literature review and on comparative analysis using gene expression studies by quantitative PCR, we have highlighted a group of key proteins that play a role in food allergenicity. These include chondroitin sulfates, zinc finger proteins, C-type lectins, retinoic acid-binding proteins, heat shock proteins, membrane proteins such as myosin and cytokines such as IL-17F, certain mast cell expressed proteins, and signal transduction molecules such as the family of MAP kinases. Almost all of these proteins play a role in immune tolerance and hypersensitivity [18–22], and they modulate cytokine patterns that favor a Th2 response that typically results in IgE-mediated allergic response. The following are the gene expression levels of the key proteins that were significantly lower in RAST$^+$ subjects as compared to RAST$^-$ subjects. Heat shock protein-40 homolog (DJC15) (19% decrease,p=0.001); C-type lectins (15% decrease;p=0.006); MAPK (24% decrease;p=0.0001); retinoic acid-binding protein (18% decrease;p=0.002); chondroitin transferase (15% decrease;p=0.006); zinc finger protein (10% decrease;p=0.05); IL-17F protein (12% decrease;p=0.02); and MCEM1 mast cell expressed membrane protein 1 (29% decrease;p=0.0001). The gene expression levels of cytokine IL-10 were significantly higher (50% increase;p=0.0001) in the RAST$^+$ subjects as compared to the RAST$^-$ subjects. No significant differences in the gene expression levels of myosin and Ikaros were observed between the 2 patient groups. Figures 3, 4, and 5 show the fold regulation in the gene expression levels of these key proteins. Although the fold changes in gene and protein changes in these key molecules are not of similar magnitude, the trends in their expression are in concordance and therefore suggests that these proteins may play a significant role in the development of food allergies.

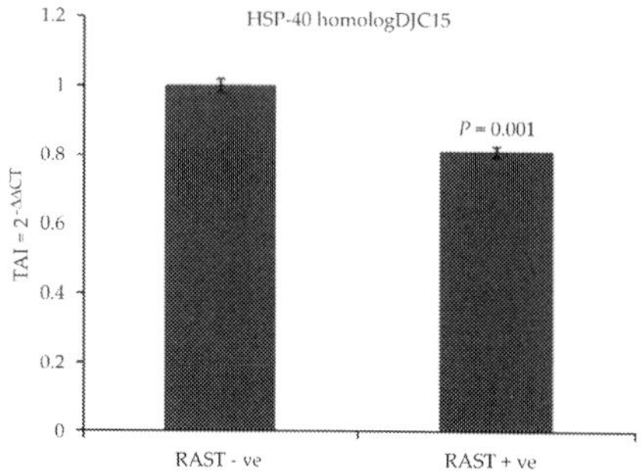

(a)

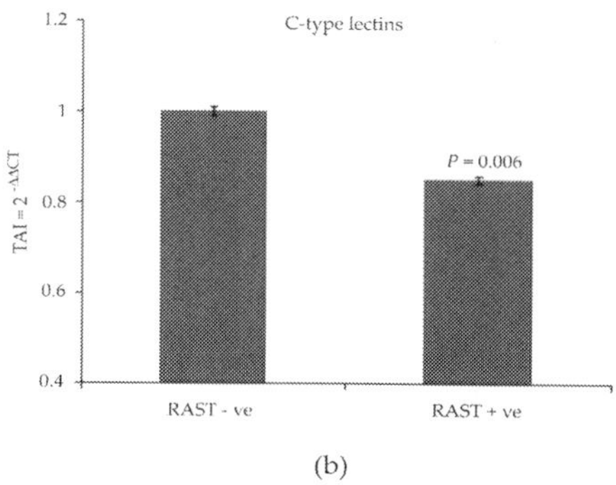

(b)

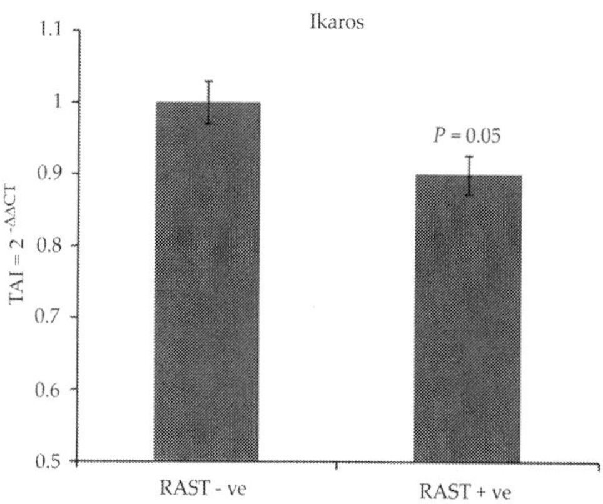

(c)

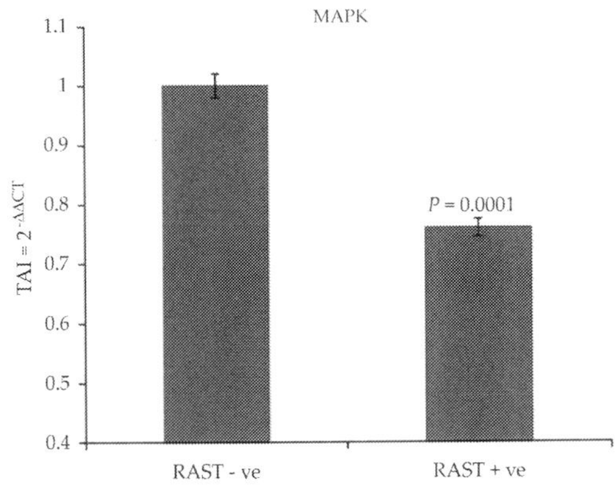

(d)

Figure 3: Relative gene expression of heat shock protein-40 homolog (DJC15); C-type lectins, Ikaros; MAPK in RAST$^+$ versus RAST$^-$Patients as quantitated by QPCR. Gene expression analysis of heat shock protein-40 homolog; C-type lectins, Ikaros; MAPK were done using RNA extracted from PBMC from RAST$^+$ and RAST$^-$ subjects. RNA was reverse transcribed and mRNA expression levels of the above genes were quantitated using real time QPCR. Relative expression of mRNA species was calculated using the comparative C_T method. Data are the mean ± SD of 3 separate experiments done in duplicate. Data are presented as TAI or the transcript accumulation index. Statistical significance was determined by Student's t-test based on comparison between the RAST$^+$ versus RAST$^-$ patients.

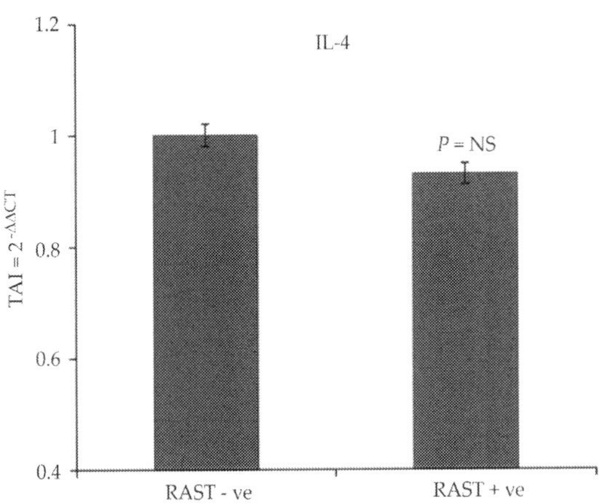

(a)

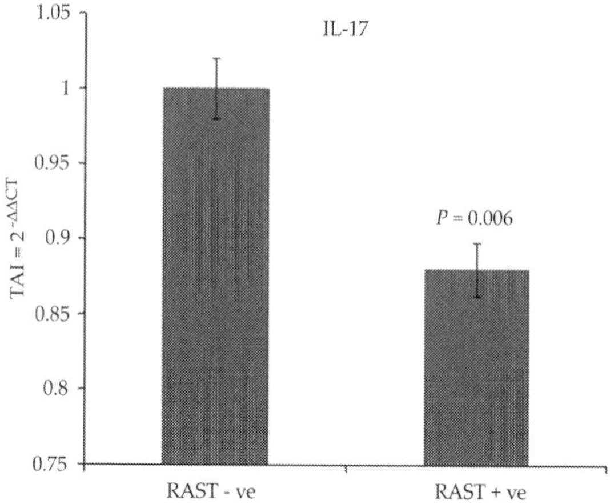
IL-17
1.05
1
0.95
0.9
0.85
0.8
0.75
TAI = 2 -ΔΔCT
P = 0.006
RAST - ve
RAST + ve

(b)

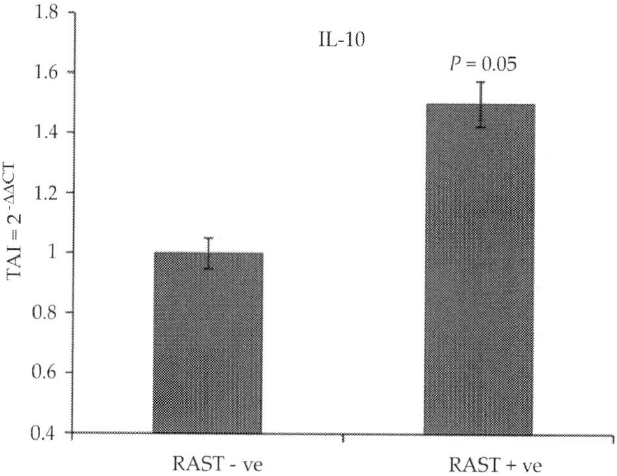
IL-10
1.8
1.6
1.4
1.2
1
0.8
0.6
0.4
TAI = 2 -ΔΔCT
P = 0.05
RAST - ve
RAST + ve

(c)

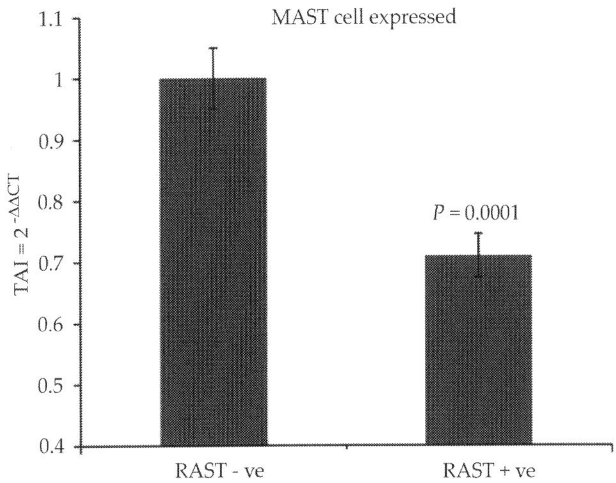

(d)

Figure 4: Relative gene expression of retinoic acid-binding protein; chondroitin transferase; myosin; zinc finger protein; in RAST$^+$ versus RAST$^-$ patients as quantitated by QPCR. Gene expression analysis of retinoic acid-binding protein; chondroitin transferase; myosin; zinc finger protein; was done in RNA extracted from PBMC from RAST$^+$and RAST$^-$ subjects. RNA was reverse transcribed and mRNA expression levels of the above genes was quantitated using real time QPCR. Relative expression of mRNA species was calculated using the comparative C_T method. Data are the mean ± SD of 3 separate experiments done in duplicate. Data are presented as TAI or the transcript accumulation index. Statistical significance was determined by Student's t-test based on comparison between the RAST$^+$ and RAST$^-$patients.

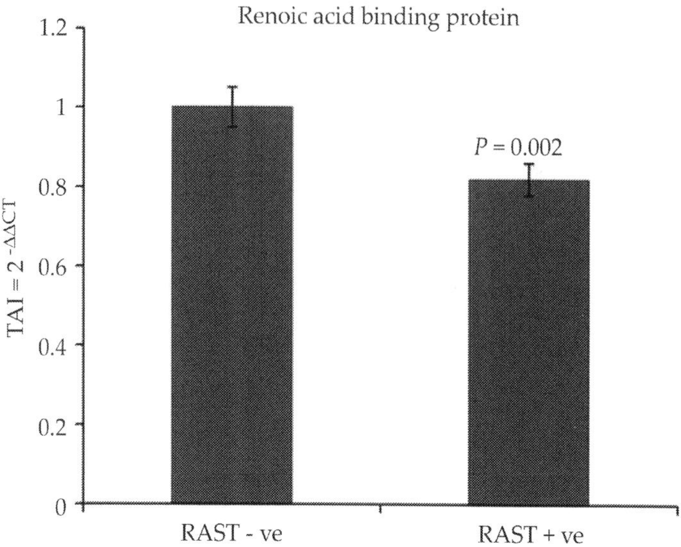
Renoic acid binding protein
1.2
1
0.8
0.6
0.4
0.2
0
TAI = 2 -ΔΔCT
P = 0.002
RAST - ve
RAST + ve

(a)

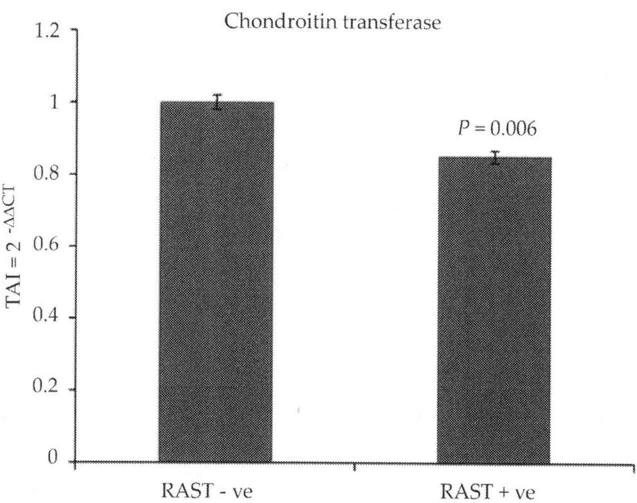
Chondroitin transferase
1.2
1
0.8
0.6
0.4
0.2
0
TAI = 2 -ΔΔCT
P = 0.006
RAST - ve
RAST + ve

(b)

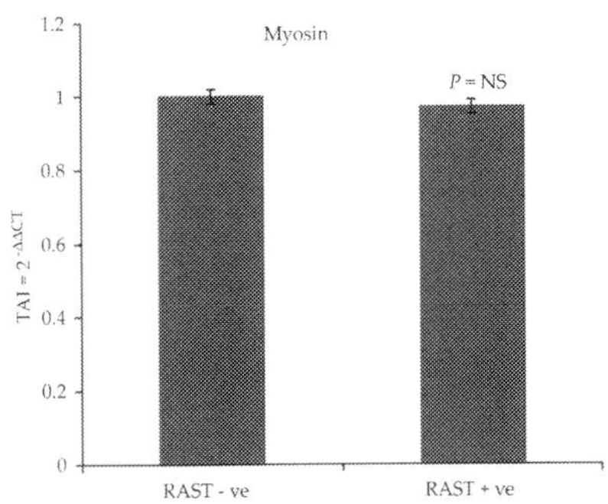

(c)

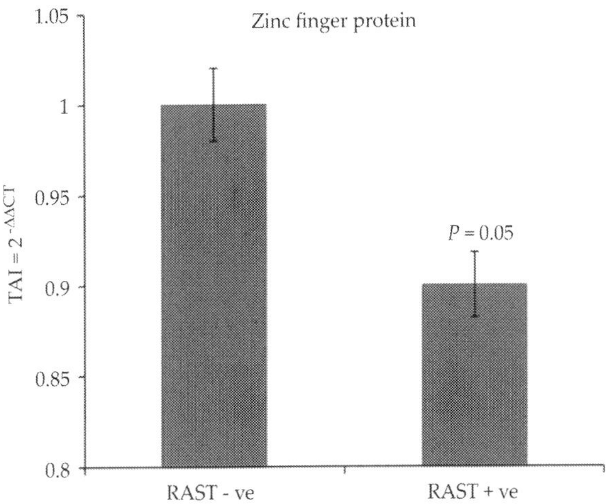

(d)

Figure 5: Relative gene expressions of IL-4, IL-17, IL-10, and MCEM1 in RAST$^+$ versus RAST$^-$ patients as quantitated by QPCR. Gene expression analysis of IL-4, IL-17, IL-10, and MCEM1 was done in RNA extracted from PBMC from RAST$^+$ and RAST$^-$ subjects. RNA was reverse transcribed and mRNA expression levels of the above genes were quantitated using real time QPCR. Relative expression of mRNA species was calculated using the comparative C_T or the ΔΔCt method. This method calculates fold changes between each normal sample and its paired affected sample. ΔCt represents normalized gene expression level (normalized to its own internal control housekeeping gene β-actin). ΔCt for test and control samples were averaged and compared to obtain the ΔΔCt value (an average value). CV, or coefficient of variation, is calculated using the formula STDEV/average. Data are the mean ±

SD of 3 separate experiments done in duplicate. Data are presented as TAI or the transcript accumulation index. Statistical significance was determined by Student's t-test based on comparison between the $RAST^{+}$versus $RAST^{-}$ patients.

Biological Network Analysis Using MetaCore

Given the high through put data that emerges from proteomic analysis, additional data analysis using novel software tools, can be used to interrogate the biological context of lists of genes and proteins. We used one such software tool, MetaCore, which uses a proprietary, manually curated database of human protein-protein, protein-DNA, and protein compound interactions, metabolic and signaling pathways which is supported by proprietary ontologies. The transcription regulation networks of differentially expressed proteins in the $RAST^{+}$ and $RAST^{-}$ groups were generated by MetaCore software using a transcription regulation algorithm. The networks were ranked by a p value and interpreted in terms of GO. The list of significantly regulated proteins emerged after 2D-DIGE/MALDI-TOF analysis of the $RAST^{+}$and $RAST^{-}$ samples, highlighted in Table 2, was entered in to MetaCore. The MetaCore software has in-built algorithms that create biological networks from the uploaded protein list and assign biological processes to the generated networks. We built these various networks using "Analyze Network" algorithm of MetaCore which is useful in providing additional information that is not present in the original list of root nodes. 37 differentially expressed proteins identified on comparison between the $RAST^{+}$ and $RAST^{-}$ groups were mapped into the software and subsequently analyzed. Networks were generated as per p value, Z-score and G-score. p value is based on the saturation of each sub-network with proteins from the user selected list. Z-score ranks the subnetworks according to saturation with the objects from the initial list of seed objects. The Z-score ranks the sub-networks of the analyze network algorithm with regard to their saturation with proteins from the experiment. The G-score modifies the Z-score based on the number of canonical pathways used to build the network. If a network has a high G-score, it is saturated with expressed proteins (from Z-score) and it contains many canonical pathways. We chose the top 3 networks from the search results as ranked by both G-score and p values. The significance of association between the biological processes and the differentially expressed proteins were represented by p values. The list of most significant GO processes associated with the network is shown in Table 3 along with the respective p values. The top 3 processes, the percentage of target nodes involved and the corresponding p values are positive regulation of immune system process (35%, p-value 2.2e-50), MyD88-independent toll-like receptor signaling pathway (19.32%, 2.51e-51) and toll-like receptor 3 signaling pathway (19.32%, 4.1e-49). Figure 6 illustrates the close clustering and interconnectedness of biochemical process networks of the identified proteins which are differentially regulated.

Table 3: List of the most significant gene ontology (GO) processes associated with the biological network shown in Figure 6 as determined by MetaCore analysis along with the corresponding values which indicates that the key protein identified by proteomic analysis are associated with the listed biological processes in a prioritized manner.

Number	Biological process	% involvement in the biological process	*P* value
1	Positive regulation of immune system process	35.23	2.280 *e* − 50
2	MyD88-independent toll-like receptor signaling pathway	19.32	2.516 *e* − 49
3	Toll-like receptor 3 signaling pathway	19.32	4.103 *e* − 49
4	Toll-like receptor signaling pathway	19.89	5.703 *e* − 48
5	Signal transmission via phosphorylation event	32.95	8.113 *e* − 48
6	Intracellular protein kinase cascade	32.95	8.113 *e* − 48
7	Positive regulation of response to stimulus	33.52	3.663 *e* − 47
8	Pattern recognition receptor signaling pathway	19.89	8.743 *e* − 47
9	Activation of innate immune response	19.89	2.639 *e* − 46
10	Innate immune response-activating signal transduction	19.89	2.639 *e* − 46
11	Positive regulation of biological process	63.64	4.550 *e* − 46

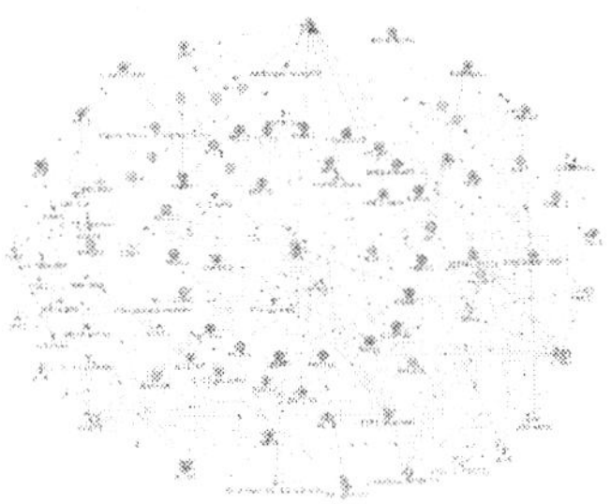

Figure 6: The biochemical processes networks of the differentially expressed proteins generated by MetaCore software using a transcription regulation algorithm. The biochemical processes regulation networks were ranked by a p value and interpreted in terms of GO. Gene network illustrates proteins and interactions of the differentially expressed proteins in $RAST^{+}$ and $RAST^{-}$ patients. The network was obtained using analyze network algorithm of MetaCore and il-

lustrates the close clustering and interconnectedness of biochemical process networks of the identified differentially regulated proteins in food allergy. Colored highlighted symbols (nodes) represent proteins (enzymes (yellow arrows), transporters (purple shapes), receptor ligands (green shapes), transfactors (red shapes), and other proteins (blue shapes)). Red solid circles correspond to differentially expressed proteins in RAST$^+$ samples identified by MALDI analysis. Blue circles show the proteins in the network within the immediate vicinity of those identified proteins. The small colored hexagons on vectors between nodes describe positive interaction (green), unspecified interactions (black), or logical relationships (blue). Each connection (vector) represents a direct interaction between proteins.

DISCUSSION

Allergic responses to food antigens involve a state of immediate hypersensitivity to certain food proteins. The mechanism underlying the initiation and development of allergic responses involves cytokine activation which is believed to directly induce the differentiation of effector Th2 lymphocytes; these Th2 responses play a pivotal role in the development of allergic responses to specific food antigenic triggers. In this paper, we utilized a high throughput proteomic approach, 2D-DIGE, coupled with MALDI-TOF protein identification methods to study the differential expression of proteins in two groups of RAST$^+$ and RAST$^-$ subjects with the goal to identify key proteins that may play a role in the development of food allergenicity. In contrast to traditional biochemical approaches that can only study one or a few particular protein species, proteomics analysis allows the simultaneous analysis of thousands of proteins. This is the first comprehensive differential proteomic analysis of PBMC obtained from RAST$^+$ and RAST$^-$ subjects. Over 1,000 distinct proteins were identified of which about 52 proteins displayed significant differential expression between the RAST$^+$ and RAST$^-$ groups. Of these 52 significantly regulated proteins, 37 were successfully identified (Table 2). A large proportion of these proteins have direct or indirect biological relevance to food allergy.

Food allergenicity is characterized by abnormal IgE production, peripheral eosinophilia, mast cell activation, and induction of Th2 lymphocytes expressing cytokines such as IL-4 and IL-10 [23, 24]. Even though we could not observe any differential expression of IL-4 and IL-10 in the 2D-DIGE analysis, Metacore analysis indicated a very active involvement of these molecules in the networks. A further QPCR analysis showed a significant increase in IL-10 levels in RAST$^+$ patients as compared to RAST$^-$ patients. IL-10 is a cytokine that downregulates both Th1 and Th2 cytokine production [25], and the fact is that its levels are significantly elevated in RAST$^+$ patients and may be useful in identifying human clinical tolerance to foods. IL-10 is also believed to help differentiation of T helper cells into the Th2 phenotype, thereby contributing to the development of allergic responses to food antigens. IL-4 levels in RAST$^+$ patients were lower but not statistically significant as compared to RAST$^-$ patients. A decrease in IL-4 is believed to impair the development of food allergy and the aversion to antigen. Production of IL-4 and levels of specific IgE/IgG1 antibodies correlate with aversion to antigen induced by food allergy in mice [26]. The mechanism underlying the initiation and development of allergic responses involves IL-4 that directly induces the differentiation of committed effector Th2 lymphocytes, and Th2 responses play a pivotal role in the development of allergic responses. IL-4 inhibits the induction of Foxp3 and the generation of inducible regulatory T cells and may also function as anti-inflammatory cytokine [27].

IL 17F is reported to be involved in inflammation and a study on effect of IL-17/IL-17F deficiency in an allergic asthma IL-17 KO mice model, exhibited reduced Th2 cytokine expression, whereas IL-17F KO mice showed elevated type 2 cytokines and eosinophil functions compared with WT mice. This indicates that IL-17 and IL-17F may have opposite functions in chronic allergic airway diseases [28]. The precise role of IL-17 and a new population of IL-17-producing Th cells (Th17) in allergic inflammation is unclear, but Th17 cells are characterized by the production of inflammatory cytokines such as IL-17A, IL-17F, IL-22, and IL-26 [29]. The highest percentage of IL-17-producing cells has been found in severe atopic dermatitis, suggesting a direct correlation between the presence of Th17 cells and severity of the disease [30], indicating that there is possible cross-talk between Th17 cells and eosinophils. Our observation of a significant decrease in IL-17 in RAST$^+$ patients is contradictory to what we expected based on its role as a promoter of allergic inflammation. This observation leads us to believe that IL-17 may be associated with patterns of cytokine dysregulation that are associated with specific atopic phenotypes, though the precise mechanism of its modulation is unclear.

Mast cells are believed to be the key effector cells of IgE-mediated anaphylactic reaction, therefore, a modulation in the expression of the mast cell expressed protein MCEM1 in $RAST^+$ patients is anticipated. We observed a significant decrease in MCEM1 expression in $RAST^+$ patients. The MAPK pathway and its downstream effectors play an important role in the regulation of cytokine expression in mast cells. Signals emanating from many cell surface antigenic receptors and environmental stimuli can converge on the MAPK pathway, which in turn phosphorylates and activates various transcriptional factors and molecular effectors, thereby modulating allergic response. Proteins such as zinc finger proteins, retinoic acid-binding proteins, chondroitin transferase, C-type lectins and heat shock proteins in the soluble microenvironment are believed to be involved in the conditioning of antigen presenting cells, thereby promoting tolerance to dietary antigen primarily via the induction of regulatory T cells.

Yet another unique protein that is significantly modulated in $RAST^+$ patients is Ikaros. This family of proteins plays an important role in hematopoietic development and is believed to modulate the levels of an immunosuppressive compound called bis(tributyltin)oxide (TBTO) that is present in the diet and which can contribute significantly to the development of food allergy to peanut or ovalbumin. TBTO can decrease allergen-specific Th2 cytokine production, number of eosinophilic and basophilic granulocytes in the blood and production of mast cell proteases after an oral food challenge.

To understand the relevance of these proteins and their role in allergy or overall inflammation and to ascertain their interaction with other proteins in known networks the identified proteins were analyzed using Metacore network building tools. Of the total 37 identified proteins imported to Metacore, 20 were successfully mapped in the networks (Figure 6). The unconnected proteins in the network were removed. The proteins differentially regulated in $RAST^+$ subjects were shown in red circles in the network. In protein network generated using the uploaded, differentially expressed proteins in this study showed significant links to inflammatory and immunological responses. Among the upregulated proteins all except Globin-1 could be mapped on to the network.

CONCLUSION

The current investigation has helped to identify key proteins involved in food allergenicity that provide evidence for a relationship between Th2 immune response and food allergy and suggests that these responses play an important role in the development of food allergy. This study provides important information about global protein expression in RAST^{+} patients with food allergies. Although we could not subdivide the patients with food allergies into specific food type, namely, peanut, egg, shell fish, and so forth, the identification of these key proteins provides clues to elucidate mechanisms of the structures that contribute to allergenicity, which thus, in turn, would help alleviate food allergens. Further, these key proteins can be utilized as diagnostic markers that will not only help in the diagnosis and management of food allergy but also may be used in future immunotherapy.

ACKNOWLEDGMENTS

This paper was supported by grants from the National Institute of Health NIAID-RO1LM009726-01 (SAS); NIDA K01 DA024577 (JR); NIDA 1R21DA030108-01 (SM); the Kaleida Health Foundations. The authors would also like to acknowledge intellectual and logistical support from Drs. Mark F. Sands and Maya Srivastava. The authors would also like to acknowledge Mr. Gary Rich for his help with the RAST tests and Sujith Alphy for the technical help with MALDI analysis.

REFERENCES

1. S. H. Sicherer and D. Y. Leung, "Advances in allergic skin disease, anaphylaxis, and hypersensitivity reactions to foods, drugs, and insects in 2008," Journal of Allergy and Clinical Immunology, vol. 123, no. 2, pp. 319–327, 2009.
2. M. P. Ross, M. Ferguson, D. Street, K. Klontz, T. Schroeder, and S. Luccioli, "Analysis of food-allergic and anaphylactic events in the National Electronic Injury Surveillance System," Journal of Allergy and Clinical Immunology, vol. 121, no. 1, pp. 166–171, 2008.
3. H. A. Sampson, "Update on food allergy," Journal of Allergy and Clinical Immunology, vol. 113, no. 5, pp. 805–819, 2004.
4. S. H. Sicherer and H. A. Sampson, "Food allergy: recent advances

in pathophysiology and treatment," Annual Review of Medicine, vol. 60, pp. 261–277, 2009.

5. H. A. Sampson, S. H. Sicherer, and A. H. Birnbaum, "AGA technical review on the evaluation of food allergy in gastrointestinal disorders. American Gastroenterological Association," Gastroenterology, vol. 120, pp. 1026–1040, 2001.
6. H. A. Sampson, "Fatal food-induced anaphylaxis," Allergy, vol. 53, no. 46, pp. 125–130, 1998.
7. R. Aebersold, B. Rist, and S. P. Gygi, "Quantitative proteome analysis: methods and applications," Annals of the New York Academy of Sciences, vol. 919, pp. 33–47, 2000.
8. J. L. Harry, M. R. Wilkins, B. R. Herbert, N. H. Packer, A. A. Gooley, and K. L. Williams, "Proteomics: capacity versus utility," Electrophoresis, vol. 21, no. 6, pp. 1071–1081, 2000.
9. L. Tonella, B. J. Walsh, J. C. Sanchez et al., "'98 Escherichia coli SWISS-2DPAGE database update," Electrophoresis, vol. 19, no. 11, pp. 1960–1971, 1998.
10. A. Pandey and M. Mann, "Proteomics to study genes and genomes," Nature, vol. 405, no. 6788, pp. 837–846, 2000.
11. M. Unlu, M. E. Morgan, and J. S. Minden, "Difference gel electrophoresis: a single gel method for detecting changes in protein extracts," Electrophoresis, vol. 18, no. 11, pp. 2071–2077, 1997.
12. R. P. Tonge, J. Shaw, B. Middleton et al., "Validation and development of fluorescence two-dimensional differential gel electrophoresis proteomics technology," Proteomics, vol. 1, no. 3, pp. 377–396, 2001.
13. G. Zhou, H. Li, D. DeCamp et al., "2D differential in-gel electrophoresis for the identification of esophageal scans cell cancer-specific protein markers," Molecular & Cellular Proteomics, vol. 1, no. 2, pp. 117–124, 2002.
14. P. Chomczynski and N. Sacchi, "Single-step method of RNA isolation by acid guanidinium thiocyanate-phenol-chloroform extraction," Analytical Biochemistry, vol. 162, no. 1, pp. 156–159, 1987.
15. R. Higuchi, C. Fockler, G. Dollinger, and R. Watson, "Kinetic

PCR analysis: real-time monitoring of DNA amplification reactions," Biotechnology, vol. 11, no. 9, pp. 1026–1030, 1993.

16. A. Radonić, S. Thulke, I. M. Mackay, O. Landt, W. Siegert, and A. Nitsche, "Guideline to reference gene selection for quantitative real-time PCR," Biochemical and Biophysical Research Communications, vol. 313, no. 4, pp. 856–862, 2004.
17. D. Wessel and U. I. Flugge, "A method for the quantitative recovery of protein in dilute solution in the presence of detergents and lipids," Analytical Biochemistry, vol. 138, no. 1, pp. 141–143, 1984.
18. S. Sakai, H. Akiyama, Y. Sato et al., "Chondroitin sulfate intake inhibits the IgE-mediated allergic response by down-regulating Th2 responses in mice," Journal of Biological Chemistry, vol. 281, no. 29, pp. 19872–19880, 2006.
19. W. I. Weis, M. E. Taylor, and K. Drickamer, "The C type lectin superfamily in the immune system," Immunological Reviews, vol. 163, pp. 19–34, 1998.
20. A. J. Currie, G. A. Stewart, and A. S. McWilliam, "Alveolar macrophages bind and phagocytose allergen-containing pollen starch granules via C-type lectin and integrin receptors: implications for airway inflammatory disease," Journal of Immunology, vol. 164, no. 7, pp. 3878–3886, 2000.
21. J. J. Fredberg, "Molecular mechanisms in allergy and clinical immunology. Frozen objects: small airways, big breaths, and asthma," Journal of Allergy and Clinical Immunology, vol. 106, no. 4, pp. 615–624, 2000.
22. H. Ishigame, S. Kakuta, T. Nagai et al., "Differential roles of interleukin-17A and-17F in host defense against mucoepithelial bacterial infection and allergic responses," Immunity, vol. 30, no. 1, pp. 108–119, 2009.
23. C. K. Wong, C. Y. Ho, F. Ko et al., "Proinflammatory cytokines (IL-17, IL-6, IL-18 and IL-12) and Th cytokines (IFN-γ, IL-4, IL-10 and IL-13) in patients with allergic asthma," Clinical & Experimental Immunology, vol. 125, no. 2, pp. 177–183, 2001.
24. E. A. Wierenga, M. Snoek, H. M. Jansen, J. Bos, R. Van Lier, and M. Kapsenberg, "Human atopen-specific types 1 and 2 T helper cell

clones," Journal of Immunology, vol. 147, no. 9, pp. 2942–2949, 1991.

25. G. Del Prete, M. De Carli, F. Almerigogna, M. G. Giudizi, R. Biagiotti, and S. Romagnani, "Human IL-10 is produced by both type 1 helper (Th1) and type 2 helper (Th2) T cell clones and inhibits their antigen-specific proliferation and cytokine production," Journal of Immunology, vol. 150, no. 2, pp. 353–360, 1993.
26. L. P. A. Dourado, J. C. D. S. Saldanha, D. L. Gargiulo et al., "Role of IL-4 in aversion induced by food allergy in mice," Cellular Immunology, vol. 262, no. 1, pp. 62–68, 2010.
27. C. R. Cardoso, P. R. Provinciatto, D. F. Godoi et al., "IL-4 regulates susceptibility to intestinal inflammation in murine food allergy," American Journal of Physiology, vol. 296, no. 3, pp. G593–G600, 2009.
28. X. O. Yang, S. H. Chang, H. Park et al., "Regulation of inflammatory responses by IL-17F," Journal of Experimental Medicine, vol. 205, no. 5, pp. 1063–1075, 2008.
29. S. C. Liang, X. Y. Tan, D. P. Luxenberg et al., "Interleukin (IL)-22 and IL-17 are coexpressed by Th17 cells and cooperatively enhance expression of antimicrobial peptides," Journal of Experimental Medicine, vol. 203, no. 10, pp. 2271–2279, 2006.
30. A. Di Cesare, P. Di Meglio, and F. O. Nestle, "A role for Th17 cells in the immunopathogenesis of atopic dermatitis?" Journal of Investigative Dermatology, vol. 128, no. 11, pp. 2569–2571, 2008.

CHAPTER

11

A NOVEL PROTEOMIC ANALYSIS OF THE MODIFICATIONS INDUCED BY HIGH HYDROSTATIC PRESSURE ON HAZELNUT WATER-SOLUBLE PROTEINS

Nuria Prieto[1], Carmen Burbano [2], Elisa Iniesto [1], Julia Rodríguez [3], Beatriz Cabanillas [3], Jesus F. Crespo [3], Mercedes M. Pedrosa [2], Mercedes Muzquiz [2], Juan Carlos del Pozo [4], Rosario Linacero [1] and Carmen Cuadrado [2,]

[1]Genetics Department, Biology Faculty, Complutense University of Madrid, Madrid 28040, Spain

[2]Food Technology Department, National Institute for Agricultural and Food Research and Technology (INIA), Ctra. La Coruña Km. 7.5, Madrid 28040, Spain

[3]Allergy Service, Research Institute Hospital 12 de Octubre (i+12), Avenida de Córdoba s/n, Madrid 28041, Spain

[4]Center for Biotechnology and Plant Genomic, Polytechnic University of Madrid-National Institute for Agricultural and Food Research and Technology (UPM-INIA), Montegancedo Campus, Boadilla del Monte, Madrid 28660, Spain

ABSTRACT

Food allergies to hazelnut represent an important health problem in industrialized countries because of their high prevalence and severity. Food allergenicity can be changed by several processing procedures since food proteins may undergo modifications which could alter immunoreactivity. High-hydrostatic pressure (HHP) is an emerging processing technology used to develop novel and high-quality foods. The effect of HHP on allergenicity is currently being investigated through changes in protein structure. Our aim is to evaluate the effect of HHP on the protein profile of hazelnut immunoreactive extracts by comparative proteomic analysis with ProteomeLab PF-2D liquid chromatography and mass spectrometry. This protein fractionation method resolves proteins by isoelectric point and hydrophobicity in the first and second dimension, respectively. Second dimension chromatogram analyses show that some protein peaks present in unpressurized hazelnut must be unsolubilized and are not present in HHP-treated hazelnut extracts. Our results show that HHP treatment at low temperature induced marked changes on hazelnut water-soluble protein profile.

Keywords: hazelnut; high hydrostatic pressure; immunoreactivity; ProteomeLab PF-2D

INTRODUCTION

High-hydrostatic pressure (HHP) is considered an emerging processing technology used to develop novel and high-quality foods. This novel-processing technique even renders harmless foods which would be of considerable benefit to consumers. HHP treatment of foods can be used to create new products (new texture or taste) or to obtain analogue products with minimal effect on flavor, color and nutritional value and without any thermal degradation. It is well established that higher pressure has a disruptive effect on the tertiary and quaternary structure of most globular proteins, with relatively little influence on the secondary structure. Therefore, higher hydrostatic pressure can unfold proteins. The typical pressure needed for the unfolding is around 500 MPa but it varies from protein to protein, in the range from 100 MPa to 1 GPa or reaching even higher pressures in special cases. The effect of HHP on immunoreactive proteins is being currently investigated through changes in protein structure [1]. Such effects have been studied in: beef [2,3], apple [4,5], celery [4] and in nuts such as peanut [5]. However, there is scarce information on the effects of such food processing techniques on hazelnut (*Corylus avellana* L.) immunoreactive proteins.

Food allergies to hazelnut represent an important health problem in industrialized countries because of their high prevalence and severity [6]. Several hazelnut allergens are well characterized being Cor a 1 (18 KDa, Bet v 1 family) the major one. Other allergenic proteins are Cor a 2 (profilin), Cor a 8 (lipid transfer protein, LPT), Cor a 9 (11S globulin), Cor a 11 (vicilin-like protein) and Cor a 12, Cor a 13 and Cor a 14 belonging to the 2S albumins [7].Understanding how food processing affects the allergenic proteins could be important to control food allergenicity risk.

Bioinformatics tools for database searching enable the quick identification of sequences of interest. Tools for sequence comparison, motif searching or sequence profiling assist researchers to identify biologically relevant sequences similarities. The guidelines to assess potential allergenicity of proteins using bioinformatics in a step-by-step procedure are well established [8,9].

Plant comparative proteomics is becoming increasingly attractive as the rapidly expanding plant genomic and Expressed Sequence Tags (EST) databases provide new opportunities for protein identification. A partial automation of this procedure consisting of a robotic lift of protein spots embedded in the gel, followed by extraction, destaining and protein digestion, has been finalized with reasonable success to the further protein characterization and identification by mass spectrometry (MS) [10]. Proteome analyses have also been performed in a "gel free" condition by using protein fractionation procedures based entirely on liquid chromatography [11,12]. ProteomeLab PF-2D introduced a two dimensional liquid chromatography based on a high-performance chromatofocusing in the first dimension followed by high-resolution reversed-phase chromatography in the second dimension [13,14]. The ProteomeLab PF-2D has become available for sample fractionation and was more resolutive at extreme pHs, both acid and basic. It offers automation of the fractionation processes and resolves proteins by isoelectric point and hydrophobicity in the first and second dimension, respectively [15].

In this study, we have undertaken a comparative proteomic analysis of the effect of HHP at low temperature on water-soluble protein profile of hazelnut immunoreactive extracts using a novel proteomic approach that combines ProteomeLab PF-2D liquid chromatography and mass spectrometry analysis.

EXPERIMENTAL SECTION

Samples and High Hydrostatic Pressure (HHP) Treatments

The hazelnut (*Corylus avellana* L.) var Negreta used in this work was provided by the hazelnut collection of Institut de Recerca i Tecnología Agroalimentàries (IRTA-Mas de Bover, Tarragona, Spain) [16].

Hazelnuts were ground and defatted with *n*-hexane (34 mL/g of flour) for 4 h and air-dried after filtration of the *n*-hexane. High-pressure experiments conditions were carried out according to Omi *et al.* [17] and Kato *et al.* [18]. Hazelnut defatted flours were dissolved in distilled water (1:4 w/v) 20 h before HHP treatment. The flours were subjected to HHP, using pressures of 300, 400, 500 and 600 MPa for 15 min in a multivessel high-pressure equipment (HHP, ACB, France) at 20 °C (Figure 1).

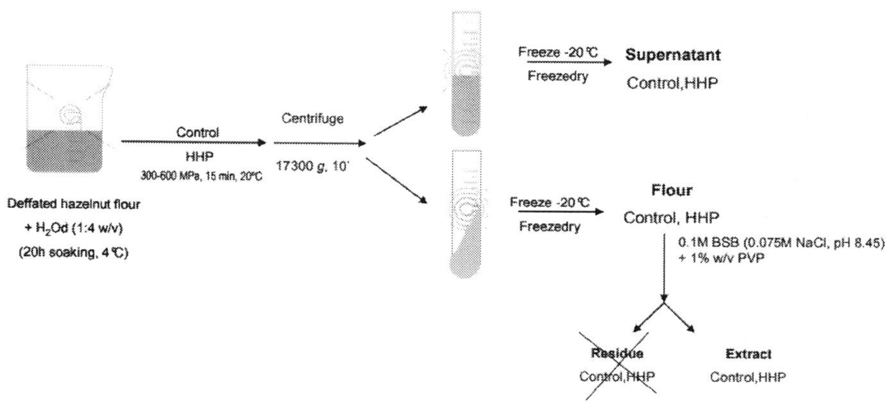

Figure 1. Scheme of high hydrostatic pressure (HHP) treatment of hazelnut samples.

Protein Electrophoresis and Immunoblotting

1D SDS-PAGE was performed according to Laemmli [19]. Protein extracts of hazelnut unpressurized (control) and HHP-treated (300–600 MPa) of supernatant and flour were mixed with XT Sample Loading Buffer (Bio-Rad, Hercules, CA, USA) heated at 95 °C for 5 min, electrophoresed in 12% Bis-Tris precast gel (Bio-Rad, Hercules, CA, USA). Proteins were visualized with Coomassie brilliant blue R250 staining.

Western blotting was performed by electrophoretic transfer to polyvinylidene difluoride (PVDF) membranes at 250 mA for 100 min, at room temperature, essentially according to the method of Towbin *et al.* [20]. After blocking with 5% bovine serum albumin (BSA) (w/v) in phosphate buffered saline (PBS), membranes were incubated overnight with the serum pool from fifteen patients sensitized to hazelnut (serum specific IgE > 0.35 kilounits/L quantified by the CAP-FEIA assay) (1:10 dilution), washed, and then treated with mouse anti-human IgE mAb HE-2 ascitic fluid (1:3000 dilution for 2 h) [21]. After washing, a rabbit anti-mouse IgG peroxidase-conjugated antibody (1:5000 dilution for 1 h; DAKO, Glostrup, Denmark) was added. Detection of IgE-binding components was achieved by means of enhanced chemiluminescence, according to the manufacturer's instructions (Amersham Biosciences, Little Chalfont, UK).

Protein Sample Preparations and 2D-Liquid Chromatography Analysis

Total protein extracts of unpressurized and HHP-treated hazelnut previously tested for IgE-reactivity were used for comparative proteomic analysis with ProteomeLab PF-2D liquid chromatography (Beckman Coulter, Fullerton, CA, USA).

Unpressurized and HHP-treated hazelnut proteins were extracted with 0.1 M borate saline buffer (BSB, 0.075 M NaCl, pH 8.5) plus 1% (w/v) PVP at a 1:10 w/v ratio for 1 h at 4 °C by stirring. The extract was clarified by centrifugation at 27,000× *g* for 30 min at 4 °C, and the supernatants were dialyzed against distilled H_2O for 48 h at 4 °C using a dialysis membrane with a cut-off of 3.5 kDa and freeze-dried. The protein content of each sample was measured by the Bradford dye binding assay (Bio-Rad, Hercules, CA, USA) using bovine serum albumin (Sigma, St.

Louis, MO, USA) as a standard. The nitrogen contents of the samples were determined by LECO analysis according to standard procedures based on Dumas method [22]. The total protein content was calculated as N × 5.3 [22]. The analyses were carried out in duplicates.

Protein extracts (2.5 mg of protein) from unpressurized and HHP-treated hazelnut were subjected to 2-D LC analysis using a ProteomeLab PF-2D instrument (Beckman-Coulter, Fullerton, CA, USA) and the protocol recommended by the manufacturer (ProteoSep®, Chemistry Kit, Sigma, St. Louis, MO, USA). The first-dimension separation was carried out by chromatofocusing on a High Performance ChromatoFocusing (HPCF) 1-D column (250 mm × 2.1 mm internal diameter, 300 Å pore size). The column was equilibrated at pH 8.5 with CF Start Buffer for 250 min at 0.2 mL/min. The pH gradient began after 20 min of sample injection when the CF Eluent Buffer at pH 4.0 moved through the column, gradually decreasing the pH from 8.5 to 4.0. Proteins were eluted according to their isoelectric points (pI) and, in the final step, the most acidic ones were eluted with 1 mol NaCl, 0.2% *n*-octylglucoside. All fractions were collected in 96 well plates using an automated collector.

All the different pH fractions collected from the first dimension were resolved on a reverse phase C18 column (HPRP column: 4.66 mm × 3.3 mm, 1.5 μm particle size,). Of each fraction, 200 μL was run through the column in solvent A (0.1% v/v TFA in water), and the proteins were then eluted with a linear gradient (0%–100%) of solvent B (0.08% v/v trifluoroacetic acid in acetonitrile) for 35 min. Separation was performed at 0.75 mL/min, and the temperature column was maintained at 50 °C. Eluted proteins were monitored by ultraviolet light at 214 nm of absorbance. The different fractions of the first dimension were collected in 12 plates (each 96 well) using an automated collector. All CF profiles were elaborated and compared using 32 Karat V1.01 software (Beckman-Coulter, Fullerton, CA, USA). Quantitative analysis of unpressurized and HHP-treated hazelnut protein peak areas and heights were performed using the Mapping tools software V1.0 (Beckman-Coulter, Fullerton, CA, USA).

Fractions from the second dimension were analyzed by 1D SDS-PAGE as above described and selected bands were manually excised for protein identification by mass spectrometry and database search.

Protein Identification by MS and Data Base Search

Immunoreactive hazelnut proteins modified by high hydrostatic pressure were analyzed by MS and data base search in order to determine their identification. Proteins were in-gel digested with trypsin (Sequencing Grade Modified, Promega, Madrid, Spain) in the automatic Investigator ProGest robot of Genomic Solutions. Briefly, excised gel bands were washed sequentially with 50 mM ammonium bicarbonate (NH_4HCO_3) buffer and acetonitrile. Eluted fractions were evaporated to a final volume of 10 μL. Protein digestions were carried out by incubating the samples in 50 mmol/L NH_4HCO_3 and 10 mmol/L dithiothreitol at 60 °C for 1 h. The alkylation of the reduced sulfhydryl groups was performed by adding 55 mmol/L iodoacetamide at 25 °C for 30 min in the dark. Proteins were digested by adding 1.5 μL of trypsin (125 μg/mL) and incubating at 37 °C overnight. The reaction was stopped with 1% of formic acid. Tryptic peptides were desalted and concentrated with ZipTipC18 columns according to the manufacturer's recommendation. Peptides were eluted in 0.1% trifluoroacetic acid , 50% acetonitrile for matrix-assisted laser desorption ionization time-of-flight (MALDI-TOF)-MS analysis, and with 1% formic acid, 50% methanol for electrospray MS analysis. To increase salt removal, samples were washed with 3–5 cycles of 0.1% triflouroacetic acid as wash solution. The solution was spotted directly onto a MALDI target and analyzed by MALDI-TOF/TOF off-line coupled LC/MALDI-MS/MS. MS analyses were performed automatically with a 4700 Analyzer MALDI-TOF/TOF instrument (Applied Biosystems, Carlsbad, CA, USA). First, MS spectra of all spotted fractions were acquired in the positive reflector mode for peak selection (S/N > 20, excluded precursor with 200 resolution), and further MS/MS spectra acquisition was done using the Collision Induced Dissociation of selected peaks. The search of filtered peptides was performed in batch mode using GPS Explorer V 3.5.0 software with a licensed version of MASCOT, in the Swiss-Prot Database. The MASCOT search parameters were: (1) species, *Coryllus avellana*; (2) allowed number of missed cleavages (only for trypsin digestion), 1; (3) considered modifications, cysteine as carboamidomethyl derivate and methionine as oxidized methionine; (4) peptide tolerance, ±50 ppm; (5) MS/MS tolerance, ±0.3 Da; and (6) peptide charge, +1 [23].

RESULTS AND DISCUSSION

1D Analysis and Immunoblottting

The protein migration patterns of hazelnut before (control) and after high-pressure treatments are shown by SDS-PAGE (Figure 2a). Samples were also analyzed for differences in IgE binding by immunoblot using pool sera from 15 patients sensitized to hazelnut (Figure 2b). Defatted flours that were directly solubilized in sample buffer were used for the immunoassays carried out in this study. The results showed that electrophoretic migration patterns of high-pressure treated flour hazelnut proteins were similar to the control hazelnut.

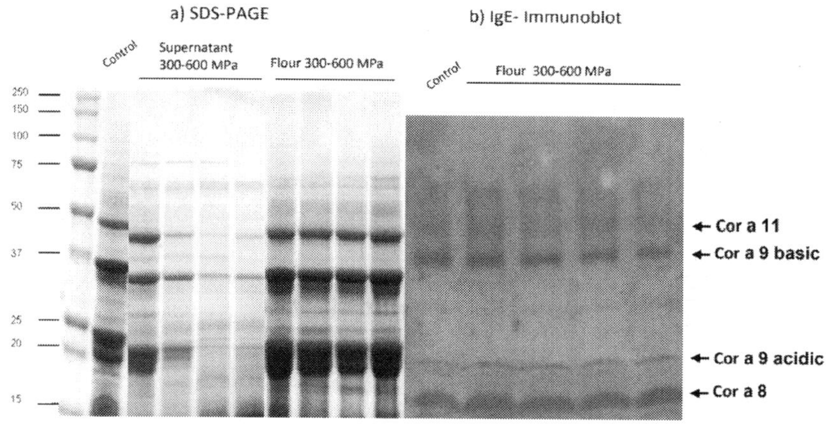

Figure 2. (**a**) SDS-PAGE patterns of supernatant and flour protein extract from control and processed hazelnuts samples and (**b**) IgE immunoblot analysis of control and processed hazelnut samples with a serum pool from subjects with specific IgE to hazelnut (31.3 kU/L).

However, the band intensity of supernatants was diminishing according the pressure increase from 300 to 600 MPa (Figure 2a). The possibility that HHP resulted in variable aggregation depending on the applied pressure could explain these findings. This is in agreement with Somkuti and Smeller [1].

The IgE-immunoblot patterns are similar in control and pressure hazelnuts treated samples from 300 to 600 MPa, showing multiple immunoreactive proteins (Figure 2b). In these samples, most immunoreactive proteins fall in the central and lower area of the gel covering 15–60 kDa. The IgE immunoblotting showed that IgE of pool sera from sensitized to hazelnut recognized bands in all the samples at 50 kDa, 40 kDa, 20 kDa and 9 kDa which might correlate with Cor a 11, Cor a 9, Cor a 1 and Cor a 8 allergens [7].

Hazelnut Proteins Analyzed by Liquid Chromatography

In the ProteomeLab PF-2D system, 2.5 mg of untreated (control) and HHP (600 MPa)-treated hazelnut protein extracts were injected into the HPCF column and were recovered as follow: 40.0% in the 8.0–4.0 pH gradient, 28.6% before the initiation of the pH gradient, and the remaining 31.4% eluted as different peaks when the column was finally washed with high salinity buffer (Figure 3a). The fractions showing higher differences between untreated and HHP treated hazelnut were collected. We collected seven fractions in the control sample and 10 fractions in de HHP-treated hazelnut sample with high salinity buffer and they were subjected to a second dimension using the high performance reverse-phase chromatography column (Figure 3b).

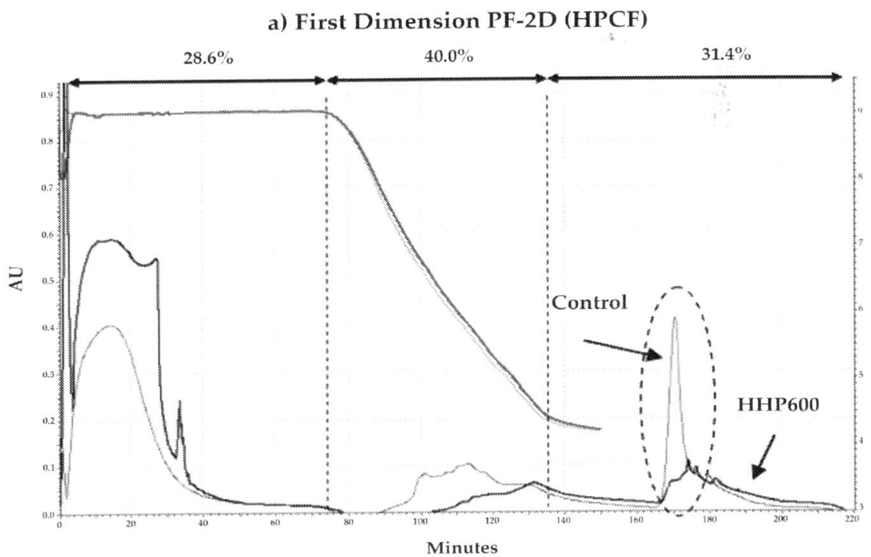

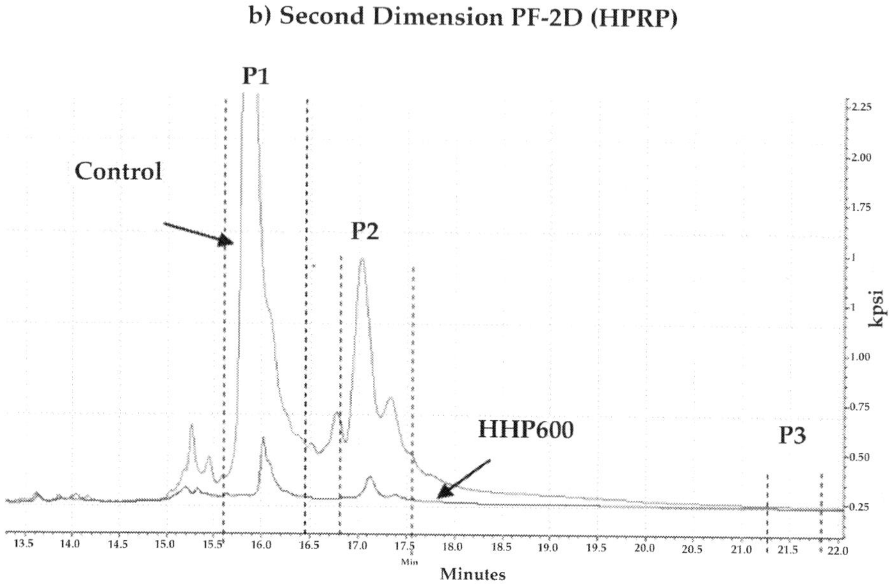

Figure 3. Chromatographic analysis of hazelnut proteins using ProteomeLab PF-2D: (**a**) Chromatofocusing of total proteins separated by pH gradient 8.0–4.0. Fraction number 29 of control and 34 of HHP hazelnut were fractionated in the second dimension. (**b**) High performance reverse phase chromatography, fractions are separated by hydrophobicity. Proteins contained in peaks marked P1, P2 and P3 were separated by SDS-PAGE for subsequent MS analyses to determine their identity.

The protein profiles obtained showed important differences between the two samples (control and HHP 600 samples) with much lesser and smaller peaks in HHP 600 hazelnut. The fractions collected after the second dimension were concentrated in three ones, namely P1 (from 23 to 26), P2 (from 27 to 29) and P3 (from 34 to 35) in control as well as HHP treated hazelnut.

Proteins Identified in Raw and HHP Hazelnuts Samples

P1, P2 and P3 fractions from raw hazelnuts (control) and HHP600 hazelnut samples were further analyzed by 1D SDS-PAGE (Figure 4a,b). Control and HHP 600 hazelnut had a distinct protein band pattern, in band number as well as band intensity, in agreement with the differences showed in the second dimension absorbance profile between both hazelnut samples (Figure 3b).

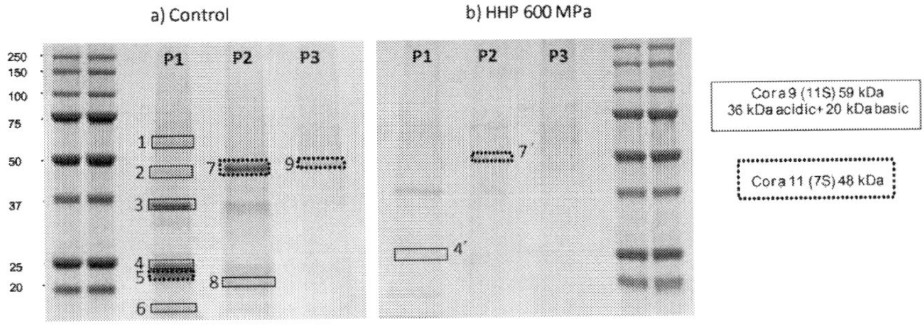

Figure 4. SDS-PAGE of the different fractions collected and concentrated (P1, P2 and P3) from the second dimension chromatography column (HPRP) of control hazelnut **(a)** and HHP 600 MPa hazelnut **(b)**. Proteins were visualized by Coomassie Blue. The rectangles indicate the protein bands that were identified by MS analysis.

Finally, nine of these polypeptide bands of control hazelnut and two bands in HHP 600 hazelnut were manually excised for MALDI-TOF-TOF and LC-ESI-MS/MS identification. Table 1 summarizes the polypeptide identification data of raw hazelnut sample. The major bands were tryptic digested in order to carry out the analysis by MALDI-TOF/MS. A peptide mass fingerprint search allowed us to identify the bands 1, 2, 3, 4, 6 and 8 of raw hazelnut as a 11S globulin-like of *C. avellana* (59 kDa) (Table 1). The bands 5, 7 and 9 of raw hazelnut were identified as a 7S vicilin-like of *C. avellana* (48 kDa). The identification carried out in HHP600 treated sample (Table 2) showed only one band (4′) corresponding to the 11S globulin-like and another one (7′) identified as the 7S vicilin-like. According to the hazelnut allergen description [7], the bands identified as 11S globulin-like could correspond to the putative Cor a 9 allergen (59 kDa) which is composed of a 36 kDa acid subunit and a 20 kDa basic subunit. The 7S vicilin-like bands could correlate with the Cor a 11 allergen (48 kDa).

Proteome analysis is a tool that can be used both to visualize and compare complex mixtures of proteins and to gain a large amount of information about the individual proteins involved in specific biological responses. In this work, proteomic analysis by bidimensional chromatography (PF-2D) shows that the high hydrostatic pressure (HHP) induces significant changes in the proteome of hazelnut extracts in agreement with results on differential solubility after HHP process reported by other authors in several foods [1,4,18,24]. Our results show that the protein solubility became different among the fractions from 300 to 600 MPa. The identification by MALDI-TOF/MS of the proteins affected by high pressure indicated that among the proteins which are insolubilized by high pressure, there are some legumins and vicilins that could correspond to the allergen Cor 9 (11S) and Cor 11 (7S) according to data base search.

Table 1. Proteins separated by ProteomeLab PF-2D from raw hazelnuts and identified by MALDI-TOF/MS.

Band No.	**No. access**	**Protein identification**	**Mascot score ***	**Mass (Da)**	**Matched peptides**
1	gi|18479082	11S globulin-like protein (*C. avellana*)	73	59,605	12
2	gi|18479082	11S globulin-like protein (*C. avellana*)	161	59,605	19
3	gi|18479082	11S globulin-like protein (*C. avellana*)	136	59,605	6
4	gi|18479082	11S globulin-like protein (*C. avellana*)	187	59,605	20
5	gi|19338630	48-kDa glycoprotein precursor (*C. avellana*)	199	51,110	21
6	gi|18479082	11S globulin-like protein (*C. avellana*)	218	59,605	13
7	gi|19338630	48-kDa glycoprotein precursor (*C. avellana*)	256	51,110	24
8	gi|18479082	11S globulin-like protein (*C. avellana*)	65	59,605	11
9	gi|19338630	48-kDa glycoprotein precursor (*C. avellana*)	199	51,110	21

* Protein scores greater than 42 are significant ($p < 0.05$). Protein score is $-10 \times \text{Log}(P)$, where P is the probability that the observed match is a random event.

Table 2. Proteins separated by ProteomeLab PF-2D from hazelnuts HHP 600 MPa and identified by MALDI-TOF/MS.

Band No.	No. access	Protein identification	Mascot score *	Mass (Da)	Matched peptides
4′	gi|18479082	11S globulin-like protein (*C. avellana*)	205	59,605	10
7′	gi|19338630	48-kDa glycoprotein precursor (*C. avellana*)	194	51,110	21

* Protein scores greater than 42 are significant ($p < 0.05$). Protein score is $-10 \times \text{Log}(P)$, where P is the probability that the observed match is a random event.

CONCLUSIONS

The proteomic approach described here allows a deeper and more detailed study of the modifications induced by high pressure. The HHP treatment employed in this study is not effective to alter the immunoreactivity to hazelnut proteins but the protein solubility became different after HHP processing. Although at present there is not an accepted method to reduce the allergenicity of foods, a combination of treatments including high hydrostatic pressure with others, such as protease digestion, could be a successful strategy towards hypoallergenization. To reach this, we have to collect more information about the pressure behavior of these proteins at various environmental conditions and in presence of food additives. There is a definite need for further studies at basic scientific level, which could report the effect of pressure on the 3D structure of the allergenic proteins. The complexity of food processing demonstrates the importance of understanding its impact at the molecular level if risk assessors are to move towards knowledge-based ways of managing allergen risks.

ACKNOWLEDGMENTS

This work was supported by BIO2008-00639, AGL2008-03453 and AGL2012-39863 research projects.

CONFLICTS OF INTEREST

The authors declare no conflict of interest.

REFERENCES

1. Somkuti, J.; Smeller, L. High pressure effects on allergen food proteins. *Biophys. Chem.* 2013, *183*, 19–29.
2. Han, G.D.; Matsuno, M.; Ikeuchi, Y.; Suzuki, A. Effects of heat and high-pressure treatments on antigenicity of beef extract. *Biosci. Biotechnol. Biochem.* 2002, *66*, 202–205.
3. Yamamoto, S.; Mikami, N.; Matsuno, M.; Hara, T.; Odani, S.; Suzuki, A.; Nishiumi, T. Effects of a high-pressure treatment on bovine gamma globulin and its reduction in allergenicity. *Biosci. Biotechnol. Biochem.* 2010, *74*, 525–530.
4. Husband, F.A.; Aldick, T.; van der Plancken, I.; Grauwet, T.; Hendrickx, M.; Skypala, I.; Mackie, A.R. High-pressure treatment reduces the immunoreactivity of the major allergens in apple and celeriac. *Mol. Nutr. Food Res.*2011, *55*, 1087–1095.
5. Johnson, P.E.; van der Plancken, I.; Balasa, A.; Husband, F.A.; Grauwet, T.; Hendrickx, M.; Knorr, D.; Mills, E.N.; Mackie, A.R. High pressure, thermal and pulsed electric-field-induced structural changes in selected food allergens. *Mol. Nutr. Food Res.* 2010, *54*, 1701–1710.
6. Piknová, L.; Pangallo, D.; Kuchta, T. A novel real-time polymerase chain reaction (PCR) method for the detection of hazelnuts in food. *Eur. Food Res. Technol.* 2008, *226*, 1155–1158.
7. Garino, C.; Zuidmeer, L.; Marsh, J.; Lovegrove, A.; Morati, M.; Versteeg, S.; Schilte, P.; Shewry, P.; Arlorio, M.; van Ree, R. Isolation, cloning, and characterization of the 2s albumin: A new allergen from hazelnut. *Mol. Nutr. Food Res.* 2010, *54*, 1257–1265.
8. European Food Safety Authority (EFSA). Scientific Opinion on the Assessment of Allergenicity of GM plants and microorganisms and derived food and feed. EFSA Panel on Genetically Modified Organisms (GMO Panel). Available online: http://www.efsa.europa.eu/en/efsajournal/doc/1700.pdf (accessed on 28 Jaunuary 2014).
9. Kulkarni, A.; Ananthanarayan, L.; Raman, K. Identification of putative and potential cross-reactive chickpea (*Cicer arietinum*) allergens through an in silico approach. *Comput. Biol. Chem.* 2013, *47*, 149–155.

10. Rose, J.K.; Bashir, S.; Giovannoni, J.J.; Jahn, M.M.; Saravanan, R.S. Tackling the plant proteome: Practical approaches, hurdles and experimental tools. *Plant J.* 2004, *39*, 715–733.

11. Agrawal, G.K.; Yonekura, M.; Iwahashi, Y.; Iwahashi, H.; Rakwal, R. System, trends and perspectives of proteomics in dicot plants part I: Technologies in proteome establishment. *J. Chromatogr. B Analyt. Technol. Biomed. Life Sci.* 2005, *815*, 109–123.

12. Lambert, J.P.; Ethier, M.; Smith, J.C.; Figeys, D. Proteomics: From gel based to gel free. *Anal. Chem.* 2005, *77*, 3771–3787.

13. Soldi, M.; Sarto, C.; Valsecchi, C.; Magni, F.; Proserpio, V.; Ticozzi, D.; Mocarelli, P. Proteome profile of human urine with two-dimensional liquid phase fractionation. *Proteomics* 2005, *5*, 2641–2647.

14. Pirondini, A.; Visioli, G.; Malcevschi, A.; Marmiroli, N. A 2-D liquid-phase chromatography for proteomic analysis in plant tissues. *J. Chromatogr. B Analyt. Technol. Biomed. Life Sci.* 2006, *833*, 91–100.

15. Wu, W.W.; Wang, G.; Yu, M.J.; Knepper, M.A.; Shen, R.F. Identification and quantification of basic and acidic proteins using solution-based two-dimensional protein fractionation and label-free or 18O-labeling mass spectrometry. *J. Proteome Res.* 2007, *6*, 2447–2459.

16. Batlle, I.; Aletà, N.; Clavé, J.; Ninot, A.; Romero, M.; Rovira, M.; Tous, J.; Vargas, F.J. Bancos de germoplasma de especies de frutos secos y desecados: Avellano, nogal, pistachero y algarrobo. *Fruticultura Profesional* 1999, *104*, 6–12.

17. Omi, Y.; Kato, T.; Ishida, K.-I.; Kato, H.; Matsuda, T. Pressure-induced release of basic 7S globulin from cotyledon dermal tissue of soybean seeds. *J. Agric. Food Chem.* 1996, *44*, 3763–3767.

18. Kato, T.; Katayama, E.; Matsubara, S.; Omi, Y.; Matsuda, T. Release of allergenic proteins from rice grains induced by high hydrostatic pressure. *J. Agric. Food Chem.* 2000, *48*, 3124–3129.

19. Laemmli, U.K. Cleavage of structural proteins during the assembly of the head of Bacteriophage t4. *Nature* 1970, *227*, 680.

20. Towbin, H.; Staehelin, T.; Gordon, J. Electrophoretic transfer of proteins from polyacrylamide gels to nitrocellulose sheets-procedure and some applications. *Proc. Natl. Acad. Sci. USA* 1979, *76*, 4350–4354.
21. Sanchez-Madrid, F.; Morago, G.; Corbi, A.L.; Carreira, J. Monoclonal-antibodies to 3 distinct epitopes on human IgE: Their use for determination of allergen-specific IgE. *J. Immunol. Methods* 1984, *73*, 367–378.
22. AOAC International. *Official Methods of Analysis of AOAC International*, 16th ed.; Association of Official Analytical Chemists: Washington, DC, USA, 1995.
23. Bairoch, A.; Apweiler, R. The SWISS-PROT protein sequence database and its supplement TrEMBL in 2000. *Nucleic Acids Res.* 2000, *28*, 45–48.
24. Peñas, E.; Gomez, R.; Frias, J.; Baeza, M.L.; Vidal-Valverde, C. High hydrostatic pressure effects on immunoreactivity and nutritional quality of soybean products. *Food Chem.* 2011, *125*, 423–429.

CITATIONS

CHAPTER 1

Peter Feist and Amanda B. Hummon, Proteomic Challenges: Sample Preparation Techniques for Microgram-Quantity Protein Analysis from Biological Samples, doi:10.3390/ijms16023537

CHAPTER 2

Perez-Riverol Y, Alpi E, Wang R, Hermjakob H, Vizcaíno JA, Making proteomics data accessible and reusable: current state of proteomics databases and repositories, doi: 10.1002/pmic.201400302.

CHAPTER 3

Mayer, G., Jones, A. R., Binz, P.-A., Deutsch, E. W., Orchard, S., Montecchi-Palazzi, L., … Eisenacher, M. (2014). Controlled vocabularies and ontologies in proteomics: Overview, principles and practice. Biochimica et Biophysica Acta, 1844(1), 98–107. http://doi.org/10.1016/j.bbapap.2013.02.017

CHAPTER 4

Sameh Magdeldin, Shymaa Enany, Yutaka Yoshida, Bo Xu, Ying Zhang, Zam Zureena, Ilambarthi Lokamani, Eishin Yaoita and Tadashi Yamamoto, Basics and recent advances of two dimensional- polyacrylamide gel electrophoresis, doi:10.1186/1559-0275-11-16

CHAPTER 5

Ignacio Ortea, Gavin O'Connor, Alain Maquet, Review on proteomics for food authentication, https://doi.org/10.1016/j.jprot.2016.06.033

CHAPTER 6

Piotr Minkiewicz, Małgorzata Darewicz, Anna Iwaniak, Jolanta Sokołowska, Piotr Starowicz, Justyna Bucholska and Monika Hrynkiewicz, Common Amino Acid Subsequences in a Universal Proteome—Relevance for Food Science, doi:10.3390/ijms160920748

CHAPTER 7

Hong C, Jiang H, Lü E, Wu Y, Guo L, Xie Y, et al. (2012) Identification of Milk Component in Ancient Food Residue by Proteomics. PLoS ONE 7(5): e37053. https://doi.org/10.1371/journal.pone.0037053

CHAPTER 8

Francesco Di Girolamo, Andrea Masotti, Guglielmo Salvatori, Margherita Scapaticci, Maurizio Muraca and Lorenza Putignani, A Sensitive and Effective Proteomic Approach to Identify She-Donkey's and Goat's Milk Adulterations by MALDI-TOF MS Fingerprinting, doi:10.3390/ijms150813697

CHAPTER 9

Romina Pedreschi, Jørgen Nørgaard and Alain Maquet, Current Challenges in Detecting Food Allergens by Shotgun and Targeted Proteomic Approaches: A Case Study on Traces of Peanut Allergens in Baked Cookies, doi:10.3390/nu4020132

CHAPTER 10

Bindukumar Nair, John C. Wheeler, Donald E. Sykes, et al., "Proteomic Approach to Evaluate Mechanisms That Contribute to Food Allergenicity: Comparative 2D-DIGE Analysis of Radioallergosorbent Test Positive and Negative Patients," International Journal of Proteomics, vol. 2011, Article ID 673618, 13 pages, 2011. doi:10.1155/2011/673618

CHAPTER 11

Nuria Prieto, Carmen Burbano, Elisa Iniesto, Julia Rodríguez, Beatriz Cabanillas, Jesus F. Crespo, Mercedes M. Pedrosa, Mercedes Muzquiz, Juan Carlos del Pozo, Rosario Linacero and Carmen Cuadrado, A Novel Proteomic Analysis of the Modifications Induced by High Hydrostatic Pressure on Hazelnut Water-Soluble Proteins, doi:10.3390/foods3020279

INDEX

T

U

V